Laboratory Studies in Earth History

Ninth Edition

HAROLD L. LEVIN
Washington University

MICHAEL S. SMITH
*University of
North Carolina - Wilmington*

**McGraw-Hill
Higher Education**

Boston   Burr Ridge, IL   Dubuque, IA   New York   San Francisco   St. Louis
Bangkok   Bogotá   Caracas   Kuala Lumpur   Lisbon   London   Madrid   Mexico City
Milan   Montreal   New Delhi   Santiago   Seoul   Singapore   Sydney   Taipei   Toronto

LABORATORY STUDIES IN EARTH HISTORY,
NINTH EDITION

Published by McGraw-Hill, a business unit of The McGraw-Hill Companies, Inc., 1221 Avenue of the
Americas, New York, NY 10020. Copyright © 2008 by The McGraw-Hill Companies, Inc. All rights reserved.
Previous editions 2004, 2001, 1997, 1993, and 1989. No part of this publication may be reproduced or distributed
in any form or by any means, or stored in a database or retrieval system, without the prior written consent of The
McGraw-Hill Companies, Inc., including, but not limited to, in any network or other electronic storage
or transmission, or broadcast for distance learning.

Some ancillaries, including electronic and print components, may not be available to customers outside the
United States.

 This book is printed on recycled, acid-free paper containing 10% postconsumer waste.

4 5 6 7 8 9 0 RMN/RMN   15   14   13   12   11

ISBN 978–0–07–305072–0
MHID 0–07–305072–5

Publisher: *Thomas D. Timp*
Executive Editor: *Margaret J. Kemp*
Senior Developmental Editor: *Margaret B. Horn*
Senior Marketing Manager: *Lisa Nicks*
Senior Project Manager: *Kay J. Brimeyer*
Lead Production Supervisor: *Sandy Ludovissy*
Associate Design Coordinator: *Brenda A. Rolwes*
Cover Designer: *Studio Montage, St. Louis, Missouri*
(USE) Cover Image: *Mammoth Skeleton:* © *PhotoLink/Getty Images;*
Hills of John Day Fossil Beds: © *Royalty-Free/CORBIS*
Lead Photo Research Coordinator: *Carrie K. Burger*
Compositor: *Aptara*
Typeface: *10/12 Times Roman*
Printer: *Quebecor World Dubuque, IA*

www.mhhe.com

Contents

15

Canadian Shield and Basement Rocks of North America

16

Mountain Belts of North America

17

The Interior Plains and Plateaus

18

Identification of Minerals

19

Igneous Rocks

20

Metamorphic Rocks

List of Tables

Colorplates*

Plates

Preface

*The past history of our planet is recorded in the rocks.
We only need to learn how to read the message. What
a joy to know, and sad to not ever see, or even ask
about the rocks beneath our feet.*

Arthur R. Green
Chief Geoscientist, ExxonMobil Exploration Co.

Instructors of Historical Geology courses have at least two important objectives in mind. One of these is to help students learn about the extraordinary events in the geologic history of the earth. The second is to give students an understanding of *how* geologists discover that these events occurred. The laboratory experience is particularly effective in promoting the second objective. As in previous editions, questions and exercises are designed to encourage use of observations as a basis for making inferences about events and conditions in the geologic past. Some of the studies do not require laboratory materials and can be assigned as homework to be discussed in subsequent laboratory meetings. This has the advantage of allowing students to work at their own pace.

As in previous editions, *Laboratory Studies in Earth History* is designed for the laboratory component of a college level course in Historical Geology. We have included information that will not only satisfy the needs of the student who will go no further in geology, but also those preparing for the advanced courses required for an academic major. The studies are arranged in developmental order so that each can build upon earlier information. However, they are sufficiently self-contained to permit rearrangement required by the instructor's preferred schedule of lecture topics.

Information and exercises about common rock-forming minerals, as well as igneous and metamorphic rocks, are provided so that the book can be used either for the laboratory of a stand-alone first course, or for the second course of a two semester sequence (Physical Geology followed by Historical Geology). For the stand-alone course, instructors may choose to begin with minerals, igneous rocks, and metamorphic rocks.

We continue to employ color, but only where it is essential so as to minimize the student's textbook expenses. The interpretation of black and white photographs has obvious limitation, but they open the way for interpretation of color slides and observations on field trips. Chapters are self-contained so as to allow for coordination with the lecture schedule.

New to the Ninth Edition

- The first four labs in the manual have been extensively revised. The new introductory material that accompanies each of these labs now focuses more on basic information about the geologic setting being studied.

- A new discussion about paleogeography as well as new study questions on this topic have been added to Chapter 9.

- The new Chapter 18, *Identification of Minerals,* is a brief lab focused on the basics of mineral identification that was added at the request of reviewers.

- Many drawings and photographs have been improved or replaced in this edition.

- We use the new 2005 Geologic Map of North America.

- The *Materials* section of the labs is now entitled *What You Will Need* and is more descriptive in its listing of lab needs.

Instructor's Manual

An instructor's manual containing exercise materials, lists, and answers resides on a password-protected site at www.mhhe.com/levin9e. Contact your McGraw-Hill sales representative for information.

Acknowledgments

Laboratory Studies in Earth History owes much to the guidance and suggestions of reviewers and colleagues who have used previous editions of this manual. We extend our thanks and sincere appreciation to all of these earth scientists, including:

Callan Bentley, *Northern Virginia Community College*

James Carew, *College of Charleston*

Melinda Hutson, *Portland Community College*

Amanda Julson, *Blinn College*

David T. King, Jr., *Auburn University*

Gary Rosenberg, *Indiana University—Purdue University, Indianapolis*

Thomas Yancey, *Texas A&M University*

1
Introduction to Sedimentary Rocks

1. Two sets of minerals that are common in sedimentary rocks. Each set should include quartz, calcite, dolomite, gypsum, anhydrite, biotite and muscovite mica, garnet, plagioclase, and potassium feldspar (orthoclase). Minerals in the first or "demonstration set" will be examined together with the instructor. The other set ("study set") is for identification by the student.

2. Demonstration set and study set of common clastic, chemical, and bioclastic sedimentary rocks: sandstones (quartz sandstone, arkose, graywacke, submature sandstone), loess, chert and flint, limestone (fossiliferous and crystalline), dolostone, coquina, shale, siltstone, claystone (or mudstone), rock salt, rock gypsum, and coal.

3. Materials to test for hardness: glass plate, steel nail, porcelain streak plate, and set of Mohs hardness minerals (see Table 1.2).

4. 10X magnification hand lens or binocular microscope.

5. Dropper bottle containing dilute (10%) hydrochloric acid (HCl).

BASIC INFORMATION

Sedimentary rocks originally came from chemically decomposed and mechanically disintegrated older rocks (igneous, metamorphic, or sedimentary). Chemical weathering can dissolve calcium, magnesium, sodium, and other elements, which can then be precipitated to form such rocks as limestone, dolostone, and rock salt.

Detrital material that composes sediments may be rock fragments, mineral fragments, organic matter, or the shells of ancient marine creatures (Fig. 1.1). These materials may be transported by moving water or wind to places where the sediment accumulates. In these areas of sediment deposition, once loose sediment may be compacted and cemented over time to form sedimentary rock. The process of forming solid sedimentary rock from loose sediment is called **lithification.**

Layers of sedimentary rocks, like pages in a history book, provide information about the earth's past. Layers of sedimentary rock cover 75% of the world's total land area. However, in order to obtain information about conditions and events on our planet long ago, it is necessary first to identify the rocks themselves. In this study, you will develop these skills.

THE TEXTURE OF SEDIMENTARY ROCKS

Sedimentary rocks are classified by their composition and texture. **Texture** refers to the *size, shape,* and *arrangement* of the individual grains that make up the rock. In hand samples, texture can be used to separate sedimentary rocks into three general categories: **clastic, chemical,** or **bioclastic.**

Clastic Sedimentary Rocks

A sedimentary rock composed of fragments of rocks, minerals, or broken fossil shells has a **clastic texture.** Individual grains that make up the rock are called **clasts** (Fig. 1.1). The *grain size* of the clasts is determined by many factors, including the nature of the transporting and depositing medium, the source material, and the environment of deposition.

For convenience, we can divide clasts into three categories based on size (see also Table 2.1 in the next chapter).

1. **Coarse** Particles larger than 2 mm (called *gravel*). The general term gravel includes pebbles (2–64 mm), cobbles (64–256 mm), and boulders (greater than 256 mm in average diameter).

2. **Medium** Particles from 1/16 mm (0.063 mm) to 2 mm (called *sand size*). This size range is similar to that in granulated sugar.

3. **Fine** Particles finer than 0.063 mm (called silt and clay size).

Individual grains are too small to be visible to the unaided eye.

If you are undecided as to whether a sedimentary rock is of medium- or fine-grain size, estimate the size as follows: If most of the grains are easily visible, consider the rock to have medium-grain size. If, however, the grains are too small to be distinguished with the unaided eye, consider the rock to be fine-grained. With a 10X hand lens, you will be able to see the fine-grained clasts in this rock.

The size of grains in a clastic rock and the shape and composition of the grains are all used, not only to identify the rock (Fig. 1.2), but also to evaluate its origin and depositional history.

Figure 1.1 Examples of clast materials that become compacted or cemented together to form clastic or bioclastic sedimentary rocks. Sandstone would be formed from the clasts in A and B, oölitic limestones from C and D, clastic limestone from E, and fossiliferous limestone or coquina from F.

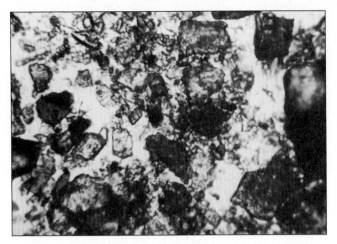

A Immature sand, artificially crushed granite. Magnification 40X.

B Mature quartz sand from the St. Peter sandstone (Ordovician). Magnification 50X.

C Oöid sand. Magnification 15X.

D Replacement of oöids by microcrystalline silica preserves the fine concentric laminae. Field of view about 4 mm.

Photo: G. Ross.

E Carbonate beach sand from the Bahamas, composed of broken shells. Magnification 15X.

F Coquina. This rock is composed almost entirely of broken bivalve (mollusk) shells cemented together by a calcite ($CaCO_3$) cement. The cement was either derived from the dissolution of the shell material or precipitated from seawater.

Figure 1.2 Four categories of sandstone as seen in thin section under the microscope. Note that although the quartz sandstone is texturally mature, its compositional maturity is only moderate because of small amounts of feldspar. Subgraywackes show characters that are transitional between quartz sandstones and graywackes. Rounding and sorting in subgraywackes are better than in graywackes but poorer than in quartz sandstones. Subgraywackes also contain less feldspar than graywackes and may exhibit mineral cements. Diameter of field about 4 mm.

From Levin, H. L., *The Earth Through Time,* 8th ed. Philadelphia: John Wiley & Sons, 2006.

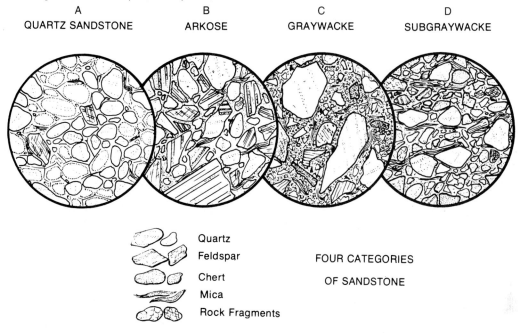

A
QUARTZ SANDSTONE

B
ARKOSE

C
GRAYWACKE

D
SUBGRAYWACKE

Quartz
Feldspar
Chert
Mica
Rock Fragments

FOUR CATEGORIES
OF SANDSTONE

Roundness is a measure of the shape of the grains and refers to the degree to which sharp corners and edges of the grains have been worn away. An angular grain released by weathering from its parent rock will become increasingly more rounded during subsequent long periods of transportation by wind or water. The composition of the grains also influences the rate at which they will be rounded. Less durable minerals may become rounded faster than durable minerals.

Sorting refers to the uniformity or lack of uniformity of particle sizes in a clastic sedimentary rock. Sandstones composed of sands that have been moved about and washed by waves are often well sorted (grains are within a small range of sizes), whereas poor sorting results when sediment is rapidly deposited without being selectively separated into sizes. Conditions leading to poorly sorted sediment occur at the base of mountains where stream gradients abruptly decrease and sediment is quickly deposited.

Maturity of Sandstones is a term used in geology to describe the amount of transportation and erosion sand grains in a sandstone have experienced. For example, a well-sorted sandstone composed mainly of rounded, very durable quartz grains would be called *mature* (Fig. 1.2A). A poorly sorted sandstone retaining angular grains of more readily destroyed minerals would be considered *immature (Fig. 1.2C).* Thus, both texture and mineral composition are important in estimating the maturity of a sedimentary rock.

Figure 1.3 is a photomicrograph of a sandstone with prominent grains of feldspar (the striped grains) and quartz. Is the sandstone mature or immature and why?

Figure 1.3 Photomicrograph of a sandstone largely composed of feldspar (striped) of a quartz grains in a clay matrix.

Chemical Sedimentary Rocks

Some sedimentary rocks are not composed of accumulations of clastic particles. They have a crystalline rather than clastic texture. The mosaic of interlocking crystals may form either by growth of developing crystals precipitated from solution or as a recrystallization of older grains in the rock. The most common minerals precipitated from solution are calcite (calcium carbonate ($CaCO_3$), hematite (iron oxide (Fe_2O_3)) or limonite ($FeO \cdot OH \cdot nH_2O$), or minerals composed largely of silica such as chalcedony or chert ($SiO_2 \cdot nH_2O$). Chemical sedimentary rocks include many types of **limestone, dolostone,** rocks formed by evaporation of seawater (such as **rock salt** and **gypsum**), and rocks formed from hot spring deposits (such as **travertine** or **sinter,** found around the geysers of Yellowstone National Park).

The crystalline texture seen in many chemical sedimentary rocks may have formed by recrystallization of sediment grains during or after the process of lithification. For example, the tiny skeletons of certain microorganisms like diatoms and radiolaria are composed of opaline silica. Their skeletons are major components of the siliceous ooze that covers many areas of the deep sea floor. During burial and lithification, the silica may recrystallize as a very fine-grained crystalline sedimentary rock called **chert.**

The shells of marine fossils are often abundant in chemical sedimentary rocks like limestone. If present in your study specimens, include a description of them in your identification notes.

The process of chemical precipitation can also produce sedimentary particles called **oöids** (Fig. 1.1C, D). They form in warm, shallow seas by precipitation of concentric layers of calcium carbonate to form tiny spheres. The individual oöids may then be cemented together to produce a limestone having **oölitic texture.** Before being incorporated into a limestone, oöids are really clastic particles. Hence an **oölitic limestone** is both clastic and crystalline.

Bioclastic Sedimentary Rocks

Bioclastic rocks are composed mainly of the shells or fragments of shells that have been transported and deposited much as if they were grains of sand or pebbles (Fig. 1.1E). The fragments of organic origin are termed **bioclasts.** They may consist of a variety of materials composed of shell, bone, fecal pellets, or plant debris.

The presence of fossils in a sedimentary rock often confuses students when they attempt to classify the rock as clastic, chemical, or bioclastic. A "rule of thumb" is that if the rock is composed almost entirely of fossil materials that are loosely held together, you should call it a bioclastic rock (Fig. 1.1F, coquina). If fossils are scattered throughout an otherwise chemical rock, it is best to simply designate the rock a **fossiliferous limestone.**

Table 1.1 will be helpful as you learn to identify the common sedimentary rocks in your demonstration and study sets. The table divides sedimentary rocks into three categories according to their overall major texture. These are *clastic,* mainly composed of fragments of minerals or *lithic* (meaning rock) fragments; *crystalline,* formed mainly by chemical or biochemical precipitation; and *bioclastic,* composed of the transported and deposited durable remains of organisms.

Figure 1.4 Thin section of sandstone composed of poorly sorted, angular grains of quartz (clear), feldspars (thin stripes), and rock fragments. Note that this sandstone lacks cement. Spaces between grains are filled with a matrix of clay and silt, which holds the grains together.

From Levin, H. L., *The Earth Through Time,* 5th ed. Philadelphia: Saunders College Publishing, 1996.

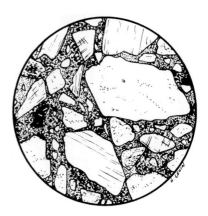

MATRIX AND CEMENTS

When a sedimentary rock is formed, the unconsolidated sediment can be bound together by compaction or cementation. Compaction occurs as the sediment is buried under the weight of overlying sediment. It causes grains to pack closely together, reducing the size of the pore spaces between grains. The finer grained detrital material, such as clay and fine silt, in which the larger grains are embedded is called **matrix** (Fig. 1.4). It serves effectively as the binding material in compacted sediments.

Cementation occurs when minerals such as calcite ($CaCO_3$), quartz (SiO_2), hematite (Fe_2O_3), or limonite ($FeO \cdot OH \cdot nH_2O$) are precipitated in the pore spaces between particles of sediment (Fig. 1.5). These cements bind the particles together (generally more strongly than a rock having only matrix). The strongest of the above cements is quartz.

In a hand sample, the kind of cement may not be easily recognized. To determine if the cement is calcium carbonate, geologists often use the "acid test." A drop of *dilute* (10%) hydrochloric acid (HCl) will cause the calcium carbonate to effervesce (form bubbles). (The acid test can also be used to distinguish calcite from dolomite. The former mineral will effervesce readily when dilute HCl is applied to its surface. Dolomite normally will not effervesce unless scratched to produce a powdery surface).

HARDNESS

The hardness of sedimentary rocks is largely determined by their composition. Limestone, for example, is composed of the mineral calcite and thus is relatively soft. A sandstone composed of quartz grains is relatively hard, especially if it has a silica cement.

Table 1.1A Key to Identification of Some Common Sedimentary Rocks

Texture	General Appearance	Diagnostic Features	Sedimentary Rock Name	
C L A S T I C	Boulders, cobbles, pebbles or coarse (2–4 mm size) particles embedded in a matrix of sand grains.	Angular rock/mineral fragments.	Breccia	
		Rounded rock/mineral fragments.	Conglomerate	
	Coarse sand and rock/mineral fragments.	Angular feldspar fragments mixed with coarse sand. Color: pink, reddish-brown, buff.	S A N D S T O N E	Arkose
	Sand-size particles.	With <25% rock/mineral fragments and sand-size matrix.		Coarse sandstone
		With few to no rock/mineral fragments - mainly quartz grains. Color: buff, white, pink, brown.		Sandstone
	Coarse to fine sand-size particles with clay-size matrix.	Fine to coarse, angular to subangular rock fragments, poorly sorted. Color: dark gray to gray-green.		Graywacke
	Fine-grained clay and silt-sized particles.	Show fissile nature, soft enough to be scratched with fingernail. Color: varies - mostly dark colored.	Shale	
		Not fissile, look like hardened mud, silt or clay. Color: varies - dark to light colored.	Siltstone, mudstone or claystone (Siltstones are grittier than mudstone, claystones are very smooth).	
	Very fine-grained silt-sized particles.	Have a silty feel between fingers, yellowish appearance, softer than a fingernail but some particles will scratch glass.	Loess	
C R Y S T A L L I N E	Very fine-grained, interlocked crystals. Grains are uniform in size.	Effervesce strongly with dilute HCl. Hardness greater than fingernail, will not scratch glass.	Limestone	
		Sample effervesces only when the rock is powdered. Hardness greater than fingernail, will not scratch glass.	Dolostone	
		Scratches glass, conchoidal fracture (looks like a broken glass bottle). Does not effervesce with dilute HCl. Color: white to gray to black (black variety called flint).	Chert	
		Same hardness as a fingernail, salty taste, greasy to waxy luster. Color: white to gray.	Rock salt	
		Hardness greater than fingernail, less than calcite. Crystals often platy to tabular in appearance. Color: varies - usually pink, buff, white.	Rock gypsum	
	Grains uniform in size and very spherical. Grains appear to have crystalline material or cement holding them together.	Spherical grains effervesce readily in dilute HCl. Color: very light (white to cream).	Oölitic limestone	
	Fossils of various sizes are present, yet most of the rock is either fine-grained matrix or cement.	Matrix/cement effervesce readily in dilute HCl.	Fossiliferous limestone	
		Matrix/cement requires powdering prior to weakly reacting to dilute HCl.	Fossiliferous dolostone	
B I O C L A S T I C	Composed mainly (>90%) of fragments of fossils (invertebrate skeletal remains, shells or other hard parts of organisms).	Whole or nearly whole shells, show abrasion on surfaces, weakly held together by matrix.	Coquina	
		Effervesce with dilute HCl, powder easily rubbed off with fingers. Minor fossils may be present. Color: white to off-white to buff.	Chalk	
		Soft, crumbles easily, but particles scratch glass. Does not react with dilute HCl. Color: gray-white (composed of microscopic siliceous algal remains).	Diatomite	
	Composed mainly of plant material.	Fibrous, brown plant fibers, soft, very porous.	Peat	
		Sooty feel, may contain wood fragments or plant impressions, from dull, dark brown to a shiny black in appearance.	Coal (Lignite - brown; Bituminous - sooty black, blocky fracture; Anthracite shiny, dense metallic black, conchoidal fracture)	

Table 1.1B Sedimentary Rock Identification Form: Use in Conjunction with Table 1.1

Sample Number	Texture	Grain/Particle Size (from Table 2.1)	Composition	Other Distinguishing Features	Rock Name

Student Name _____ Class/Section Number _____

Figure 1.5 Two common types of cement in sandstones. (A) Quartz sandstone composed of well-sorted, rounded quartz grains tightly cemented by quartz (SiO_2) overgrowths. (B) Sandstone composed of quartz, feldspar, and rock fragments cemented by coarse, sparry calcite. Both are drawings of thin sections as viewed under the microscope. Diameter of areas each 1.0 mm.

From Levin, H. L., *The Earth Through Time*, 5th ed. Philadelphia: Saunders College Publishing, 1996.

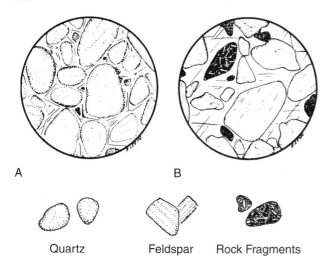

A   B

Quartz   Feldspar   Rock Fragments

When used along with other observable physical properties, a simple hardness test will help you identify sedimentary rocks and their constituents. It is especially useful in distinguishing limestone and dolostone from sandstones.

A scale of mineral hardness, based on common minerals, was developed by Frederich Mohs in 1822. The *Mohs hardness scale* (Table 1.2) tests the ability of one mineral to scratch another. On the Mohs hardness scale, quartz, which is almost always present in sandstone, has a hardness of 7. Quartz will scratch glass or a steel nail, indicating these objects are softer than 7 on the Mohs scale. Limestone, which is composed of calcite and has a hardness of 3, is much too soft to scratch these objects.

Chalk, as well as rocks composed of gypsum or fine-grained and/or loosely consolidated sediment (some shale, mudstone, claystone), are soft and can often be scratched by

Table 1.2 Mohs Scale of Hardness

Mineral	Scale #	Common Object
Talc _____	1	(SOFTEST)
Gypsum _____	2	Fingernail
Calcite _____	3	Copper wire/penny
Fluorite _____	4	
Apatite _____	5	Pocket knife
Orthoclase _____	6	Window glass
Quartz _____	7	
Topaz _____	8	
Corundum _____	9	
Diamond _____	10	(HARDEST)

your fingernail. Loess is easily scratched, but slide it forcefully across a glass plate and the tiny constituent quartz grains may produce scratches.

A word of caution. If you are testing for hardness by scratching a rock against glass, be sure the glass is placed flat on a table and the rock held firmly against it. More than one edge or corner of the rock must be tested because some rocks that are otherwise soft contain a few scattered quartz grains. If the rock is really scratching the glass plate, you will feel it "bite" into the glass surface.

COLOR

The color of sedimentary rocks can provide information about the environment in which the rock was deposited. The most common colors are black (signifying the presence of carbon) and shades of brown, red, or pale green (signifying the presence of iron compounds. In red and brown rocks, iron oxide (Fe_2O_3) is the common pigment.

The oxidizing conditions required for the development of ferric compounds are more typical of continental than marine environments. Thus, red and brown colored rocks are commonly derived from floodplain, alluvial fan, or deltaic deposits. Sometimes brown or red sediment that was formed on land is carried into the sea, and the original color is retained.

The high carbon content of black or dark gray sedimentary rocks indicates an abundance of organic matter in the depositional environment and a deficiency of oxygen so that the organic matter is not destroyed and lost by oxidation. These conditions occur in areas of the sea with poor circulation and in many lakes, swamps, estuaries, and tidal flats.

For convenience, you can use the following terms to describe the color of the sedimentary rocks in your study set.

1. For rocks in the white to black color range:

   white — light gray — medium gray — dark gray — black

2. For rocks having a red shade or hue:

   light grayish pink — medium red — dark red

3. For rocks having a brown or orange hue:

   light grayish orange — medium brown — dark brown

4. For rocks having a yellow hue:

   light grayish yellow — olive gray — dark olive gray

5. For rocks having a green hue:

   light grayish green — greenish gray — dark greenish gray

BEDDING

Bedding (or *stratification*) is the most obvious feature of sedimentary rocks when you encounter them in the field. Bedding results from changes in texture, composition, color, or rock type from one bed to the next. It is usually more obvious on a weathered outcrop than on a fresh one, such as a recently excavated road cut.

Although stratification can occur in some igneous rocks (e.g., lava flows), it is particularly characteristic of sedimentary rocks. How do you account for this?

_____

_____

Where individual beds or layers are thicker than 1 cm, the bedding is not likely to be seen in a hand sample. However, beds less than 1 cm in thickness, called **laminae,** can be observed in some hand samples.

Thin bedding should not be confused with **fissility,** a property of shales that permits one to separate the rock into thin, flat pieces. Fissility is mainly caused by the parallel alignment of platy clay minerals or flat, platy minerals like mica.

Exercise

With the information provided in Table 1.1A, along with the study set of demonstration minerals, sedimentary rocks, and materials for testing hardness, examine the unknown samples provided by your instructor and determine their texture and identity. Use the Sedimentary Rock Identification Form that accompanies Table 1.1 to record your observations.

TERMS

arkose   A feldspar-rich sandstone, typically coarse-grained and pink or reddish in color. It is composed of angular to subangular grains, either poorly or moderately sorted. Arkoses generally reflect a terrestrial depositional environment where an uplifted granitic rock body undergoes rapid erosion, transportation deposition, and burial.

bed   One of the layers of rock in a stratified sequence of rocks, having well-defined boundaries with the overlying and underlying layers. A bed is generally greater than 1 cm in thickness.

bioclastic texture   A texture composed mainly of organic remains that are weakly cemented together. The bioclasts may range from shell fragments (see **coquina**) to plant fragments (see **peat** or **coal**).

bioclasts   Fragmental organic remains usually consisting of shell or skeletal material of marine invertebrates or calcareous algae.

breccia   A coarse-grained clastic rock, composed of angular broken rock and mineral fragments held together either by a mineral cement or in a fine-grained matrix.

cementation   The precipitation of minerals (such as calcite, silica, or iron oxides) in pore spaces between particles of sediment. Cement is added to a sediment *after* deposition.

chert   A hard, dense sedimentary rock composed of microcrystalline silica (silicon dioxide; SiO_2) and characterized by conchoidal fracture.

clast   An individual grain of a detrital sedimentary rock or sediment produced by the disintegration of a larger rock mass through the processes of erosion and weathering.

clastic texture   A texture consisting of the broken fragments of rock, minerals, or skeletal remains held together by either a cement or a matrix. The individual clasts are described based on identity, grain size, degree of rounding, and sorting.

coal   A rock formed from the compaction and induration of variously altered plant remains similar to those in peat. Formed as a result of the compaction of peat over long periods of time.

compaction   Reduction in the pore spaces between sediment grains as a result of burial or overlying pressure. In compaction, the sediment grains are bound together by finer-grained sediment particles, called the **matrix.**

compositional maturity   A measure of the amount of weathering, erosion, and transport experienced by a sediment as indicated by the variety of mineral constituents ultimately deposited to form the sedimentary rock. Examples: Feldspars weathered to clay; ferromagnesian minerals (olivine or pyroxene) weathered into iron oxides such as the minerals limonite or hematite.

conglomerate   A coarse-grained clastic sedimentary rock, composed of rounded to subangular rock and/or mineral fragments greater than 4 mm in diameter. These components are either set together in a fine-grained matrix or held together by a mineral cement.

coquina   A bioclastic sedimentary rock composed largely of shells of marine invertebrates, usually mollusks, that are held weakly together with little to no matrix present and have high porosity.

crystalline texture   A term that describes a sedimentary rock composed of crystals (rather than clasts) precipitated from a saturated solution. The mineral crystals are fine to very fine in grain size and form an interlocking mosaic. Rocks formed by chemical precipitation are called **chemical sedimentary rocks.**

detrital   A term applied to *any* particles of minerals or rocks (clasts) that have been derived from preexisting rocks by processes of weathering or erosion.

fissile   A property of some sedimentary rocks that separate into thin, flat layers, usually along bedding planes.

graywacke   A general term for a dark gray, firmly indurated, coarse-grained sandstone that consists of poorly-sorted, angular to subangular grains of quartz and feldspar along with a variety of dark-colored rock fragments embedded in a fine-grained matrix.

immature sand   A poorly sorted sand that contains abundant, relatively unstable (to weathering), often angular grains. An immature sandstone would be similarly composed.

lamina   A thin layer of sediment or sedimentary rock in which the planes of stratification are 1 cm or less apart.

lithification   The process (or processes) that convert(s) unconsolidated rock-forming materials into a coherent rock mass.

loess   A soft, crumbly clastic sediment formed from accumulations of wind-blown silt.

matrix   Clastic, fine-grained particles (often clay) that are deposited at the same time as larger grains and help to hold (or bind) the grains together. Also called the *groundmass.*

mature sand   A well-sorted sand consisting primarily of subrounded to rounded grains of very stable minerals, usually quartz. Most quartz sandstones are derived from mature sands.

oöids   Spherical particles of sand size that are mostly composed of concentric laminae of calcium carbonate.

A limestone made up of cemented oöids has an **oölitic texture.**

peat   An unconsolidated deposit of semicarbonized plant remains in a water-saturated environment, such as a bog or fen. It is considered an early stage in the development of coal.

pore space   The open spaces (or voids) between sediment grains.

precipitation   Process whereby materials that are carried in solution (dissolved) are deposited as a crystalline solid.

roundness (of sedimentary particles)   The degree to which sharp corners or edges of a particle are worn away. Roundness is commonly expressed as the ratio of the average radius of the corners to the radius of the maximum inscribed circle for the particle (see Fig. 2.3).

shale   A fine-grained clastic (detrital) sedimentary rock, formed by the compaction and consolidation of clay, silt, or mud that is characterized by finely laminated structures (very thin layers).

sinter   A chemical sedimentary rock formed from the precipitation of calcite- (or silica-) saturated hot water from hot springs or geysers. Calcareous sinter is also called *travertine.* Siliceous sinter is often only called *sinter.*

sorting   A measure of the uniformity of particle sizes in a sediment or a sedimentary rock.

subgraywacke   A "dirty" sandstone with more quartz and less feldspar than a graywacke and more rock fragments. The quartz and feldspar grains are more rounded than those in a graywacke, indicating more textural maturity.

textural maturity   A measure of size and shape variation of the constituents of a sedimentary rock. Increased textural maturity is characterized by greater rounding of the grains and decreased grainsize variation.

2

Textural Clues to the History of Sediment

1. Demonstration specimens of sediment grains showing various degrees of rounding and angularity.

2. Tray containing eleven or more pebbles ranging in size (diameter) from 8 to 64 mm.

3. Vials of coarse sand and fine gravel ranging in size from 1 to 8 mm.

4. Millimeter scale, graph paper, and Table 2.1 in this chapter.

5. 10X magnification hand lens or binocular microscope.

6. Hand calculator.

PARTICLE SIZE AND SORTING

The texture of a clastic sedimentary rock such as a sandstone can often be used to infer the manner in which sediment was transported and deposited. We are all aware that a stronger current of water is required to move large pebbles than grains of sand. Thus the grain size of the particles in a clastic rock serve as an indicator of the approximate velocity and density of the transporting medium (such as running water or wind). Water in a swiftly flowing stream can transport much larger grains than desert winds. Winds may have greater velocity than water in a stream, but they lack the density of water.

Information about conditions of deposition can also be obtained from the distribution or sorting of grain sizes in a clastic rock. **Sorting** is a measure of the uniformity or lack of uniformity of particles in a sediment. It is more specifically defined as the range of particle sizes that deviate from the average size. Poorly sorted sediments have a wide range of particle sizes as in the sandstone depicted in Figure 2.1A. Rocks composed of grains that are all about the same size are said to be well sorted (Fig. 2.1B).

Water or wind currents may **winnow** out finer particles in an originally poorly sorted sediment, leaving behind a better sorted sediment composed mostly of larger grains. The fine sediment removed by winnowing may then be transported to another location and deposited as a fine-grained sediment like siltstone or shale.

In general, poor sorting is characteristic of sediment that has been rapidly deposited under conditions that prevent

the sorting action of waves and currents. Such conditions exist at the foot of mountain ranges or along the margins of glaciers.

To provide uniformity in the use of terms that describe the size of sedimentary particles, geologists use a table of particle sizes called the Wentworth grain-size scale (Table 2.1). As you examine Wentworth's scale, you will see that the size

Figure 2.1 Two sandstones as seen in thin section under the microscope. (A) is a poorly sorted, immature sandstone composed of angular to subangular grains of quartz, feldspar, and rock in a matrix of clay and micaceous minerals. (B) is a well-sorted, mature quartz sandstone with chemical cement.

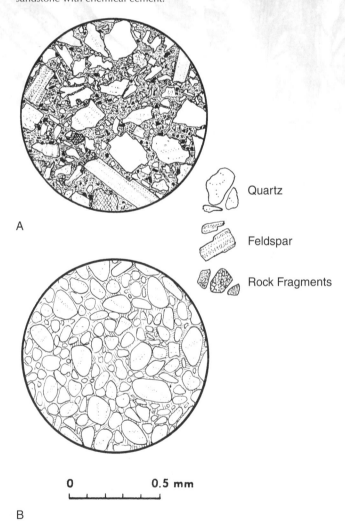

Table 2.1 The Wentworth and Phi (Ø) Grain-Size Classification for Sediment Grains

Size (m)	Class Boundary (mm)	Size Classes		Phi (Φ) Units
	2048	Gravel — Boulders	very large	−11
	1024		large	−10
1	512		medium	−9
	256		small	−8
	128	Cobbles	large	−7
10⁻¹	64		small	−6
	32	Pebbles	very coarse	−5
	16		coarse	−4
10⁻²	8		medium	−3
	4		fine	−2
	2		very fine	−1
	1	Sand	very coarse	0
10⁻³	500µm		coarse	1
	250µm		medium	2
	125µm		fine	3
10⁻⁴	63µm		very fine	4
	31µm	Mud — Silt		5
	16µm			5
10⁻⁵	8µm			6
	4µm			7
	2µm	Clay		8
10⁻⁶				9

groups increase geometrically by a factor of two. This is desirable because a given arithmetic increment of increase, for example one millimeter, makes no difference in the behavior of big grains but makes a large difference in the behavior of small grains. To illustrate, if a sediment composed of many grain sizes is poured into a water-filled glass cylinder, grains having a diameter of 2 mm fall about twice as fast as those having a diameter of 1 mm, but a difference of 1 mm makes no appreciable difference in the settling velocities of pebbles with diameters from 50 to 60 mm.

The millimeter-based Wentworth grain-size scale can be transformed into a logarithmic scale, which can then be plotted on ordinary graph paper. This technique evolved into the **Phi (Ø) grain-size scale** (Table 2.1). In this scale, $Ø = -\log_2 (D)$, with $Ø$ as the Phi number and D as the particle diameter in millimeters. Phi units are helpful in taking into account the large range of grain diameters that exist in clastic rocks. As is evident from Table 2.1, the more negative the Phi number, the coarser grained the sediment, and the more positive the Phi number, the finer grained the sediment.

GRAIN SHAPE

In the previous chapter we learned that the roundness of clastic grains can provide information about the history of sediment before it became sedimentary rock. The angular grains in the sandstone shown in Figure 2.1A have probably not been subjected to substantial wear resulting from extensive transportation and reworking. The sandstone in Figure 2.1B, however, is more likely to have experienced long episodes of transport and wear by waves, water currents, or wind. For example, pebbles in the downstream portion of streams like the Colorado River are more rounded than their upstream counterparts because of the larger number of impacts they have been subjected to over the longer span of transport in the swirling waters of the river.

Two terms useful in describing the shape of clastic particles are roundness and sphericity. **Roundness** refers to the degree to which sharp corners and edges of rock or mineral have been worn away. The term should not be confused with **sphericity,** which is a measure of how closely a particle approaches the shape of a sphere.

Roundness may be expressed either as an approximation based on comparison with grains of known roundness (as in Figure 2.2), or it may be measured and expressed quantitatively as *the average radius of the corners of a grain (or pebble) divided by the radius of the maximum inscribed circle* (Fig. 2.3).

1. In *qualitative* terms (using Figure 2.2) what is the *roundness* of the pebbles drawn enlarged in Figure 2.3?

    Pebble A _____

    Pebble B _____

2. Determine the numerical value of roundness of pebbles A and B in Figure 2.3. Assume all edges and corners are shown. Measure the radii of edges (r_1, r_2, r_3, etc.) and the radius of the maximum inscribed circle with a millimeter scale.

    Pebble A _____

    Pebble B _____

Figure 2.2 Drawings of sand and gravel particles for estimating approximate roundness.

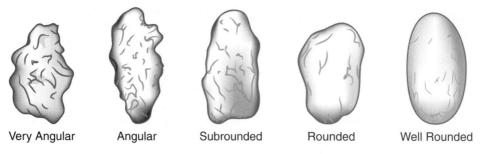

Very Angular   Angular   Subrounded   Rounded   Well Rounded

Figure 2.3 Grain profiles to be used in calculating the numerical value of roundness.

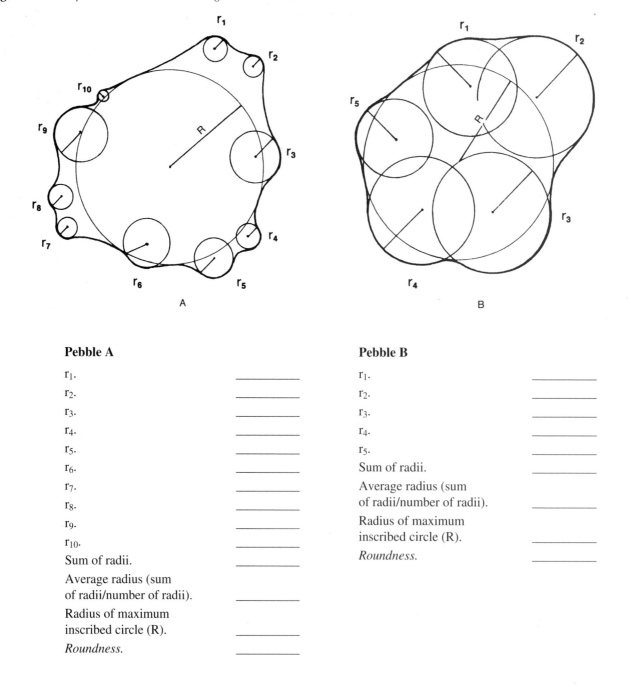

A

B

Pebble A

r_1.	_____
r_2.	_____
r_3.	_____
r_4.	_____
r_5.	_____
r_6.	_____
r_7.	_____
r_8.	_____
r_9.	_____
r_{10}.	_____
Sum of radii.	_____
Average radius (sum of radii/number of radii).	_____
Radius of maximum inscribed circle (R).	_____
Roundness.	_____

Pebble B

r_1.	_____
r_2.	_____
r_3.	_____
r_4.	_____
r_5.	_____
Sum of radii.	_____
Average radius (sum of radii/number of radii).	_____
Radius of maximum inscribed circle (R).	_____
Roundness.	_____

SIZE DISTRIBUTION OF COARSE GRAVEL

Most natural particles are not perfectly spherical. Thus, a sandstone or a pebble will have more than a single diameter. Actually, three diameters can be measured as shown in Figure 2.4. When geologists refer to the diameter of a grain, they are usually referring to the **b** (intermediate) diameter. Note that a pebble ordinarily rests on a surface such that its **c** (shortest) diameter is in a vertical orientation. When viewed from above, the shorter of the two visible diameters is usually the **b** (intermediate) diameter.

Pebble Size Analysis

Arrange the pebbles from your study set in a row according to increasing intermediate *(b)* diameters. You may make your measurements by placing each pebble on the millimeter-scale graph paper (Fig. 2.4). Record these measurements on the table accompanying Figure 2.4 in order from the smallest to the largest intermediate *(b)* diameter. Complete the calculations and record the results in the appropriate blanks.

1. What is the **median** of the diameters? (The median is the middle term in a series of items when they are arranged in order of magnitude. In this analysis, it is the diameter of the pebble in the middle of the row.)

    _____

    _____

2. What is the arithmetic **mean** of the pebble diameters? (An arithmetic mean is the sum of the diameters of the pebbles divided by the total number of pebbles.)

    _____

    _____

3. What is the modal diameter of the pebbles? (A **mode** is the item in a group that occurs most often. If no two pebbles in the group have the same diameters, then there is no mode. With more samples, the probability of two or more pebbles having the same diameter increases.)

    _____

    _____

Questions 1, 2, and 3 refer to the "average," or the *central tendency,* of the pebble population. Another set of values is needed to express the "spread" or the *dispersion,* of the population. These mathematical values are called the **range,** the **mean deviation,** and the **standard deviation** and allow the investigator to gauge whether a group of objects are similar *or* dissimilar to one another. For example, if you had a group of pebbles with a small range of sizes and of the same type of rock, you would suggest that they all were from the same source. However, if you had a group of pebbles that were not the same type of rock but did have a small range in sizes, could you make the same

interpretation? The statistical concepts of mean deviation and standard deviation are techniques that will allow you to say if the grains are similar or dissimilar to one another in this exercise.

4. What is the **range** of the pebble diameters? (The range is the numerical difference between the largest and smallest diameters in the sample population.)

    _____

    _____

5. What is the **mean deviation** of the pebble diameters? The mean deviation (MD) is obtained by first subtracting each pebble diameter from the arithmetic mean for all of the pebbles in the study. Some of these differences will be negative values, so you will use the absolute value of the number. Add the absolute values of these differences and then divide this total by the total number of pebbles in the sample population. For example, if X_{mean} is the arithmetic mean, N is the total number of pebbles, and each individual pebble diameter is X_{pebble}, the mean deviation is as follows:

    $$MD = \frac{\Sigma \left| X_{pebble} - X_{mean} \right|}{N}$$

    _____

    _____

    _____

6. What is the **standard deviation** of the pebble diameters?

    The standard deviation (SD) is obtained by squaring the mean deviation computed above and adding them together. This sum is then divided by the total number of pebbles and the square root is taken of the resulting number. For example, if the sum of the squares of the individual deviations is 1600 mm^2 for a population of eleven pebbles, the standard deviation is this:

    $$SD = \sqrt{\frac{\Sigma \left(\left| X_{pebble} - X_{mean} \right| \right)^2}{N}} = \sqrt{\frac{1600}{11}} \text{ or } 12.06 \text{ mm}$$

    _____

    _____

    _____

7. Assume the largest pebble in the population is replaced by a pebble twice as large.

    a. What would be the new value of the median?

    _____

    _____

    b. What would be the new value of the mode?

    _____

    _____

Figure 2.4 The diameters of a nonspherical particle.

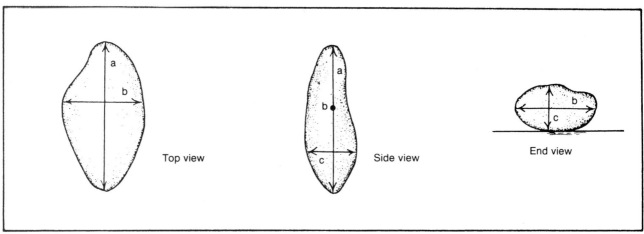

Intermediate Diameter in mm	Roundness: A, SA, SR, or R	Deviation from the Mean $[X_{pebble} - X_{mean}]$	Square of the Deviation from the Mean
_____	_____	_____	_____
_____	_____	_____	_____
_____	_____	_____	_____
_____	_____	_____	_____
_____	_____	_____	_____
_____	_____	_____	_____
_____	_____	_____	_____
_____	_____	_____	_____
_____	_____	_____	_____
_____	_____	_____	_____
Totals _____	_____	_____	_____

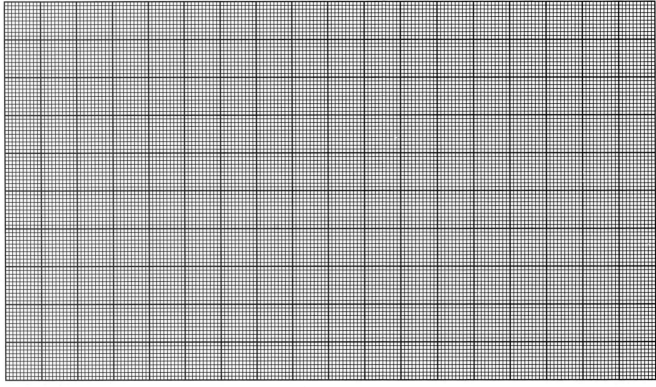

c. Which value would be influenced the most, the mean deviation or the standard deviation?

_____

_____

8. Which would be the best sorted, a gravel with a standard deviation of 30 mm or a gravel with a standard deviation of 15 mm?

_____

_____

Size Analysis of Coarse Sand and Fine Gravel

The size distribution of sand is commonly measured by using graduated-size meshes called **sieves,** but a few dozen grains may be measured for this study by spreading the grains on millimeter-scale grid paper. The grains used here have been sieved for greater convenience in the study such that their maximum diameters lie between 8 mm and 1 mm. Because of the variability in the wire mesh used in sieves, a sieve measures the intermediate diameter rather than the maximum diameter of grains. Thus, some of the study grains will be a little larger than 8 mm and some will be a little smaller than 1 mm.

Spread the grains from the vial onto the millimeter-scale paper (Fig. 2.4) and separate them into the size groups listed below. Count the grains in each group and record the count for each group in the second column of the table below. You will find it easier to separate and move the small grains by using either a toothpick or the sharpened end of your pencil.

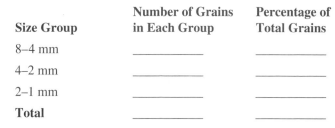

Size Group	Number of Grains in Each Group	Percentage of Total Grains
8–4 mm	_____	_____
4–2 mm	_____	_____
2–1 mm	_____	_____
Total	_____	_____

The determination of the median and mean diameters of a large population would be tedious and not very feasible by the method discussed above. However, the modal diameter may be determined approximately by construction of a **histogram** and **frequency curve** on graph paper as shown in Figure 2.5. Such a frequency curve is similar to the grading curve constructed by some teachers who follow a procedure known to students as "grading on the curve." The size distribution of natural sediments commonly describes, when plotted on a geometric scale as is used in Figure 2.5, a rather symmetrical bell-shaped curve. The sand in your vial *may not* plot in this manner, as it has been assembled from many different sources so as to illustrate variation in roundness and composition. Histograms, as in Figure 2.5, should be plotted so that the *area* enclosed by each bar represents the approximate percentage. The frequency curve, shown by the dotted line, is drawn from the midpoints of the histogram bars.

Study the composition of the grains with a hand lens while they are spread out on a sheet of paper. Most of the grains that are clear, transparent, or translucent are probably quartz. Feldspar grains are commonly opaque, pink, or gray and may show cleavage faces. Shiny spherical white grains are **oöids** of calcium carbonate ($CaCO_3$), and rounded or

Figure 2.5 Grids for the construction of histogram and frequency curves.

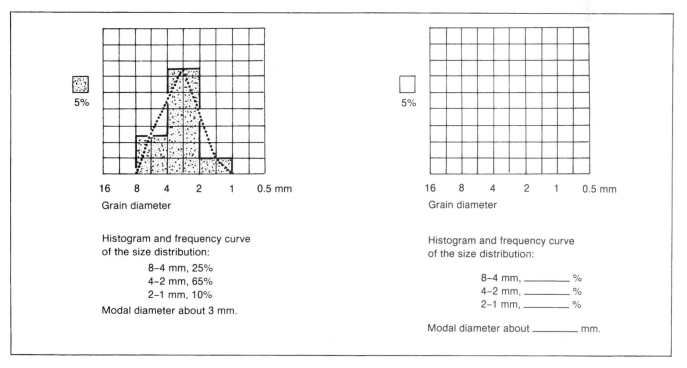

Histogram and frequency curve
of the size distribution:
    8–4 mm, 25%
    4–2 mm, 65%
    2–1 mm, 10%
Modal diameter about 3 mm.

Histogram and frequency curve
of the size distribution:
    8–4 mm, _____ %
    4–2 mm, _____ %
    2–1 mm, _____ %

Modal diameter about _____ mm.

angular shell fragments may also be present. However, many grains are not easily recognized by means of a hand lens.

1. Of the grains you can identify, what is the approximate percentage of quartz?

_____

_____

2. What percentages of the grains are either rounded or subrounded (see Fig. 2.2)?

_____

_____

_____

Small grains require a great deal more transport (by either water or the wind) to become well rounded than do larger pebbles. This is because the impacts that cause rounding are more efficiently cushioned by the thin film of water that surrounds the sand grains. If the transport of sand grains is continued for a long period of time, the less durable grains of feldspar and carbonate will eventually be mechanically destroyed or chemically dissolved. The residue will consist of the more durable quartz and **chert** grains.

QUESTIONS FOR DISCUSSION

1. How would you expect a sand derived directly from the weathering of igneous rocks in a mountainous region to differ in composition and roundness from a sand derived primarily from the weathering of preexisting sedimentary rocks?

_____

_____

_____

_____

2. How might one account for the presence of both angular and rounded grains within a given quartz sandstone stratum?

_____

_____

_____

3. What type of sandstone is characterized by poor sorting and angularity of grains (refer to Table 1.1)?

_____

_____

_____

4. Imagine that you have been given the opportunity to purchase only one of two pieces of land known to be underlain by a sandstone that contains sporadic accumulations of petroleum. Test drilling indicates that beneath land parcel A the mean particle diameter in the sandstone is 0.125 mm, while in parcel B the mean particle diameter is 0.250 mm. Which parcel of land has the greater probability for the discovery of oil? Why?

_____

_____

_____

_____

5. Why did Wentworth base his grain-size scale on a geometric increase in size rather than a simple arithmetic increase? For what reasons are the Phi (Ø) grain-size scale used in addition to the Wentworth grain-size scale?

_____

_____

_____

_____

_____

_____

6. Why are grains of windblown sand, such as those found in desert or barrier island dunes, usually well sorted?

_____

_____

_____

_____

7. How do you explain the observation that finer-grained sediments are often deposited farther away from a coastline than coarser-grained sediments?

_____

_____

_____

TERMS

chert   A hard, dense sedimentary rock composed of microcrystalline silica (silicon dioxide; SiO_2) and characterized by conchoidal fracture.

clast   An individual grain of a detrital sedimentary rock or sediment produced by the disintegration of a larger rock mass through the processes of erosion and weathering.

clastic   Pertaining to a rock or sediment composed of broken fragments of preexisting rocks, minerals, or skeletal matter (called **clasts**). The individual clasts are described based on identity, grain size, degree of rounding, and sorting.

histogram   A graph depicting the relative abundance of items, in which equal intervals of values are marked on a horizontal axis and the **frequency** corresponding to each interval is indicated by the height of a bar having the interval at its base.

mean (arithmetic)   The arithmetic average of a series of values. An arithmetic mean is calculated by adding all of the items and then dividing that sum by the total number of items.

mean deviation   For a sample of N measurements, the mean deviation is the sum of the absolute values of the differences from the **arithmetic mean** divided by the number of measurements (N).

median   The value of the middle term in a set of data that has been arranged in order of magnitude or rank.

mode   The value or group of values that occur most frequently in a set of data. For example, the diameter that is the most frequent in a particle size distribution.

oöids   Spherical particles of sand size that are mostly composed of concentric laminae of calcium carbonate. A limestone made up of cemented oöids has an **oölitic texture.**

Phi (Ø) grain-size scale   A logarithmic transformation of the Wentworth grain-size scale, where the negative logarithm to the base two of the particle diameter (in mm) is substituted for the diameter value classification. The Phi grain-size scale was developed to permit the direct application of conventional statistical practices to sedimentary data.

range   The numerical difference between the largest and smallest values in a given series of numbers or measurements.

roundness (of sedimentary particles)   The degree to which sharp corners or edges of a particle are worn away. Roundness is commonly expressed as the ratio of the average radius of the corners to the radius of the maximum inscribed circle for the particle.

sieve   A device used to separate sediment (or soil) into its component sizes. Sieves are usually brass cylinders that have wire or cloth mesh (of a particular mesh size) stretched across the base.

sorting   A measure of the uniformity of particle sizes in a sediment or a sedimentary rock. Sorting is often expressed as the standard deviation (spread) in particle sizes on either side of a mean particle size for the population.

sphericity   A measure of the shape of a particle and also of the deviation of its shape from an equivalent sphere.

standard deviation   A measure of the dispersion of data for a population from a mean value. Also called the *standard error*.

winnow   To separate free particles from coarser ones by the action of wind or water currents.

3

Sedimentary Rocks Under the Microscope

1. Hand specimens of carbonate rocks and well cemented, non-porous sandstones.

2. Rock thin sections prepared from each of the hand specimens (Fig. 3.1).

3. Petrographic microscope or a non-polarizing microscope fitted with pieces of polarizing film as shown in Figure 3.2. (Polarizing film can be purchased from Edmund Scientific, 60 Pearce Ave., Tonawanda, N.Y., www.scientificsonline.com.)

1. Dropper bottle, 2-ounce, filled with acetone.

2. Rectangular pieces of the above hand specimens cut into a shape similar to Figure 3. 1B and polished on at least one surface.

3. Squares of cellulose acetate film, 0.0015 or 0.002 inch thickness, cut somewhat larger than the prepared rock surface.

4. Small wooden or cardboard _tray_ containing a few facial tissues.

5. Paper towels.

6. Two glass plates hinged with tape along one side. Cover glasses from 35 mm slide binders are good for this purpose.

BASIC INFORMATION

Thin Sections

Because of small grain size, it is often difficult to identify the mineral components of sedimentary rocks. Identification is easier when a _thin section_ of the rock is examined with the aid of a microscope. A rock thin section (Fig. 3.1) is prepared by first cutting a thin slab of rock with a diamond saw. The small slab of rock is then cemented to a glass slide and ground with abrasive powders to a uniform thickness of about 0.03 mm (30 microns). At this thickness, most minerals are nearly transparent to visible light. A glass cover slip is cemented over the thin slice of rock, and the completed thin section is placed under the microscope for viewing.

The mineral grains in a sedimentary rock thin section can be seen more distinctly when viewed between crossed polarizers (Fig. 3.2). Most minerals are crystalline substances that transmit light between crossed polarizers because each mineral grain has a different orientation and transmits light to

Figure 3.1 Preparation of a thin section involves first cutting a thin chip of rock from the hand specimen (A and B) with the use of a rock saw. The sawed chip (B) is then polished on one side and mounted with the polished side down on a glass slide (C). Canada balsam is usually used as a mounting medium. The chip is then ground (D) to a thickness of about 30 microns (0.03 mm or 0.001 inch). After all grinding debris is washed away, a thin cover glass is placed upon the completed thin section (E).

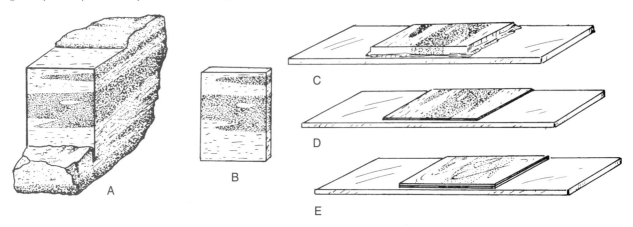

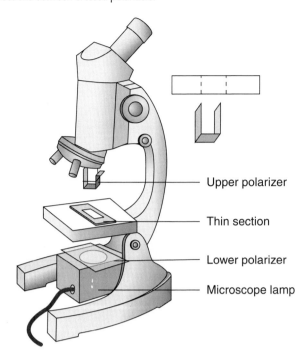

Figure 3.2 Wide field microscope prepared for viewing rock thin sections between crossed polarizers.

Upper polarizer

Thin section

Lower polarizer

Microscope lamp

a different degree. Quartz and feldspar grains—the common constituents of sandstone—range in color from light gray to black when viewed between crossed polarizers. Plagioclase feldspar often displays a striped pattern in black and gray.

The identification of mineral components of a sedimentary rock requires a good deal of practice; however, the textural relationships of the grains are easy to discern. In Chapter 2, we defined texture as the size, shape, and arrangement of mineral grains in a rock. In this study, textures seen in thin sections or acetate peels can be examined and evaluated by comparing them with diagrams and photographs (Figs. 3.4 to 3.8).

Acetate peels

An acetate peel is made by wetting the polished surface of a rock chip with acetone. A piece of acetone film is placed on the polished wet surface. The acetone softens the film, which then records an accurate impression of the texture of the rock. The peel technique is widely used in the study of texture of carbonate rocks like limestone and dolostone as well as nonporous sandstones.

Preparation of Acetate Peels

1. For sandstone, your instructor will provide a rectangular cut of the hand specimen rather like that shown in Figure 3.1B but thicker. This chip will have been polished with very fine aluminum oxide (Al_2O_3) powder. Many peels can be made from a single chip.

2. For carbonate rocks, the smooth surface of the chip is etched lightly with dilute hydrochloric acid (HCl).

Several peels can be made from this surface, but eventually it will have to be resurfaced and re-etched.

3. Use crumpled tissues to wedge the rock chip in one corner of the tray so that the prepared surface is *facing up* and is *level.*

4. Flood the level surface with acetone from the dropper, and *within a few seconds* place the square of acetate film on the wet surface.
    a. If the surface is not covered with a continuous layer of acetone, the film will not adhere to it and the peel will not be a success. Because of the speed required, it is easier if two persons work together, one applying the acetone and the other applying the film.
    b. Opposite edges of the film should be held by the thumb and forefinger and the film bent downward into a U-shape. Quickly place the bottom of the "U" on the surface and simultaneously release the two sides. The film will "recline" onto the wet surface, creating a successful peel. Do not press down on the film or touch it while it sets. Some practice may be required. If you do not succeed in laying the film smoothly on the surface on your first try, simply wait several minutes for the film to dry, pull it off, and try again with a fresh piece of acetate film.

5. Allow the film to dry for about five minutes, then grasp a corner and peel it off slowly, taking care not to tear it. The peel will warp on drying. It can be flattened by placing it between two glass plates such as the cover glasses provided with 35 mm slide holders. Otherwise it must be held flat while it is being observed under the microscope.

TEXTURES OF SANDSTONE

The possible *textural* components of sandstone are (1) sand-size grains, (2) **matrix** (which consists of silt and clay), and (3) **cement** (as discussed in Chapter 1). In *composition,* the grains of most sandstone range between quartz (or chert), feldspar (either plagioclase or potassium feldspar), and lithics (rock fragments). Figure 3.3 shows a common classification scheme for sandstone (or sand-size grains) modified from McBride (1963). Note that a quartz sandstone (containing 90% or more quartz grains) is also called a **quartz arenite,** and this type of sandstone is both texturally and compositionally mature.

The grains are held together by either cement or matrix. However, some varieties of sandstone have little or no matrix while others have little or no cement. When sandstones are examined, the textural and compositional maturity changes from *immature* to *submature* to *mature sandstone* are marked by a gradual increase in **roundness** of grains, improved **sorting** of grains, and decrease in amount of matrix, as shown in Figure 3.4. In general, immature sandstones

Figure 3.3 A simplified mineralogical classification for sandstones modified from McBride (1963). Note that most sandstones have a variable compositional maturity even when the grain-size of the constituent grains is the same. The apices of the triangular diagram represent 100%. As you move away from an apex (for example, feldspar), you decrease the amount of that material in the sandstone and increase the amounts of the other components.

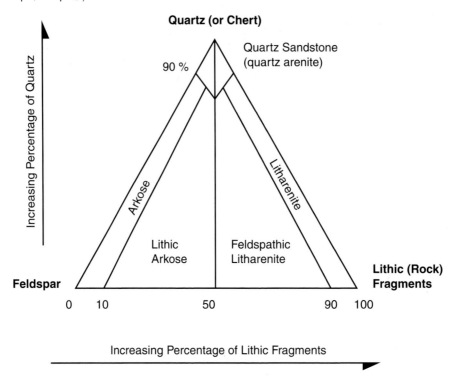

Figure 3.4 Textures of sandstones as seen in thin section under the microscope. Rock names are given at left and top, degrees of textural maturity at bottom. Sometimes the rock names are not used and a sandstone is identified simply as immature, submature, or mature. Typically, an **arkose** is derived from weathered granite. **Graywacke** is derived from a variety of rocks, including volcanic rocks, slate, and schist. Most of these rocks have mineral components (or the rock fragments themselves) that are dark in color and thus impart a dark color to graywacke. Both arkose and graywacke contain angular and poorly sorted grains, but graywacke generally has more fine-grained matrix than an arkose. *Submature sandstones* typically have only a small amount of matrix and consist of subrounded and moderately sorted grains, some of which may be dark in color. The grains in a *mature quartz sandstone* are typically subrounded to rounded. The sorting of grains is moderate to good, and few, if any, dark grains are present. Grains in a mature quartz sandstone are commonly held together by cement.

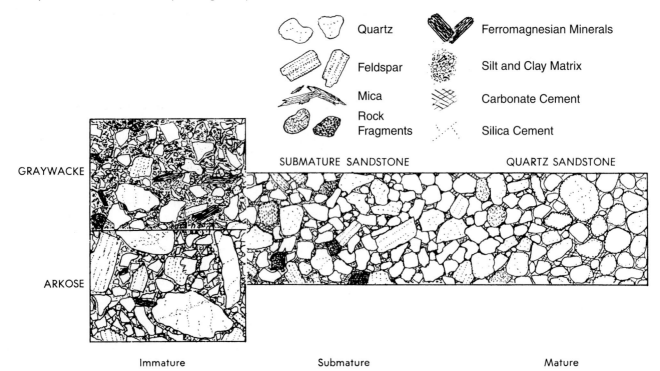

Figure 3.5 Types of contacts between grains in sandstone. As sandstone is buried to successively greater depths, grains in contact are pressed together, and one or the other grain dissolves at the contact. This is called **pressure solution.** Abundant matrix prevents pressure solution by keeping grains apart. The transition from point contacts to sutured contacts represents an increasing degree of pressure solution between grains.

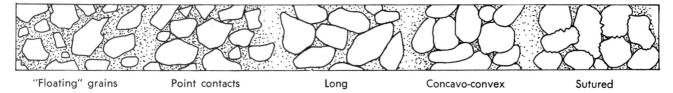

"Floating" grains   Point contacts   Long   Concavo-convex   Sutured

contain little or no cement, and the sand-sized grains are not in close contact because they are separated by matrix. The matrix, which holds together the sand-sized grains, can become very durable if the mineral chlorite or clay minerals (e.g., micas), which are often found in this material, recrystallize and grow into larger crystals. From Figure 3.5 you can see that the compositional maturity of a sandstone depends upon the amount of feldspar (plagioclase or potassium feldspar) grains as well as the amount of lithics (rock fragments). An arkose contains more feldspar and quartz grains than a litharenite, which has more rock fragments and quartz.

Submature sandstones are held together by a combination of matrix, cement, and **pressure solution** of the grains. Pressure solution occurs when the constituent mineral grains are pressed together under sufficient pressure that chemical reactions occur at the grain boundaries. At the contact, the grains may start to dissolve, and this process cements the grains together. Different degrees of pressure solution are illustrated in Figure 3.5.

Mature sandstones (quartz arenites) are held together either by cement or by pressure solution (Fig. 3.6C). If the grains of a sandstone are not closely packed, yet are not separated by matrix, cement is probably present. The most common cements, quartz (SiO_2) and calcite ($CaCO_3$), are typically clear and uniform in appearance. Iron oxide (Fe_2O_3; hematite) cement is commonly dark brown or red.

The **compositional** and **textural maturity** of a sandstone is used to interpret the geologic setting for the deposition of the sediments. In general, great thicknesses of immature sandstone (e.g., arkose), which occur in many mountain belts, are inferred to have been deposited rapidly on subsiding areas of the earth's crust. The rate of accumulation was too rapid for the grains to be rounded and the matrix to be winnowed out. In contrast, mature sandstones are inferred to have accumulated slowly on more stable parts of the crust. In their movement (by fluvial processes), there was sufficient time (and transport distance) for the sediment grains to become highly rounded and the matrix (e.g., clay minerals or micas) to be winnowed out.

In Figure 3.6 you can examine sandstones of three different maturities in thin section. Using the guidelines suggested in Table 3.1, convince yourself you can recognize the differences in maturity. Your instructor may have additional thin sections or hand samples of sandstone to interpret using Table 3.1.

TEXTURES OF CARBONATE ROCK

The term *carbonate rock* includes two very similar kinds of rock: *limestone* and *dolostone*. Limestone consists mainly of the mineral calcite ($CaCO_3$), and dolostone consists mainly of the mineral dolomite ($CaMg(CO_3)_2$); both may have a crystalline texture. Limestone and dolostone are sometimes difficult rocks for the beginner to distinguish, and some carbonate rocks are a mixture of calcite and dolomite (review the material of Chapter 1 for differences in carbonate rocks).

Table 3.1 Guide for Observing Sandstones Under the Microscope

	Specimen Number or Other Designation
1. What is the length of the largest grain, in mm?	
2. In general, are the sand-sized grains (a) angular to subangular, or (b) rounded to subrounded?	
3. By comparison with Figure 3.4, the grain contacts are best described as (a) point, (b) long, (c) concavo-convex, (d) sutured, or (e) not in contact, floating.	
4. Is the sandstone apparently held together by (a) cement, (b) matrix, (c) pressure solution contacts, or (d) some combination of the above?	
5. By comparison with Figure 3.3, is the texture (a) immature, (b) submature, or (c) mature?	

Figure 3.6 Photographs of sandstone textures as seen under the microscope. (A) Graywacke of Cretaceous age from the Coast Range of California. (B) Submature sandstone of Cretaceous age (Frontier Fm.) from Wyoming. (C) Mature quartz sandstone of Ordovician age (Roubidoux Fm.) from Missouri. A peel of each sandstone is shown in the column at left, a thin section as viewed in plain light in the center column, and a thin section as viewed between crossed polarizers in the column at right.

Peel   Thin section, plain light   Thin section, crossed polarizers

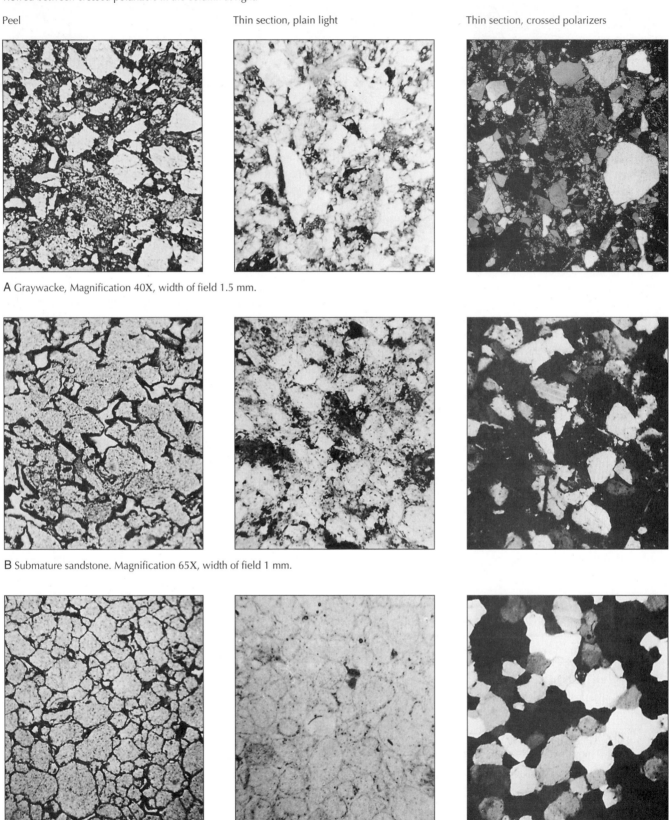

A Graywacke, Magnification 40X, width of field 1.5 mm.

B Submature sandstone. Magnification 65X, width of field 1 mm.

C Mature quartz sandstone. Magnification 40X, width of field 1.5 mm.

Figure 3.7 Textures of carbonate rocks as seen in thin section or peel under the microscope. Rock names are given at top of the figure, textural components are identified at bottom.

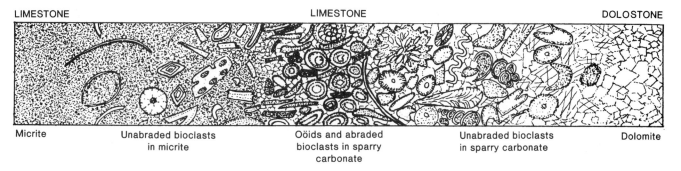

LIMESTONE   LIMESTONE   DOLOSTONE

Micrite   Unabraded bioclasts in micrite   Oöids and abraded bioclasts in sparry carbonate   Unabraded bioclasts in sparry carbonate   Dolomite

Carbonate rocks are classified based on the (1) individual grains, (2) cement, and (3) matrix that make up the rock, much like the more **clastic sandstones.** The grains, according to the common classification scheme of Folk (1959, 1962), can be separated into four major types. These are **bioclasts** (made of broken and whole skeletal parts), **oöids** (spherical grains formed by the precipitation of carbonate around a nucleus), **fecal pellets,** and **intraclasts** (which are sand or gravel-size pieces of limestone or dolostone). These grains are called **allochemical grains;** along with the amount of cement or matrix, they can be used, as we saw with sandstones, to classify most of the common carbonate rocks.

The allochemical grains of a carbonate rock are held together by either cement or matrix. The most common carbonate cement is called **sparry carbonate** (also called **sparite**). This cement is a clear crystalline carbonate that has been precipitated between **clasts** or has developed by the **recrystallization** of carbonate clasts. Matrix in most carbonate rocks is a murky, fine-grained calcium carbonate mud called **micrite.** Unlike sparite, the fine-grained carbonate mud is deposited with the clasts and, during lithification, is recrystallized. The typical appearance of these components and some of the common allochemical grains, as seen through the microscope, is shown in Figures 3.7 and 3.8.

Table 3.2 shows a modified classification scheme for carbonate rocks based on the work of Dunham (1962). In this classification scheme, the allochemical grains must be identified as discussed above. Although the classification scheme may seem a bit obtuse, it is an attempt to more completely describe carbonate rocks (and their variability). There are two key observations necessary in using this classification. The first is whether material between the allochemical grains is matrix (micrite) or cement (sparite). The second is whether the *allochemical grains form a framework that is self-supporting without the presence of matrix or cement.* If there are sufficient numbers of allochemical grains and they support themselves without the intervening matrix (or cement), they are called *grain-supported.* However, if the allochemical grains cannot support themselves without the intervening micrite, they are called *mud-supported.* Use Table 3.2 as a flow chart for examining these differences.

For example, in Figure 3.7, the nearly pure micritic limestone to the left of the figure is composed mostly of micrite with few allochemical grains. Based on the Dunham (1962) classification, this would be called a mud-supported (mudstone) limestone. By contrast, the limestone in Figure 3.8A has slightly more than 10% allochemical grains, but is mostly composed of carbonate mud. In this example, there are only a few allochemical grains and the rest is micrite. This type of limestone would be called a mud-supported limestone (Table 3.2), and with more than 10% allochemical grains, it would be called a fossiliferous wackestone. Lastly, note that if the limestone is composed of mainly allochemical grains (as in Figure 3.8C), the grains themselves support one another. This is a good example of a grain-supported limestone.

Table 3.2 Classification of Limestones Modified from Dunham (1962)

A. Does the limestone have micrite (as matrix) or sparite cement?

B. Are the allochemical grains supported by themselves to form a framework, [or] would they fall apart without the matrix or cement?

   1. If the allochemical grains are self-supporting—this is a *GRAIN-SUPPORTED* limestone.

      a. If the material between the allochemical grains is micrite—this is a *PACKSTONE* limestone.

      b. If the material between the allochemical grains is cement (sparite)—this is a *GRAINSTONE* limestone.

   2. If the allochemical grains are not self-supporting—this is a *MUD-SUPPORTED* limestone.

      a. If there are < 10% allochemical grains—this is a *MUDSTONE* limestone.

      b. If there are > 10% allochemical grains—this is a *WACKESTONE* limestone.

Figure 3.8 Photographs of carbonate rock textures as seen under the microscope. The bioclasts that look like shell fragments are of brachiopods; the rest of the bioclasts are mainly from crinoids or bryozoans (see Chapters 10 and 11 for additional details).

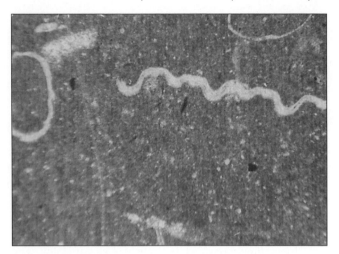

A Unabraded bioclasts in micrite. Magnification 65X, width of field 1.2 mm.

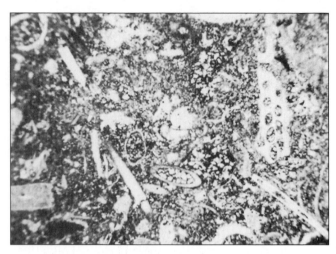

B Bioclasts in fine-grained, sparry carbonate. Magnification 20X, width of field 4 mm.

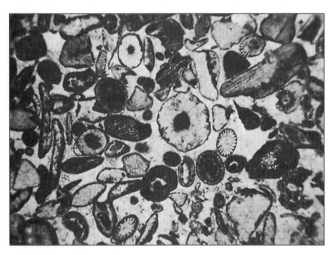

C Abraded bioclasts and oöids in sparry carbonate. Magnification 13X, width of field 6 mm.

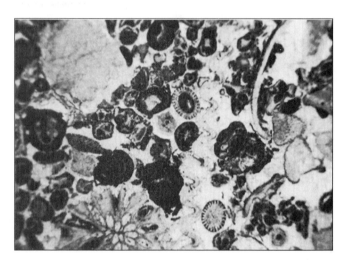

D Unabraded bioclasts and oöids in sparry carbonate. Magnification 20X, width of field 4 mm.

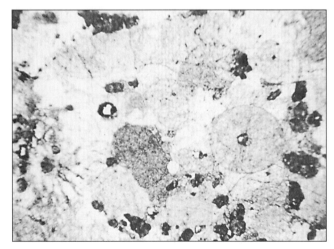

E Bioclasts (mostly recrystallized) in sparry carbonate. Magnification 13X, width of field 6 mm.

F Rhomb-shaped crystals of dolomite. Magnification 65X, width of field 1.2 mm.

Table 3.3 Guide for Observing Carbonate Rocks Under the Microscope

	Specimen Number or Other Designation
1. Is micrite (a) absent, (b) minor constituent, less than 25%, or (c) major constituent, more than 25%?	
2. Are bioclasts (a) absent, (b) minor constituent, or (c) major constituent?	
3. If present, are bioclasts generally (a) abraded, or (b) unabraded?	
4. Is sparry carbonate (a) absent, (b) minor constituent, or (c) major constituent?	
5. Are rhomb-shaped crystals of dolomite (a) absent, (b) minor constituent, or (c) major constituent?	
6. Was rock probably deposited in (a) quiet water, or (b) turbulent water, or (c) is there no basis for inference?	

Since sparry carbonate cement (and not micrite) cements the grains, this limestone would be called a grainstone.

Although most carbonate rocks originate as allochemical grains held together by micrite or sparry carbonate cement, the clastic texture is likely to be partly or wholly obliterated by recrystallization. Fine-grained calcite tends to go into solution and then recrystallizes as larger crystals, and the texture thus becomes partly or wholly **crystalline** (Fig. 3.8E). The minerals calcite and dolomite are similar in appearance under the microscope, but dolomite usually develops rhombshaped crystals, whereas calcite crystals have no regular geometric outline (Fig. 3.8F).

Primary crystallization of dolomite is extremely rare in modern sedimentary environments. Most dolomite forms through diagenetic alteration of limestone. In general, the presence of micrite suggests deposition in quiet water because fine carbonate mud would not likely settle to the bottom in turbulent water. Whole, unbroken fossil shells also indicate deposition in quiet water because they would be abraded or broken in turbulent water. Today, carbonate sediments are accumulating in quantity in the oceans and some large lacustrine (lake) systems and presumably this has been true of the geologic past.

In Figure 3.8 you can examine a variety of carbonate rocks in thin section. Using the guidelines suggested in Tables 3.2 and 3.3, convince yourself that you can classify these carbonate rocks as mud-supported or grain-supported. Your instructor may have additional thin sections or hand samples of carbonate rocks to interpret using Tables 3.2 and 3.3.

TERMS

allochemical grain (also called **allochem**)   A collective term introduced by Folk (1959) for one of several varieties of carbonate aggregates that serve as the coarser framework grains in most mechanically deposited limestones. The common allochemical grains are **bioclasts, intraclasts,** *fecal pellets,* and **oöids.**

arkose   A red to pink coarse-grained feldspar-rich sandstone composed of poorly (to moderately) sorted angular to subangular grains. Arkoses generally reflect a tenestrial depositional environment where an uplifted granitic rock body experienced rapid erosion, transportation deposition, and burial.

bioclasts   Detrital particles of sediment composed of fragments of calcareous algae, shells, plant matter, or the skeletons of marine invertebrates.

cement   A chemical precipitate such as SiO_2, $CaCO_3$, or Fe_2O_3 that crystallizes in voids between sedimentary particles following their deposition.

clastic   Pertaining to a rock or sediment composed of broken fragments of preexisting rocks, minerals, or skeletal matter (called **clasts**). The individual clasts are described based on identity, grain size, degree of rounding, and sorting.

compositional maturity   A measure of the amount of weathering, erosion, and transport experienced by sediment, as indicated by the variety of mineral constituents ultimately deposited to form the sedimentary rock. Examples: Feldspars (plagioclase or potassium feldspar) weathered to clay; ferromagnesian minerals (olivine or pyroxene) weathered into iron oxides, such as the minerals limonite or hematite.

crystalline   A term that describes a sedimentary rock composed of crystals (rather than clasts) precipitated from a saturated solution. The mineral crystals are fine to very fine in grain size and form an interlocking mosaic. Rocks formed by chemical precipitation are called chemical sedimentary rocks.

graywacke   A dark gray, firmly indurated coarse-grained sandstone that consists of poorly sorted angular to subangular grains of quartz and feldspar, with a variety of rock and mineral fragments embedded in a clayey matrix (texturally and compositionally immature). Graywackes generally reflect a marine depositional environment where rapid erosion, transportation, deposition, and burial are associated with an orogenic tectonic setting.

intraclasts   Gravel-, sand-, or silt-sized particles of limestone (or dolostone) rock fragments found as framework grains in carbonate rocks. These grains range from angular to rounded in shape.

matrix   Clastic, fine-grained particles (often clay) that are deposited at the same time as larger grains and help to hold (or bind) the grains together. Also called the *groundmass.*

maturity   A measure of the weathering and transportation of a clastic sedimentary rock by evaluation of the heterogeneity of its composition (i.e., **compositional maturity**) and the amount and degree of rounding and sorting (i.e., **textural maturity**).

micrite   Mechanically deposited lime mud with particle sizes in the 1- to 3-micron range that has been subsequently lithified (or experienced recrystallization) to form a very finely textured limestone.

oöids   Spherical particles of sand size that are mostly composed of concentric laminae of calcium carbonate. Limestone made up of cemented oöids has an *oölitic texture.*

pressure solution   The dissolution of clastic mineral grains at the points of grain contacts because of the concentration of pressure at those locations.

quartz arenite   Another term for a sandstone that has 90% (or more) sand-size quartz grains. This is texturally and compositionally mature quartz sandstone.

recrystallization   The formation of new mineral crystals in a rock after the rock has formed. This process may result from changes in pressure or temperature, or from the reorganization of the chemical elements of preexisting minerals into new minerals.

roundness (of sedimentary particles)   The degree of abrasion of a clastic particle as shown by the sharpness of its edges and corners.

sorting   The process by which sedimentary particles having some particular characteristic, such as size, shape, or specific gravity, are separated from other dissimilar particles by the agents of transportation (e.g., running water, waves, glacial ice, wind, etc.).

sparry carbonate   Clear, crystalline calcite or dolomite deposited as cement between clasts or developed by recrystallization of clasts. Also called **sparite.**

textural maturity   A measure of size and shape variation of the constituents of a sedimentary rock.

texture (sedimentary)   The general physical appearance of the rock, including the size, shape, and arrangement of the constituent mineral (or rock) components.

4

Ancient Sedimentary Environments

1. *Demonstration* or *display* specimens that reflect different environments of deposition and have different textural and compositional maturity. These may include the following:

   Limestone (both crystalline and fossiliferous).
   Chert or flint.
   Sandstones, including mature quartz sandstone to immature sandstone such as arkose.
   Shale and silty shale.
   Conglomerate and breccia.
   Anthracite and bituminous coal, lignite, peat.
   Evaporites that include gypsum and halite.

2. *Demonstration* or *display* specimens showing different sedimentary structures, including the following:

   Cross-bedding.
   Normal and reverse graded bedding.
   Plane bedding.
   Ripple marks.
   Sole marks.
   Trace fossils.
   Stromatolite algal mat.

ENVIRONMENTS OF DEPOSITION

One of the most interesting things geologists do is determine the environment in which different kinds of sedimentary rocks are deposited. With this kind of information it is possible to unravel the succession of changes in a region over an interval of geologic time. It is the way geologic history is written.

Environment, as used in this study, refers to the geographic nature of the place in which sediments are being deposited. Today, as in the geologic past, environments are very diverse. Here we will examine only a few of those that have left their imprint in sedimentary rocks.

Nonmarine environments (also called the *terrestrial* or *continental;* see Figure 4.1 and Table 4.1) include alluvial fans, river floodplains, lakes, glaciers, and eolian (windswept) environments.

Alluvial fans (Fig. 4.8) are wedges of clastic sediment deposited by rivers and mudflows emerging from mountains onto a plain or basin.

Floodplains are lowlands bordering streams composed of river deposits and dry except when streams overflow their banks during flood stages. Floodplains are **fluvial environments** of deposition.

Eolian refers to environments where the action of wind has transported, arranged, and deposited sand and silt. Such deposits collect along shorelines but mainly in deserts as dunes and blankets of sand. The silty sediment known as *loess* is an eolian deposit. (The term is derived from the name of the Greek God *Aeolus,* who controlled the winds by releasing them at will from his cave.)

Glacial environments are those in which glaciers have deposited huge volumes of rock debris, including large boulders. Characteristically, these deposits are poorly sorted mixtures of boulders, gravel, sand, and clay. Where these materials have been reworked by meltwater flowing from the glacier, they are less chaotic and may resemble stream deposits.

Lacustrine refers to lake environments. Lacustrine environments are common in nearly every nonmarine region and may even occur in desert regions. Desert lacustrine features include ephemeral **playa lakes,** formed by heavy runoff from sudden rainstorms. Playa lakes often evaporate completely, leaving dry lake beds floored by clay and evaporite minerals such as gypsum and halite.

Paludal refers to swamp and marsh environments. Such environments are likely to be coastal but may also border the floodplains of rivers.

Transitional environments are marked by the interaction of fluvial and near-shore marine processes. Here we find deltas, beaches, tidal flats, bars, and lagoons, all having distinctive structures useful in identification.

Deltas are accumulations of sediment that form when the velocity of a stream is reduced upon reaching the ocean. Without the carrying power of a fast-flowing current, sediment formerly held in suspension is deposited. The amount of sediment deposited is usually so great that deltaic deposits are usually unaffected by tides, waves, or ocean currents.

Beaches are transitional environments that are affected by tides, waves, and ocean currents. A beach is a zone of sediment that accumulates between the average low-water level and a landward change in topography (a sea cliff or other rise in land elevation). The majority of beaches are

Figure 4.1 Environments in which sediments are deposited. The diagram is not to scale, and the steepness of slope for both the continental slope and the continental rise are greatly exaggerated.

From Levin H. L., *Contemporary Physical Geology*, 2nd ed. Philadelphia: Saunders College Publishing, 1986.

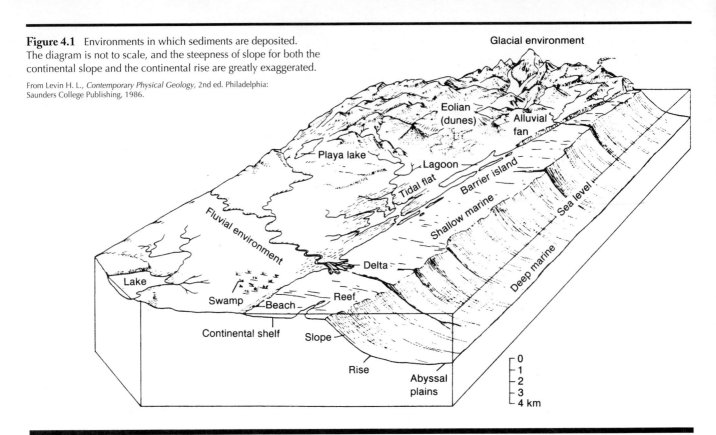

Table 4.1 Features of Major Sedimentary Environments

Major Environment	Sedimentary Features
Nonmarine	
Fluvial	*River floodplain:* overbank deposits are plane-bedded silt or clay; *channel deposits* are cross-bedded sandstone (ss.) or conglomerate (cng.). *Alluvial fan;* irregular lens-shaped beds of poorly sorted ss., and cng. Brown, red, or white.
Aeolian	*Desert:* cross-bedded ss., large-scale, steep foreset beds, usually no preserved ripple marks. Brown, red, or white. *Loess:* thick accumulations of unconsolidated and unstratified silt-sized particles. Yellow or buff.
Lacustrine	Ephemeral environment. Deposits of water-transported (some minor aeolian) sediments. Plant and animal debris. *Turbidity current* sedimentation possible in larger lakes. Beach deposits, low-energy ripple marks, and possible wave-cut terraces.
Glacial	Poorly sorted rock and mineral fragments with *low* textural and compositional maturity (called *till*). If reworked by wind or water, many exhibit some sorting and bedding forms.
Swamp (Paludal)	Associated with overbank deposits. Low-lying areas accumulating fine muds and silts. Plant and animal debris. May be confused with the late stages in the history of a lake. Gray clay or coal, plane bedding.
Transitional	
Delta	Usually cannot be interpreted from a single outcrop. Sometimes cross-bedded ss., sometimes plane-bedded ss., freshwater and marine fossils, brown or gray shale. Many have cross-cutting channels and overbank deposits.
Beach	Environment between highest and lowest levels of the tide. Well-sorted sands and/or pebbles, broken shells, rare mud or clays.
Bar	Deposits of sand and/or gravel, cross-bedded. Usually elongate in overall shape.
Tidal Flats	Fine-grained sand, silt, and clay; gravels rare. Usually cross-cut by numerous gullies and channels, variety of bedding. Covered and uncovered in each tidal cycle.
Lagoon	Associated with tidal flats. Silty and muddy sediments, highly *bioturbated,* usually interbedded with sandy layers deposited by wind or water during storms.
Marine	
Shallow marine (Water depth less than 200 m; includes continental shelf and reefs)	Limestone, dolomite (or fossils of marine animals) most diagnostic, if present. Cross-bedded clastic limestone indicates shallow turbulent water, limestone, quiet water; massive organic structures, a reef. Graded bedding uncommon. Many rock types, gray shales. Most rocks outside mountain belts were deposited in this environment.
Deep marine (Water depth 200 to 10,000 m; includes continental rise, abyssal plain, trench)	Typically gray ss. and shale, often in alternating beds, ss. likely graded. Cross-bedding rare. Bedded cherts and volcanic rocks sometimes present. Where land-derived material is scant, thin beds of dark, limy shale may form.

composed of sand, and the most abundant mineral in most beach sand is quartz.

Tidal flats are muddy or marshy areas that are repeatedly inundated and drained by the tides.

Lagoons are bodies of shallow water that are often found behind bars where they are partially protected from the waves and currents of the open ocean. As a result, lagoonal deposits usually consist of fine-grained clastics heavily burrowed by mollusks and worms. If these sediments become lithified, the borings may be preserved.

Bars in transitional environments usually consist of shallowly submerged or emergent embankments of sand and gravel built on the sea floor by ocean waves and currents.

Marine environments may be broadly separated into *shallow marine* (water depths less than 200 m) and *deep marine* (water depths from 200 m to 10,000 m). Depth alone, however, does not solely determine the type of sediment being deposited on the ocean floor. Other factors that affect marine sedimentation include distance from shore, topography of the ocean floor, and storm activity.

Shallow marine environments are often dominated by sand and silt. This is because wave action and marine currents keep finer particles in suspension and carry them farther out to sea. However, coarse sediment can be carried to great depths where topographically deep areas lie close to the shoreline. Limestones (fossiliferous or crystalline) originate mainly on shallow continental shelves where there is little terrigenous sediment input from rivers or turbidity currents. Clear, warm water of normal salinity is important for the formation of limestones.

Deep marine regions are characterized by very fine clay, volcanic ash, or the calcareous and siliceous remains of microscopic organisms. Exceptions include sporadic occurrences of coarser sediment that have been carried down continental slopes into the deeper parts of the ocean by **turbidity currents** (see Fig. 4.6).

The marine depositional environment can be divided into four large-scale topographic divisions called continental shelves, continental slopes, continental rises, and abyssal plains. Along Atlantic-type coasts (see passive margin discussion in Chapter 5), the **continental shelf** (inclination about 0.1°) extends to an average depth of about 200 m, and its outer edge is marked by a fairly abrupt change in slope (to about 4°, and is called the **shelf-slope break**), which is the beginning of the **continental slope. Reefs** may be present on the shelf, but they are not a typical feature of the shelf edge, as may be implied from Figure 4.1. The continental slope merges into the **continental rise** (inclination about 0.5°), which sweeps seaward to the **abyssal plain.**

Along mountainous Pacific-type coasts (see active margin discussion in Chapters 5 and 6), the continental shelf is likely to be narrow, and there are no features corresponding to the continental slope and rise of Atlantic-type coasts. Mountain chains or island arcs may occur in ocean basins or on a continental border and deep linear subduction trenches usually adjoin them.

BEDDING AND RELATED FEATURES

Bedding (or **stratification,** which is another term for bedding) is the most obvious feature of sedimentary rocks in outcrop. It results from a change in grain size, color, or rock type from one bed to the next, and it is usually more obvious on a weathered outcrop than a fresh one. For example, a bedding plane between apparently identical beds of limestone sometimes can be discerned by the presence (or absence if it is weathered away) of a very thin layer of shale or clay.

Often the burrowing, churning, and stirring of the original sediment may have destroyed the finer laminations and bedding of a bed by organisms that live on or within these sediments. This process is called **bioturbation** and may result in the bed being totally disrupted so that original layering is impossible to discern.

The essential properties of a bed are its *thickness* and *lateral continuity.* A bed whose thickness is greater than 61 cm is described as *thick;* from 5 to 61 cm as *medium;* and 1.5 to 5 cm as *thin.* Beds that continue for tens to hundreds of meters without much change in thickness are described as *laterally uniform.* Those that change in thickness over several to tens of meters are called *laterally variable* in thickness.

Bedding in undeformed sedimentary rocks can often be described as planar or cross-stratified. **Plane bedding** (Fig. 4.2) can apparently form in any sedimentary environment. These beds are often laterally extensive and commonly thinly laminated. Cross-stratified bedding is an arrangement of the beds (or laminations) in which one set is *inclined* relative to the others. Cross-stratified bedding is often termed **cross-bedding** (Fig. 4.3) and indicates the action of strong currents of water, as in rivers or the shallow marine environment, or of wind. Thus, the direction of the inclination of the sloping beds can be useful in determining the direction of current (water or air) flow that produced these structures.

Cross-bedding is separated into two types; *tabular-planar cross-bedding* and *trough cross-bedding* (Fig. 4.4). Tabular-planar cross-bedding (not to be confused with plane bedding) is seen in beach deposits and dunes while trough cross-bedding is often formed in river and stream channels. Tabular-planar cross-bedding formed by wind (called aeolian cross-bedding) is often confused with the tabular-planar cross-bedding formed by fluvial (running water) processes. A "rule of thumb" is that aeolian cross-bedding results in units that tend to be thick, and the slope of the **foreset beds** is steep (Fig. 4.5).

Another feature in sedimentary rocks is graded bedding. **Graded bedding** consists of repeated beds, each of which has the coarsest grains at the base of the bed and successively finer grains near the top (Fig. 4.6D). Graded beds, which range in thickness from millimeters to many meters, are attributed to **turbidity currents** (Fig 4.6), which

Figure 4.2 *Left,* plane bedding, and *right,* two processes by which it is formed. *Top right,* weak currents of water or *swash* on a beach move sand in thin sheets. Successive thin sheets build up to form an internally laminated bed. *Bottom right,* particles settling from suspension form laminae, which may accumulate to form thicker beds.

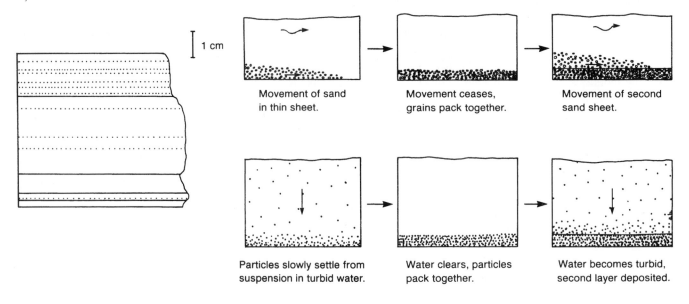

Figure 4.3 The development of cross-bedding in current-deposited sand (A and B) and in windblown sand (C and D). (A) Current fills in a depression on the river bottom or sea floor with sediment. (B) Continued sedimentation may cover the first set of crossbeds with another. (C) Sand is deposited as inclined layers **(foreset beds)** on the downwind side of a dune. (D) A second dune covers the first. Cross-bedding may change orientation if the wind direction shifts.

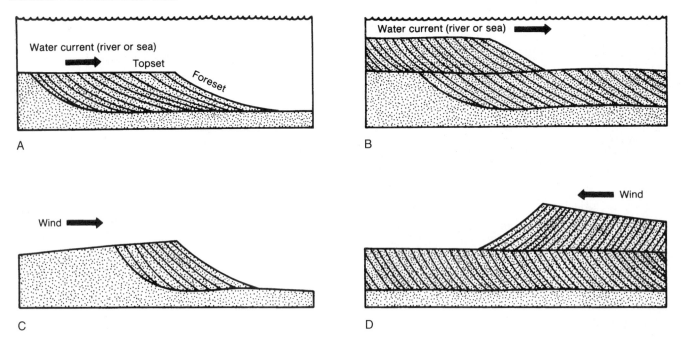

Figure 4.4 Common types of cross-bedding. Tabular-planar cross-bedding (A) is often seen in beach deposits and dunes. Trough cross-bedding (B) is frequent in stream deposits. Under certain conditions, tabular-planar cross-bedding ranging from (A) to the wedge-shaped cross-bedding (C) with its thick beds and steep foreset bed slope can be formed by wind (aeolian) processes (see Fig. 4.5).

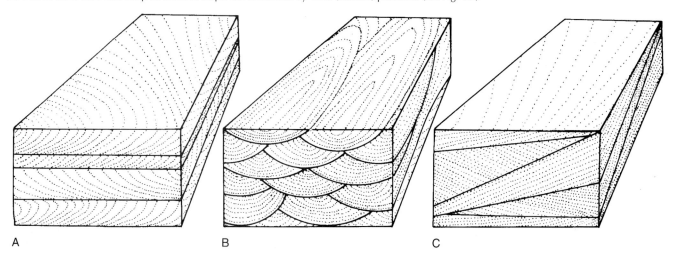

A   B   C

Figure 4.5 Aeolian sandstone with distinctive, large-scale cross-bedding. Checkerboard Mesa, Zion National Park, Utah.

Courtesy of R. L. Clouser.

are most common in the deep marine environments (turbidity currents may also occur in large lakes). Because of the great distances that turbidity currents may travel—up to tens or even hundreds of kilometers—a graded bed may have great areal extent. Most graded beds deposited by turbidity currents when observed in the field have an abrupt, non-gradational contact with the underlying stratum as a result of how they were deposited.

Most bedding surfaces are flat, but some have distinctive markings called *bedding plane markings.* These include **ripple marks,** which are formed most abundantly in shallow water, and a variety of features called **sole marks** (Fig. 4.7). Sole marks can include **scour marks,** which occur at the bottom of some graded beds, **tool marks,** and **load casts.** The direction of an ancient current (called a **paleocurrent**) is inferred from ripple marks, which trend transverse to current direction, and from sole marks, which point in the direction of turbidity current flow.

Besides these markings, a variety of other features are found on and within beds. These include tracks and burrows (often referred to as **trace fossils**), **stromatolites** (an algal mat that traps fine grains of sediment and forms moundlike shapes), desiccation cracks and raindrop prints, and the impressions of plants or other fossils. Features of this type can help in determining the environment at the time of sediment deposition and give a "This Side Up" clue (called a **geopetal indicator**) when examining a section of vertical or overturned strata.

Figure 4.6 Deposition of graded bedding by a turbidity current.

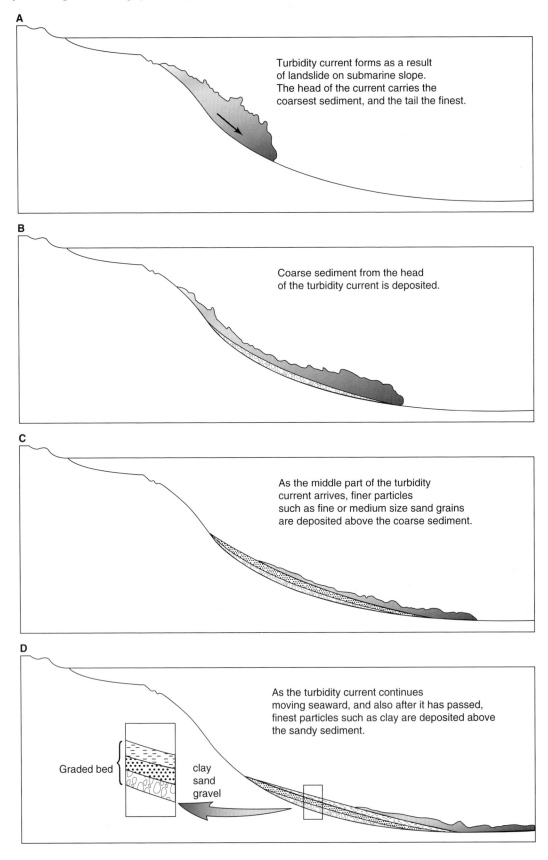

A

Turbidity current forms as a result of landslide on submarine slope. The head of the current carries the coarsest sediment, and the tail the finest.

B

Coarse sediment from the head of the turbidity current is deposited.

C

As the middle part of the turbidity current arrives, finer particles such as fine or medium size sand grains are deposited above the coarse sediment.

D

As the turbidity current continues moving seaward, and also after it has passed, finest particles such as clay are deposited above the sandy sediment.

Graded bed {

clay
sand
gravel

Figure 4.7 Bedding plane markings: (A) symmetrical ripple marks, (B) asymmetrical ripple marks, and (C) **sole marks,** which occur as raised forms on the bottom of a graded bed. These and other markings can be used to determine flow direction (arrows) and the orientation (top and bottom) of a bed.

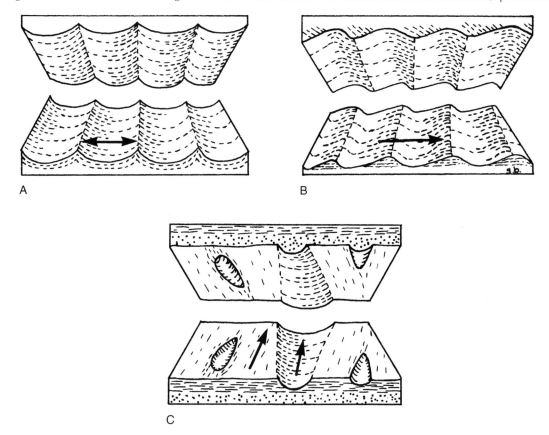

A

B

C

COLOR

Some inferences about the type of environment of sediment deposition can be made from rock color. However, it is possible that the original color of a sediment (or sedimentary rock) may have changed since the time that it was buried or that the color has changed as a result of weathering after it was exposed at the earth's surface. The color of shale or siltstone is more diagnostic than the color of sandstone or limestone. Gray or greenish-gray immature sandstones are very likely to turn brown at the weathered surface. The color of unweathered sedimentary rocks is mainly influenced by amount of iron (generally in the form of an iron oxide mineral such as hematite (Fe_2O_3) or limonite ($FeO \cdot OH \cdot n\ H_2O$)). The oxidation of iron results in colors such as red, yellow, or brown. Furthermore, oxygen is likely to be more abundant in the nonmarine than the marine environment, and the presence of iron oxide coloration is used as an indicator of terrestrial sedimentation. With these cautions in mind, sedimentary rock color can be interpreted using the criteria displayed in Table 4.2.

Table 4.2 Color and Environment of Deposition

Color	Interpretation	Environment
Red, yellow, brown	Oxidizing conditions	Probably nonmarine
Black, gray, or greenish-gray	Reducing conditions	Marine or nonmarine (swamps or swampy flood plains)
Light gray or white	Little iron present	Either nonmarine or marine

Exercise

Answer the questions associated with Figures 4.8 through 4.19.

Figure 4.8 A view of the Panamint Range, Inyo County, California.

Courtesy of U.S. Geological Survey, photograph by H. E. Malde.

1. What are the prominent depositional features formed at the mouths of the canyons?

    _____

2. Describe the probable maturity and textural characteristics of the sediment forming the features named above

    _____

    _____

    _____

3. What other non-marine environments or features exist on the flat terrain in front of the range?

    _____

    _____

    _____

Figure 4.9 Ripple marks in intertidal zone at Puerto Penasco, Sonora, Mexico. Tides in this area have the greatest range of any location in the Gulf of California (exceeding 5 m).

Courtesy of Dr. Guillermo A. Salas, Universidad de Sonora, Departamento de Geologia, Hermosillo, Sonora.

1. What is the direction of the tidal current? How can you tell? Draw an arrow on the photograph to show the direction of flow.

    _____

    _____

2. Are these ripple marks symmetrical or asymmetrical? How can you tell?

    _____

    _____

Figure 4.10 Lime mud composed of very small crystals of calcium carbonate ($CaCO_3$) from a tidal flat at Andros Island, Bahamas.

Courtesy of N. D. Newell.

1. What would be the texture of a limestone that could form from such fine limy mud?

_____

_____

2. What type of bedding would be found in this depositional environment?

_____

_____

3. What would the presence of plants (shown in photograph) do to the bedding of this depositional environment?

_____

_____

Figure 4.11 Stratified sedimentary rock of the Pumpernickel Fm., Humboldt County, Nevada. The more prominent beds are chert and they are separated by thin layers of shale. The scale bar is 15 cm in length.

Courtesy of the U.S. Geological Survey.

1. Does this represent a marine or nonmarine (terrestrial) environment of deposition? Explain your answer.

_____

_____

2. If this was a marine depositional environment, was the water shallow or deep? If this was a terrestrial depositional environment, would you predict that it was fluvial (river or stream) or aeolian? Explain your reasoning.

_____

_____

3. What do the interbeds of shale in this rock suggest about the history of deposition? What factors (tectonic, sedimentation rate, etc.) would produce the rock that you observe in this photograph? Explain your reasoning.

_____

_____

Figure 4.12 Beds in mature sandstone of the Mississippian Tar Spring Fm., southern Illinois.

1. What type of bedding does this represent?

_____

_____

2. Does this suggest a fluvial, marine, aeolian, or indeterminate environment of deposition? Explain.

_____

_____

_____

Figure 4.13 Cross-bedding, accentuated by weathering, in the Jurassic Navajo Sandstone, north of Kanab, Kane County, Utah.

Courtesy of E. D. McKee, U.S. Geological Survey.

1. What type of bedding does this represent?

_____

_____

2. Was this sandstone deposited in a nonmarine (terrestrial) or marine environment? If it was terrestrial, was the sandstone deposited by fluvial or aeolian processes? If it was deposited in a marine environment, was it shallow or deep water? Explain.

_____

_____

_____

Figure 4.14 Bedding surface on a mature sandstone of the Silurian Clinton Fm., Peters Mountain, Virginia. Rock hammer for scale. Bedding surface is oriented near vertical.

Courtesy of G. W. Stone, U.S. Geological Survey.

1. What type of markings are present?

_____

_____

2. Assuming the rock is marine in origin, what do these markings indicate about the depth of the water in which the sediment was deposited?

_____

_____

_____

Figure 4.15 Red sandstone and shale of the Eocene Colton Fm., Utah. Bold line outlines the bottom of an ancient stream channel filled with sediment (for example, channel fill).

1. What type of bedding does this represent?

_____

_____

2. Was this sandstone deposited in a nonmarine (terrestrial) or marine environment? Explain your reasoning.

_____

_____

_____

Channel fill

Figure 4.16 A highway roadcut exposing beds of the Ordovician Joachim Dolomite near Pacific, Missouri.

1. What type of bedding does this represent?

_____

2. Was the formation deposited in a nonmarine (terrestrial or continental) or a marine environment? Explain your reasoning.

_____

_____

_____

_____

Figure 4.17 Coarse grained conglomerate of the Malone Fm. near the southeastern end of the Malone Mountains, Hudspeth County, Texas. The scale is 6 inches long.

Courtesy of C. C. Albritton, Jr.

1. What type of bedding does this represent?

_____

_____

_____

2. Was this conglomerate deposited in a fluvial or aeolian environment? Explain your reasoning.

_____

_____

_____

Figure 4.18 Gray, immature sandstone of the Late Proterozoic Great Smoky Group, Great Smoky Mountain National Park, Tennessee.

Courtesy of W. B. Hamilton, U.S. Geological Survey.

1. What has caused the separation of the sandstone into layers?

_____

_____

_____

2. Account for the massive character of the thicker bed lying between sequences of thinner beds.

_____

_____

_____

Figure 4.19 Detail of bedding in steeply dipping sandstone of the Great Smoky Group. At this location in the Great Smoky Mountain National Park, strata of the Great Smoky Group are vertical. Examine the texture closely near areas labeled *a* and *b.* Rock hammer for scale.

Courtesy of P. B. King, U.S. Geological Survey.

1. In which areas or zones, *a* or *b,* are the coarsest grains visible?

_____

2. What type of bedding characterizes these sandstones?

_____

3. Was this sandstone deposited in a nonmarine (terrestrial) or marine environment? If it was a marine environment, was it deposited in shallow or deep water? Explain your reasoning.

_____

_____

_____

_____

TERMS

abyssal plains   The relatively level areas of the ocean floor that lie at depths exceeding 2,000 meters.

aeolian (also *eolian*)   Term applied to sediment that has been transported and deposited by the wind. For example, the sand of desert dunes is aeolian.

alluvial fan   A cone-shaped deposit of unconsolidated, poorly sorted sediment made by a stream where it passes from an area of steep gradient to lower gradient. This change of slope causes rapid deposition of sediment transported by the stream.

bar   A general term for any of the various elongate offshore ridges, banks, or mounds of sand, gravel, or other unconsolidated material, submerged at least at high tide, that are built up by the action of waves or currents on the water bottom.

beach   The relatively thick and temporary accumulation of loose water-borne material that is in active transit along, or deposited on the shore zone between the limits of low and high tide.

bedding (also called *stratification*)   The arrangement of a sedimentary rock in layers of varying thickness and character. The term may also be applied to the layered arrangement and structure of an igneous or metamorphic rock.

bioturbation   Reworking of existing sediments by organisms. This reworking may result in the destruction of distinctive sedimentary structures such as bedding.

continental rise   The ocean floor beyond the base of the continental slope, generally with lower gradient than the continental slope.

continental shelf The gently sloping region that extends from the edge (coast or shoreline) of a continent seaward to a line (called the **shelf-slope break**) marked by an abrupt increase in slope.

continental slope The submerged region of steep slope extending from the seaward edge of the continental shelf (the shelf-slope break) down to the upper margin of the continental rise.

cross-bedding (also called *cross-stratification*) Beds or laminations arranged at an oblique angle to the main bedding. Separated into *tabular-planar cross-bedding* (near shore, fluvial or aeolian environments) and *trough cross-bedding* (fluvial stream channel environments).

delta The low, nearly flat-lying, triangular-shaped tract of land at or near the mouth of a river resulting from the accumulation of sediment supplied by the river in such quantities that it is not removed by tides, waves, or currents.

evaporite A nonclastic sedimentary rock composed primarily of minerals produced from a saline solution as a result of extensive evaporation. Examples include halite (rock salt), gypsum, and various nitrates and borates.

floodplain The lowlands that border a stream (or river), composed of sediment deposited by the stream, except when the stream overflows its banks during flood stages.

fluvial Pertaining to rivers. For example, fluvial sediments are sedimentary materials deposited by rivers and streams.

foreset bed In a cross-bedded unit, the inclined layers of material deposited on the relatively steep frontal slope. As these materials are deposited they cover the **bottomset bed** and in turn are covered (or truncated) by the **topset bed.**

geopetal indicator Pertaining to any rock feature that indicates the relation of top to bottom at the time of the formation of the rock.

graded bedding A sequence of repeated beds, each of which has the coarsest grains at the base and successively finer grains toward the top. Often found with turbidity current deposits.

lacustrine Pertaining to, produced by, or formed in a lake.

load cast A **sole mark** often caused by soft sediment deformation prior to and during the lithification of water-rich sediments. Load casts are often irregular in shape and difficult to use for paleocurrent direction determinations.

paleocurrent An ancient current (generally of water) that existed in the geologic past, whose direction is inferred from the sedimentary structures and textures of the rocks formed at that time.

paludal Pertaining to a marsh or swamp.

plane bedding Layers of originally horizontal sediment or sedimentary rock that lie approximately parallel to the surface of deposition.

playa lake A shallow, intermittent lake (with no outlet) in an arid or semiarid region, covering or occupying a small sandy area at the mouth of a stream in the wet season and usually drying up in the summer. Commonly, the evaporation of the water results in the precipitation of evaporite minerals.

reef A ridgelike or moundlike structure, layered or massive, built from the shells or body parts of sedentary calcareous marine organisms such as corals or mollusks.

ripple marks An alternating sequence of small troughs and ridges produced on the surface of fine clastic deposits by wind or water movements.

scour marks Produced as a result of erosion of a sediment surface by the current of water flowing over it. When they are filled and preserved they are referred to as **sole marks** or *bedding plane markings*.

sole marks Casts of sedimentary structures such as cracks, tracks, or grooves, commonly revealed after the original underlying sedimentary layer has weathered away. Sole marks include **scour marks, tool marks,** and **load casts.**

stratification The layering in sedimentary rocks that results from changes in texture, color, or rock type from one layer (bed) to another.

stromatolite A sedimentary structure produced by the trapping of fine sediments as a result of the growth and metabolic activity of cyanobacteria (formerly called blue-green algae). The fine sediment trapped by the sticky filaments of the cyanobacteria often accumulates in cabbage- or mound-shaped masses that build up to form reefs.

tidal flat An extensive, nearly horizontal, marshy or barren tract of land that is alternately covered and uncovered by the tide. Consists of unconsolidated sediment (mostly mud and sand).

tool marks A form of **scour mark** resulting from objects (such as a branch, shell fragment or bone fragment) swept along by the current that is dragged or bounces along the muddy bottom.

trace fossil Indirect evidence of the biologic activities of ancient organisms, such as burrows, tracks, feeding traces, and fecal pellets, that are recorded in sedimentary rocks.

turbidity current A current of sediment-laden water that is denser than surrounding clearer water and therefore flows downslope beneath the less turbid water.

5
Tectonic Settings

WHAT YOU WILL NEED

1. Demonstration specimens representing different rock associations.

   a. Limestone (or dolostone), quartz sandstone, and shale.

   b. Immature sandstone, conglomerate containing fragments of volcanic rock and chert, "dirty limestone" (limestone with terrigenous clastic components).

   c. Arkosic sandstone and arkosic conglomerate, breccia.

   d. Coal, peat, rocks containing evaporites (gypsum, halite).

2. Demonstration specimens of

   a. Igneous and metamorphic rocks common in continental and oceanic crust.

   b. Gabbro and basalt, granite and rhyolite, volcanics such as pumice, scoria, volcanic ash, and obsidian.

   c. Metamorphic rocks such as greenstone (metamorphosed basalt), slate, schist, and gneiss.

3. Table 1.1 (Chapter 1).

TECTONIC SETTINGS

Rocks are the pages of earth history. By studying associations of rocks, we can discover where the earth was relatively quiet and stable or where conditions were dynamic, mobile, and characterized by mountain building and volcanic activity. The more stable of these two settings is called **cratonic** (from Greek *kratos,* power or strength). The more active is the **orogenic setting** (from Greek *oros,* mountain, and *genesis,* to be born).

In an orogenic setting, rapid uplift of the earth's surface is likely to be associated with rapid subsidence in adjacent areas. In contrast, vertical movements in a cratonic setting will be relatively small, and slow uplift is likely to be associated with slow subsidence. Both the orogenic and cratonic settings are two kinds of **tectonic** settings, which means that they both involve movements of the earth's rigid outer layer or lithosphere.

THE TECTONIC SETTING-SEDIMENTARY ROCK CONNECTION

Geologists are fond of saying, "The present is the key to the past." Thus, we study sediments being deposited in present tectonic settings in order to understand tectonic settings of the geologic past. Another approach, which we will use here, is to employ hypothetical or model settings. From models like the two illustrated in Figure 5.1, and from premises about how they operate, deductions can be made about the sedimentary deposits associated with each model.

Model 1 in Figure 5.1 represents an **orogenic** (Pacific-type) coast bordered by a narrow continental shelf and deep trench. Model 2 represents a **cratonic** (Atlantic-type) coast that is bordered by a wide, gently sloping continental shelf leading to a continental slope and continental rise. The tectonic and topographic characteristics of each model directly affect sedimentation.

1. The rapidly uplifted *source area* (which may also have volcanoes producing lava, cinders, and ash) supplies a large quantity of land-derived debris (called **terrigenous sediment**).

2. Rock-forming materials also originate in the depositional areas, mainly by the accumulation of the remains of shelled organisms, but partly from direct chemical precipitation of certain minerals (e.g., calcite) from seawater. The amount (and type) of chemical sediment deposition is controlled, in part, by the amount of terrigenous sediment in the water. Other controls on chemical sediment deposition are salinity and water temperature.

3. Sediment deposited in a rapidly subsiding area or in deep water tends to remain in place, but some of it will slump and be transported to deeper depths (even possibly the abyssal plains) by turbidity currents.

4. In a shallow, slowly subsiding area, sediment is continually stirred up and reworked. This may occur by currents or bioturbation from living organisms or a combination of both. The end result is that fine silts and clays are separated from the coarser sands and transported to deeper (or quieter) water.

40

Figure 5.1 Model Tectonic Settings.

Model 1 Generalized view of an *active margin,* where an oceanic plate is being subducted beneath a continental plate. Characterized by rapid uplift in the source region and rapid sedimentation (usually texturally and compositionally immature) and subsidence in the depositional area. Note the small continental shelf area, steep continental slope, and trench.

Model 2 Generalized view of a *passive margin* (also called a trailing margin). The source area generally has lower relief or is farther from the depositional area. This results in a range of compositional and textural maturities of the sediments being deposited. Deep-water sedimentation is mainly pelagic and/or fine silt/clay-size sediments. Note the wider continental shelf area and the lack of a trench.

Scale for both models: length of base approximately 200 km, vertical scale exaggerated.

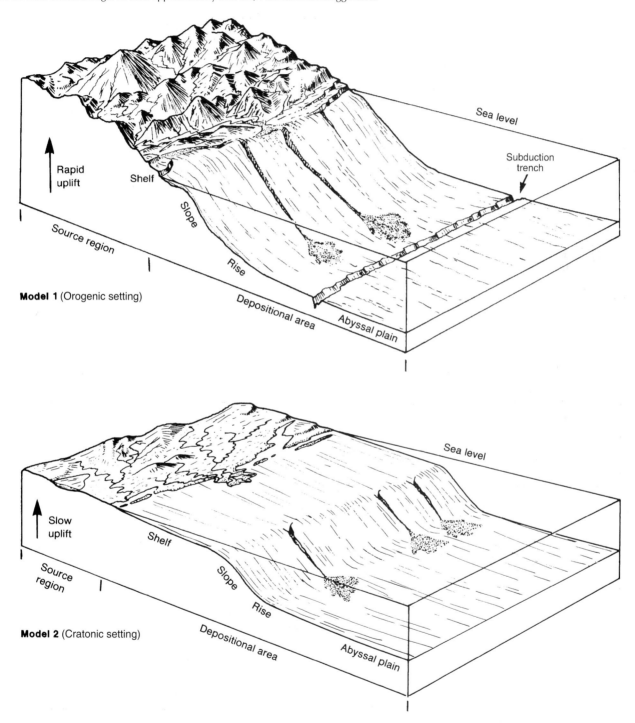

5. In general, in the absence of turbidity currents, deposits of land-derived sediment tend to get finer grained with increasing distance from land.

Study Questions

1. What kind of sandstone (Table 1.1), in terms of maturity, texture, and rock name(s), would be formed from sediments deposited on the landward side of the subduction trench in Model 1?

_____

_____

_____

2. How do the rocks formed from sediments deposited in the subduction trench differ from the rocks of Question 1 in texture and thickness?

_____

_____

_____

3. What kind of sedimentary rock would form from sediments deposited farthest from the source area in either model?

_____

_____

_____

4. What kind of sandstone would be formed from sands deposited on the shelf of Model 2?

_____

_____

_____

5. What factors would determine the relative amounts of sand and carbonate (chemical and biogenic) sediment deposited on the shelf of Model 2?

_____

_____

_____

_____

6. What kind of sedimentary rock would be formed from sediment deposited on the continental rise of Model 2?

_____

_____

_____

_____

7. During a specific period of geologic time, where in the two models would the greatest thickness of sediment be expected to accumulate? Why?

_____

_____

_____

_____

8. In some regions of the earth, very substantial thicknesses of carbonate rock (about 3000 m) are found. What modification of Model 2 would be required to account for this scale of accumulation?

_____

_____

Figure 5.2 Walther's Law stipulates that vertical and horizontal variations in facies generally are nearly identical. As an example, note that at Section B in the cross section, the nearshore silt facies is overlain by offshore shale. The same offshore shale is also found laterally at Section A. Similarly, the beach sand that lies directly beneath the nearshore silt at Section B can be found laterally at Section C. Inclination of sedimentary units shown has been greatly exaggerated.

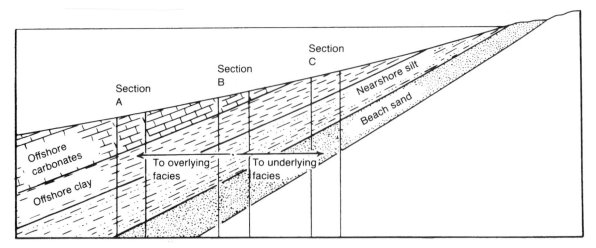

Facies and Associations of Beds

Within a particular tectonic setting, beds of different rock types are likely to be deposited in association. **Association** implies that either (1) a bed of one rock type may grade laterally into another rock type or (2) a bed of one rock type may be overlain by a bed of another rock type. It is a general rule that only rock types representing a particular tectonic setting will occur in association. For example, an association of coarse-grained (arkose) sandstone, conglomerates, and shales (termed a **molasse**) generally represents an orogenic setting where sediments are produced by the rapid erosion of a mountain range (see association 4, Fig. 5.3).

Within a particular tectonic setting, subtle changes in the environments of deposition can occur. The term **facies** is used to refer to the general aspect of a rock type from which its environment of deposition can be inferred. A *lateral* succession of sedimentary rock types that grade into one another

Figure 5.3 Diagram showing major tectonic settings (top) and a chart indicating associations of beds that are typical of these settings (bottom). Numbers on top diagram refer to the different associations. Also refer to Table 4.1.

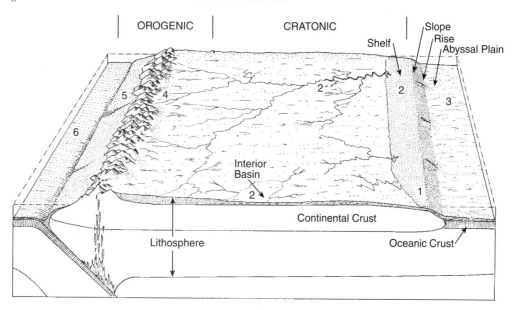

Association of Beds	Bedding of Sandstone	Color of Shale	Tectonic Setting	Sedimentary Environment
1. Limestone or dolomite, clay shale, mature sandstone.	Plane or cross-bedded	Lt. gray	Cratonic (Shelf)	Shallow marine
2. Unstable shelf: immature sandstone silty shale, limestone (w/coal).	Plane or cross-bedded	Lt. gray, red, brown	Cratonic (Shelf)	Nonmarine or shallow marine
3. Shale, fine-grained immature sandstone.	Graded or plane	Dark gray	Cratonic (Abyssal Plain)	Deep marine
4. Arkose, molasse, or clastic wedge: coarse-grained immature sandstone, conglomerates, siltstone.	Cross-bedded or plane	Red, brown or gray	Orogenic (Mountains)	Mostly nonmarine (Terrestrial)
5. Flysch or graywacke: shale and immature sandstone w/bedded cherts and volcanic rocks.	Graded or plane	Dark gray	Orogenic (Shelf)	Mostly marine, shallow/deep
6. **Ophiolite:** siliceous shale, dark limestone, dark sandstone, chert, igneous rock (pillow basalts).	Thin beds, some graded	Dark gray	Oceanic (Trench and Accretionary Wedge)	Deep marine, distant from land

Note: The subsurface layers of the earth shown on the front of the diagram are drawn according to the concepts of plate tectonics. Briefly, the lithosperic plate at left, which consists of oceanic (basaltic) crust (pattern of short vertical lines), is being subducted beneath the plate at the right, which consists of continental (granitic) crust. The term *orogenic* applies to areas where mountain building is in progress. The term *cratonic* refers to the stable part of the continent where no mountain building has occurred for a long period of geologic time.

as the environment of deposition changes is an example of a **facies change.** For example, quartz sandstone and **carbonate rocks** are a common association of facies, wherein the quartz sandstone represents deposition along the nearshore margin of a shelf environment and the carbonate rocks are deposited (or chemically precipitated) in the offshore environment. In contrast, the association of clastic limestone beds and graded immature sandstone beds is rare. This is because these rocks are deposited in highly different environments of deposition. The clastic carbonate beds occur in shallow water with low terrigenous sediment influx, and the graded immature sandstones form in a marine environment of moderate depth characterized by volcanic rocks and a tectonically active environment. Note that not only are these facies incompatible, but also their tectonics settings differ greatly.

An important generalization about facies, sometimes called **Walther's Law** (after the German geologist J. Walther), states that the **succession of facies** that occurs laterally will also be found in the vertical succession of facies, assuming there are no significant breaks in the rock record at the locality. Thus, to determine what facies are to be encountered laterally from a locality (within a particular tectonic regime), one has only to examine the vertical sequence of beds at the locality. This relationship can be seen in Figure 5.2.

At an outcrop, the cause for vertical change in rock type can be inferred with reasonable certainty in some cases but only speculatively in many others because of the large number of factors that are involved. For example, a layer of dark shale between graded sandstone beds may have been deposited by the slow settling of silt and clay during the intervals between turbidity currents. On the other hand, a layer of shale between limestone beds reflects some change in the sedimentary environment that may be difficult to specify. This shale layer could reflect either a change in water temperature or, perhaps, an influx of silt and clay from the

source area. An important cause of vertical change in rock type that can often be well established is a change in the position of the shoreline. For example, in one cratonic setting coal may be deposited in a coastal swamp while limestone is being deposited in the sea offshore. If, at an outcrop, limestone is found to overlie coal, it is highly probable that this is due to an advance of the sea over the land by [*either*] a rise in sea level [*or*] by subsidence of the land.

SUMMARY OF ASSOCIATIONS AND SETTINGS

In Figure 5.3 are shown major tectonic settings and the association of beds that are typical of each. Among the cratonic settings, the term **shelf** is used in the tectonic sense for a flat area of low to moderate subsidence that borders the coast and receives a relatively thin, well-winnowed cover of sediments (Figs. 5.3 and 5.4A). A somewhat "dirty" association (with correspondingly low textural and/or compositional maturity) is deposited on the shelf if it receives drainage from rivers flowing through an orogenic region (Fig. 5.5A). An interior **basin** is an area of moderate subsidence within the craton. Orogenic settings are characterized by relatively rapid subsidence and usually by mountainous source areas. One important setting, illustrated in Figure 5.1, involves a mountainous arc, an oceanic trench, and a depositional area between the arc and the trench. This arc-trench gap can receive large amounts of terrestrial and volcanic material and later gives rise to a variety of sedimentary associations such as **molasse** and **flysch** (Fig. 5.3). Other orogenic settings are fault troughs or basins surrounded by mountains (Fig. 5.5C) or located between the mountains and the craton.

Figure 5.4 Associations of beds deposited in some cratonic settings. Examine each setting and answer the following questions. Use Figures 5.1 and 5.3 for assistance in answering these questions.

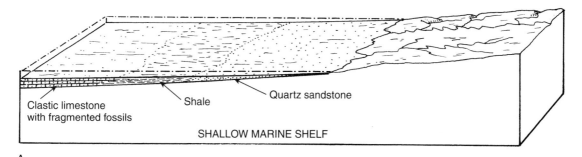

A

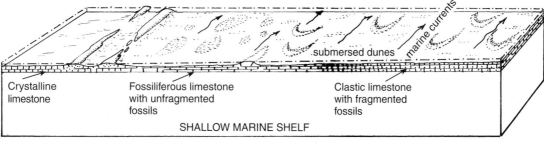

B

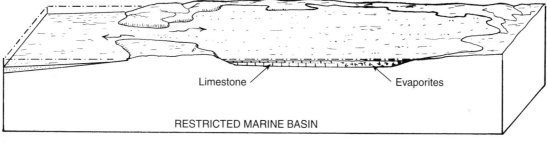

C

1. In setting A, what is the composition, texture, maturity, and kind of bedding likely to be present in the sandstone?

_____

_____

   Why is there no shale deposited near the shoreline?

_____

_____

2. In setting B, how do you account for the accumulation of fragmented fossil material in the clastic limestone, and whole fossils in the fossiliferous limestone?

_____

_____

   What type or types of bedding are likely to be present in the clastic limestone?

_____

_____

What differences may have existed in the environment of deposition to account for the deposition of clastic limestone, as opposed to crystalline limestone?

_____

_____

3. In setting C, what environmental conditions are required for the formation of evaporites?

_____

_____

   What minerals are found in evaporites?

_____

_____

   What environmental conditions might account for the concentration of limestone (rather than evaporites) in the western part of the basin?

_____

_____

Figure 5.5 Associations of beds deposited in some orogenic settings. Examine each setting and answer the following questions. Use Figures 5.1 and 5.3 to help answer these questions.

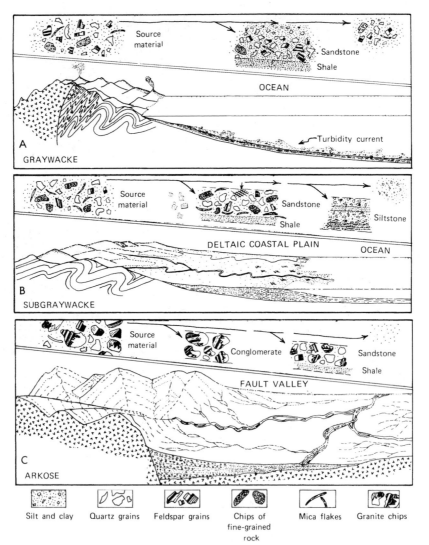

1. In setting A, what is the composition, texture, maturity, and kind of bedding likely to be present in the sandstone?

_____

_____

_____

What effect will a turbidity current have on the kind of bedding found in the offshore deposits?

_____

_____

2. In setting B, how do you account for the change in bedding between the sediments found on the deltaic coastal plain and those on the ocean floor?

_____

_____

_____

What minerals do you find in the coastal plain that are not found on the ocean-floor sediments? Explain your reasoning.

_____

_____

_____

3. In setting C, what is the composition, texture, maturity, and kind of bedding likely to be present in the sandstone?

_____

_____

_____

What are the differences in composition and maturity between the arkose and the conglomerates deposited in this setting?

_____

_____

Figure 5.6 *Right,* bluff along the Mississippi River at Alton, Illinois, composed almost entirely of limestone of Mississippian age. About 80 ft (24 m) high. *Left,* sketch of bluff where the beds are given informal names.

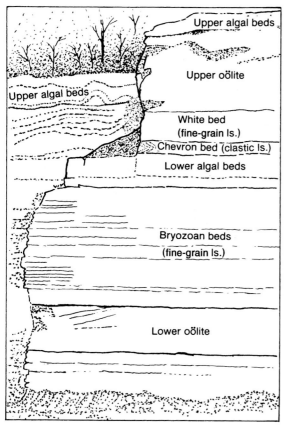

1. To which association do these beds belong? (See Fig. 5.3.)

2. Type(s) of bedding?

3. Tectonic setting?

4. Sedimentary environment?

5. Why are the bryozoan beds (in the fine-grained limestone) more likely to be separated by shale (which causes the thin bedding) than are the oölite beds?

Figure 5.7 Beds of the Pennsylvanian Kansas City Group, exposed along Interstate 70 at Kansas City, Missouri. Resistant beds at the top and near the middle of the outcrop are limestone; rilled beds are light gray shale that contain plant fragments and, locally, thin seams of coal.

1. Which association? (See Fig. 5.3.)

2. Type of bedding?

3. Tectonic setting?

4. Sedimentary environment(s)?

5. Probable reason for vertical change from mostly shale to limestone?

Figure 5.8 Coarse breccia from the Old Red Sandstone Fm., Scotland. Rock hammer is 16 in long.

Courtesy of the Institute of Geological Sciences, London.

1. Which association? (See Figures 5.3 and 5.5.)

_____

_____

2. Tectonic setting?

_____

_____

3. This breccia is often found interbedded with a more texturally mature conglomerate containing similar mineral and rock fragments. What type of sedimentary environment would you propose for the breccia? What about the interbedded conglomerate? What does this suggest?

_____

_____

_____

_____

Figure 5.9 Gray, immature sandstone (thick bedded) and gray, silty shale (darker, thin bedded) in the Pennsylvanian Pottsville Fm. near Birmingham, Alabama. The sandstone is ripple marked and contains plant fragments.

1. Which association? (See Fig. 5.3.)

_____

_____

2. Type of bedding?

_____

_____

3. Tectonic setting?

_____

_____

4. Sedimentary environment? What evidence did you use to choose your answer?

_____

_____

Figure 5.10 Beds in the late Miocene Capistrano Fm., Orange County, California. Lensing beds of boulder conglomerate interbedded with immature sandstone. Microfossils in associated shale below these beds indicate a water depth of 2000 to 4000 ft.

Courtesy of J. G. Vedder, U. S. Geological Survey.

1. Tectonic setting? (See Fig. 5.3.)

_____

_____

2. Sedimentary environment?

_____

_____

3. Provide a hypothesis for how these boulders might be transported to such depths.

_____

_____

Figure 5.11A Alternating beds of immature sandstone with graded bedding and dark, silty shale of Cretaceous age, cropping out along the California coast at San Pedro Point.

Figure 5.11B Near view of beds at same locality as figure 5.11A. Foot rule gives scale.

1. Which association? (See Fig. 5.3.)

_____

_____

2. Tectonic setting? (See Fig. 5.3.)

_____

_____

3. Reason that base of the lighter sandstone beds is irregular? See Fig. 5.11B to see the irregular base of the beds.

_____

_____

4. Sedimentary environment?

_____

_____

Figure 5.12 Fine, mature sand (from the top of the roadcut to a few feet below the man) grading downward into gray clay in the Lake Cretaceous Blufftown Fm. near Cusetta, Georgia. The clay contains marine fossils.

1. Which association? (See Fig. 5.3.)

_____

2. Tectonic setting? (See Fig. 5.3.)

_____

3. Suggest a possible reason for upward change from marine clay to marine sand.

_____

_____

_____

TERMS

associated beds   Beds composed of different rock types that are commonly found to occur together (in association) as a result of their having been deposited within a particular tectonic setting. For example, coal beds are frequently associated with silty shales deposited in cratonic, nonmarine areas; immature sandstones form an association with bedded chert and volcanic rock and reflect an orogenic, tectonically active marine depositional environment.

basin   A low area in the earth's crust, of tectonic origin, in which sediments have accumulated. The accumulation of sediments causes subsidence and allows for the deposition of thick layers. The shape of a basin may range from circular to elliptical.

carbonate rock   Rocks composed of minerals that contain the radical ion group $(CO_3)^{2-}$. Examples include chalk or limestone (composed of calcite: $CaCO_3$) and dolostone (composed of dolomite: $CaMg(CO_3)_2$).

cratonic   Pertaining to those parts of the continents that have been tectonically stable for several hundred million years.

facies   The aspect, appearance, and characteristics of a rock unit, usually corresponding to a certain environment or mode of origin.

facies change   A lateral or vertical change in the lithologic or paleontologic characteristics of contemporaneous (deposited at the same time) sediments. It is caused by, or reflects, a change in the depositional environment.

flysch   A pre- or syn-orogenic sedimentary deposit resulting from erosion of uplifting mountain ranges. Comprises thick sequences of marine sedimentary facies characterized by calcareous/sandy shales and muds interbedded with conglomerates, coarse sandstone, and graywackes.

graywacke   A dark gray, firmly indurated coarse-grained sandstone that consists of poorly sorted angular to subangular grains of quartz and feldspar, with a variety of rock and mineral fragments embedded in a clayey matrix (texturally and compositionally immature).

Graywackes generally reflect a marine depositional environment where rapid erosion, transportation, deposition, and burial are associated with an orogenic tectonic setting.

molasse   A post-orogenic sedimentary deposit resulting from the erosion of mountain ranges. Varies from marine to continental facies composed of clastic sandstone, conglomerates, and shales in very thick sequences.

ophiolite   An association of mafic and ultramafic igneous rock whose origin is related to ocean crust formation. The presence of ophiolite rocks on dry land indicates some tectonic activity that moved oceanic crust. This was another piece of evidence in the development of the hypothesis of plate tectonics.

orogenic   Pertaining to deformational crustal processes by which great mountain systems are formed. An orogenic region is one that is susceptible to earthquake activity, volcanism, and general tectonic instability.

shelf (in tectonic sense)   A stable cratonic area that was periodically flooded by shallow marine waters and received a relatively thin, well-winnowed cover of sediments.

stable tectonic setting   A region in which vertical movements of the crust are slight and usually very slow.

succession of facies (Principle of)   Observation that within a given sedimentary cycle, the same succession of facies that occurs laterally is also present in vertical succession. Also called **Walther's Law.**

tectonic   Pertaining to movements of the earth's rigid outer layer (**lithosphere**) and to the effects of these movements on the origin and deformation of rocks.

terrigenous sediment   Marine sediment consisting of material eroded from a land surface.

unstable tectonic setting   An orogenic setting characterized by rapid uplift and subsidence of the crust accompanied by deformation of rocks, earthquakes, and volcanism.

Walther's Law   (See **succession of facies.**)

6

Sea-Floor Spreading and Plate Tectonics

AN OVERVIEW OF SEA-FLOOR SPREADING AND PLATE TECTONICS

The Medieval philosopher William of Ockham (A.D. 1349?) proposed what has become an enduring ideal for finding answers to complex questions. Simply put, he stated that the simplest explanation is usually the best. His concept is now known as "Ockham's Razor." Geologists have been searching for a way to apply Ockham's Razor in order to derive one unified theory to explain the global relationship between the interaction of continents and oceans, the cause and location of mountain building, and the distribution of volcanoes and earthquakes. They have been searching for a *gestalt* view of our planet. The term *gestalt* refers to a collection of physical entities that can be brought together to creat a single unified concept.

Early thoughts on such a gestalt view were triggered by the remarkable fit of the continental shorelines on either side of the Atlantic Ocean. Like a jigsaw puzzle, the "nose" of Brazil seems to fit easily into the indented western coastline of Africa (Fig. 6.1). This and other "fits" around the globe prompted many to formulate theories involving the breakup of an ancient supercontinent.

Figure 6.1 World map outlining the continental landmasses and the locations of lithospheric plates, oceanic ridges, deep-sea trenches, and plate margins.

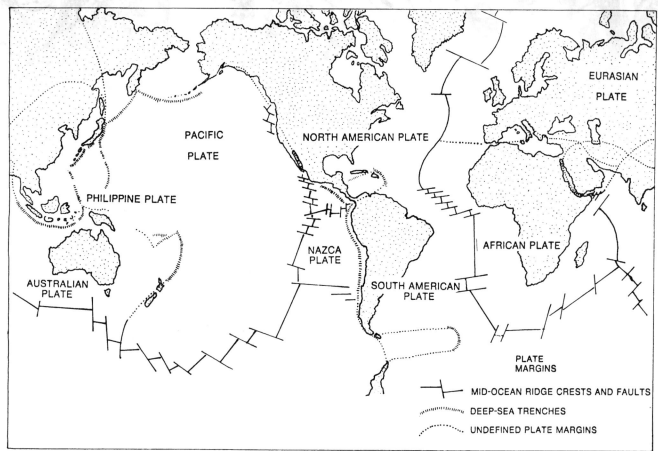

A major proponent of the supercontinent hypothesis was the German meteorologist Alfred Wegener (1880–1930). Wegener postulated that an ancient supercontinent named **Pangea** broke into two parts, **Laurasia** (which included North America and Europe) and **Gondwanaland** (which included most present Southern Hemisphere continents). In addition to the fit of the continents, Wegener found other evidence for the breaking up and drifting apart of continents in the similarity of ancient climatic conditions (as indicated by fossils) on now widely separated landmasses. The location of coal and evaporite deposits around the globe substantiated the climatic evidence supported by fossils.

Hypotheses are made to be tested, and the idea of **continental drift** espoused by Wegener failed to answer a major question; namely, what was the *mechanism* that moved the continents?

The search for a mechanism capable of transporting continents had to await the development of instruments sensitive enough to measure the earth's ancient magnetic field, as recorded in the rock itself. Lava, molten igneous rock, has certain iron-rich minerals that record the orientation of the earth's magnetic field when these minerals cool below a certain temperature (called the **Curie point**). When measurements of the **paleomagnetism** of the rock were made, the position of the earth's magnetic field appeared to be different for different continents. This implied either that the earth's magnetic pole had moved (a concept called **polar wandering**) or that the continents had moved in relation to a fixed magnetic pole. These data, and improved seismic information derived from earthquakes, led to a dusting off of the continental drift hypothesis of Wegener and its modification into the theory of **plate tectonics.**

Plate tectonics is a relatively simple concept. It involves an earth whose outer shell is a rigid layer of rock (called the **lithosphere,** Figs. 6.2 and 6.3), broken up into 7 larger and 20 or so smaller **plates.** The larger plates are approximately 75 to 125 km thick and behave as rigid, brittle units. Movement of the plates, some of which have continents attached to them, causes them to converge (move together or collide), diverge (spread apart), or slide past one another at the plate boundaries. This motion results in frequent earthquakes along the plate margins as the rigid, brittle rock is broken by the movement. When the locations of earthquakes are plotted on a world map, the boundaries of these plates are clearly defined (Fig. 6.1).

The rigid plates ride upon a partially molten (ductile) region of the upper **mantle** called the **asthenosphere** (from the Greek *asthenos,* weak). These divisions, shown in Figures 6.2 and 6.3, are determined by the studies of seismic waves caused by earthquakes passing through these regions. The mechanism for motion of the plates, and the attached continents, is thought to be **convection** within the mantle. Convection of the partially molten material in the asthenosphere is similar to the convection of heated water in a glass coffeepot on the kitchen stove. As the water is heated by the stove, it becomes

Figure 6.2 Internal structure of the earth. Notice that the mantle extends from the base of the **crust** to the top of the outer **core** and makes up about 83% of the total volume of the earth. The divisions shown here are determined by the study of seismic waves and how they travel through various materials.

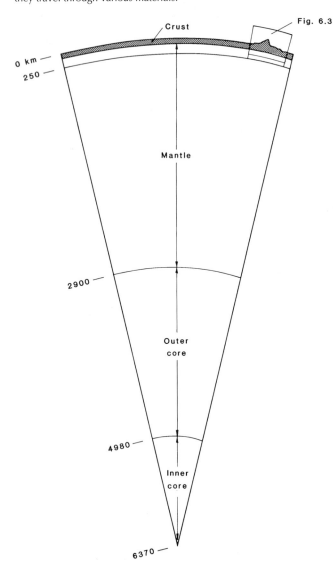

Figure 6.3 Cross section of the crust and the upper mantle. The crust of the earth is of two different types. Continental crust is about 40 km to 60 km thick, whereas oceanic crust is only about 10 km thick. The terms *lithosphere* and *asthenosphere* refer to the physical properties of the rocks in these particular regimes and are not equivalent with the terms *crust* and *mantle*. Rocks in the lithosphere deform brittlely, whereas rocks in the asthenosphere deform plastically, like putty.

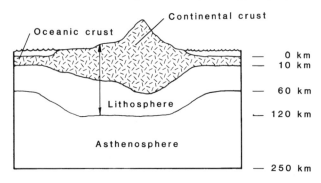

less dense and rises in the pot. As the heated water rises, it displaces the cooler water, forming a circulation cell. This upward movement of heated, less dense material and the downward flow of cooler material is analogous to what is thought to occur in the asthenosphere. As suggested in Figure 6.4, the motion of the rigid lithospheric plates results from movement in the underlying asthenosphere. A junction between lithospheric plates marked by a ridge axis and newly formed oceanic crust is called a **divergent plate boundary.**

The process by which new sea floor is produced along the oceanic ridges and conveyed slowly away from the ridges is termed **sea-floor spreading.** Thus, new oceanic crust is produced at the oceanic ridges and consumed at regions of **subduction** (called **convergent plate** boundaries) where it is pushed/pulled back into the asthenosphere, melted, and brought back to the earth's surface either as volcanic material or as new oceanic crust.

Plate tectonics provides an explanation for the different kinds of continental margins we observe today. For example, a tectonically **active margin** (both a *convergent* and *transform* plate boundary) has been formed along the western side of the North American plate by convergence with the plates underly the Pacific Ocean. In contrast, the Atlantic coast is a **passive margin** produced by the widening of the Atlantic Basin. Here, subsidence and sediment accumulation are occurring.

These concepts have revolutionized geological thought and provide a unifying framework for the earth sciences. They have not, however, invalidated the principles by which earth history is interpreted. In fact, long-established principles can be used to verify the hypothesis of sea-floor spreading, as will be seen in the following exercises.

Study Questions

The following questions should be answered with respect to Figure 6.4.

1. Is the sea floor in Figure 6.4B at "C" older or younger than the sea floor at "W"?

    _____

    _____

2. Why are the bands at "C" (and at "W") equidistant from the mid-oceanic ridge/rift?

    _____

    _____

3. In Figure 6.4B, note that at the oceanic floor/continent boundary, the basalt is covered by a thick wedge of sediment. Supply an explanation for why this sediment would build up to such a thickness in this location. *Hint:* Think about the differences in the basalt at the mid-oceanic ridge/rift and at the continent margin.

    _____

    _____

    _____

Figure 6.4 Sea-floor spreading. The block diagrams show the widening of an ocean basin and separation of continents. (A) An early stage of spreading after the initial rifting of a continental plate at a spreading center. (B) A later stage of ocean basin widening where severed continental plates have moved a considerable distance apart. The black and white bands indicated on the oceanic crust are a record of the changes in the orientation of the earth's magnetic field.

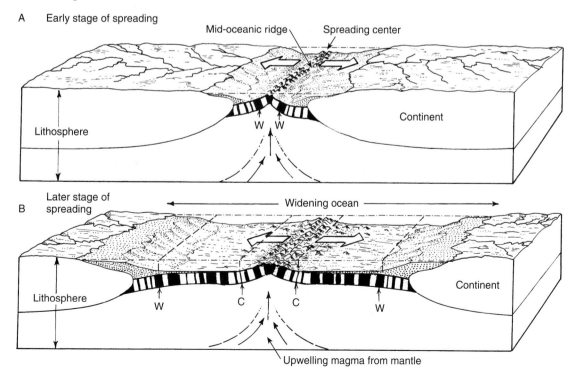

4. Can there be any sediment on the new sea floor that is older than the time of initial rifting of the ancient continent? Why or why not?

_____

_____

_____

5. Why would the thickness of sediment be expected to increase away from the ridge/rift and in the direction of a continent?

_____

_____

_____

6. Assume that Figure 6.4B represents present conditions and that the age of the basaltic strip at "C" has been accurately measured as 10 m.y. (million years). What has been the rate of separation of the continents, in cm/yr (centimeters per year), during the past 10 m.y.? *Hint:* Use the horizontal scale of 1 mm = 50 km.

_____

_____

_____

7. Assuming that this same rate of spreading has prevailed between "C" and "W," what would be the age of the sea floor at "W"?

_____

_____

_____

8. The dashed lines drawn on the sea surface above "C" and "W" are called **isochrons** (from *iso,* same, and *chrone,* time) and they follow lines of constant age on the sea floor. Using the results from Questions 6 and 7, label the isochrons with the age, in million years, that they represent.

VERIFICATION OF SEA-FLOOR SPREADING

When the sea-floor spreading hypothesis was first proposed, there seemed to be no feasible way of determining the age of large areas of the basaltic sea floor. By a lucky coincidence, geophysicists working on the magnetic properties of basalts on the continents had shown that periodic reversals of the earth's magnetic field were imprinted in successive, superimposed basalt layers, and a time scale for the magnetic reversal pattern during the past 5 m.y. had been established. If the hypothesis of Figure 6.4 is correct, the pattern of reversals ought to be repeated horizontally on the sea floor because the bands (or strips) of basalt would get successively older

Figure 6.5 Photograph of the drilling research vessel *Glomar Resolution.* The *Glomar Resolution* replaces her mothballed sister ship *Glomar Challenger* in the search for more information about the oceans. These vessels were specially designed to provide confirmation of the hypothesis of sea-floor spreading by obtaining cores drilled through sediment on the deep-sea floor. Below the drilling derrick, which stands 194 ft above the water line, is a center well, through which the drilling apparatus has been lowered to drill holes in the sea floor at a water depth of 20,000 ft.

Courtesy of Ocean Drilling Program, National Science Foundation.

with increasing distance from a mid-oceanic ridge/rift. Such a pattern of normal and reversed magnetic bands, symmetrical on either side of a ridge, was indeed found by towing sensitive *magnetometers* behind ships on courses at right angles to the ridge (Fig. 6.5). The pattern that occurs with increasing distance from a ridge (whether in the Atlantic or the Pacific) can be seen by looking down the magnetic reversal column in Figure 6.6. Absolute ages extending back 5 m.y. from present are based on correlation with the pattern of magnetic reversals for dated continental basalts. Ages for the rest of the reversal pattern are based partly on assumptions as to the rates of sea-floor spreading, and partly on the ages of fossils in sediments overlying the basalt. The information about the basal sediments and their fossils was obtained by the highly successful Deep Sea Drilling Program (DSDP), supported by the National Science Foundation and carried out by the Joint Oceanographic Institutions for Deep Earth Sampling (JOIDES). The drilling is done from a unique ship

Figure 6.6 Stratigraphic ranges of fossils in relation to the radiometric time scale and the magnetic reversal scale.

Magnetic reversal scales from Larson and Pitman (1972), *Geol. Soc. America Bull.,* **83:** 3645–3662. Time scale from Palmer (1983), *Geology,* **11:** 503–504.

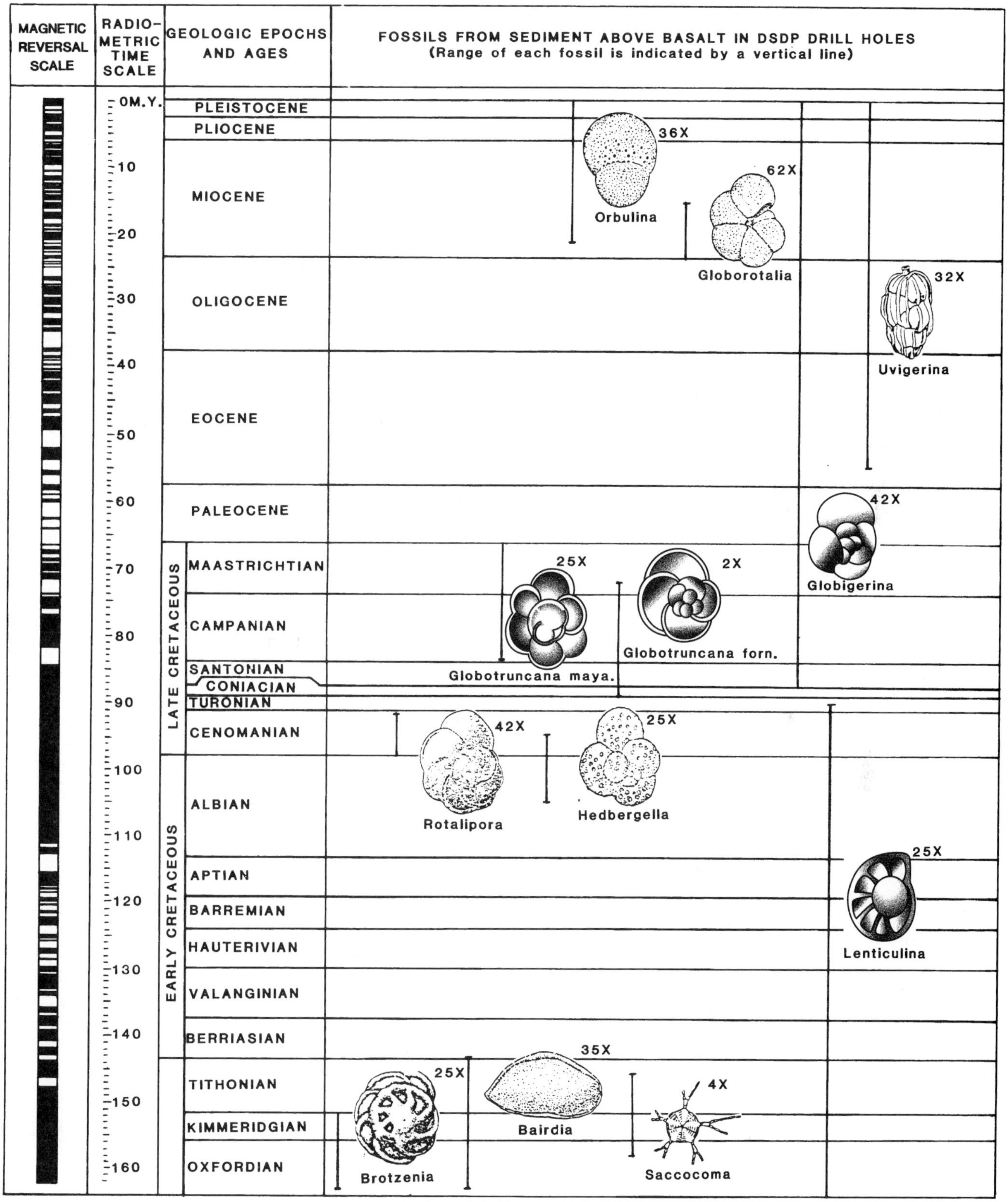

named the *Glomar Resolution* (Fig. 6.5). Cores of basaltic sea floor were also obtained, but these are not suitable for radiometric dating because the basalt was altered, when hot, by contact with sea water (this process is called *hydrothermal alteration;* see Chapter 20).

Study Questions

1. Index fossils found in the basal sediment just above the basalt in each DSDP hole are illustrated in Figure 6.6. From the ranges given in Figure 6.6, determine the geologic epoch indicated by the fossil association at each drill hole in Figure 6.7. Identify the fossils associated with the stratigraphic section at each drill hole and write them in the space provided in Figure 6.7.

2. Now write the geologic age of the epoch, or part of epoch, as indicated by the radiometric time scale, in million years (m.y.).

3. The pattern of magnetic reversals at each DSDP hole is shown in Figure 6.7. Cut off the magnetic reversal scale along the bottom margin of Figure 6.7, and by matching it with the pattern at each hole, find the age in million years of the basaltic sea floor at the hole. (In the technical paper by Pitman and Talwani [1972] on which this study is based, the available pattern of magnetic reversals extended continuously across the ocean, along the courses of ships towing magnetometers. Recognizing the pattern and correlating it from place to place was difficult and to some degree uncertain. **Isochrons** were drawn at prominent anomalies in the magnetic reversal pattern, rather than at drill sites.)

4. Compare the ages for each hole as determined by the two methods, magnetic reversals and fossils. An agreement within plus or minus 5 m.y. would be considered very good, particularly for the older parts of the sea floor. Which method is probably the more reliable at present?

_____

_____

_____

_____

_____

5. Under the assumption that the magnetic reversal age is correct, draw isochrons passing through each drill site and parallel to the ridge/rift axis. Assume that the spreading rates have been symmetrical on either side of the axis, and draw the four isochrons at the correct distance on each side of the axis. Label each isochron with its age in million years. By folding either of the pages inward and superimposing the isochron lines for a particular time, you can see the

relative positions of Africa and North America at that time.

6. What has been the separation rate between North America and Africa, in cm/yr, from the present ridge/rift axis to the isochron corresponding with DSDP Hole 11?

_____

_____

_____

_____

_____

_____

7. What has been the separation rate, in cm/yr, between the isochrons corresponding with Holes 10 and 105?

_____

_____

_____

_____

_____

8. Now measure the total distance of separation between the ridge/rift axis and the 2000-m depth line, along the dotted path of travel passing just south of Bermuda. Using the separation rate, in km/m.y., between Holes 10 and 105 as representative of the average since the beginning of separation, calculate how long ago the ancient continent began to separate.

_____

_____

_____

_____

_____

9. Note the kind of sediment found in the DSDP Holes, the deepest of which penetrated the sea bottom to a depth of 633 m (2077 ft). **Ooze** is defined as a very fine-grained sediment composed in part of the shells and fragments of pelagic microorganisms. Why was coarser land-derived sediment, such as sand, not found in the drill holes?

_____

_____

_____

10. The evidence for sea-floor spreading in the Atlantic Ocean, as represented here, has been the "clincher" that convinced most earth scientists of the validity of the hypothesis and the reality of continental drift. However,

Figure 6.7 Diagram for plotting the time sequence of sea-floor spreading at about Latitude 35° North in the Atlantic Ocean. Dotted "flow lines" connecting the continents indicate the inferred path of travel during separation. Numbered circles indicate location and designation of DSDP drill holes.

Generalized from Pitman and Talwani (1972), *Geol. Soc. America Bull.,* **83:** 619–646.

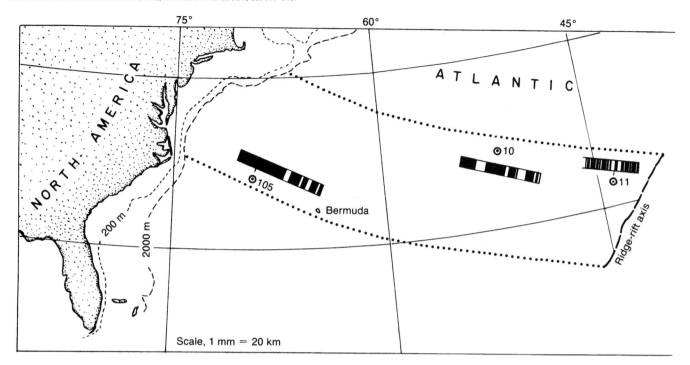

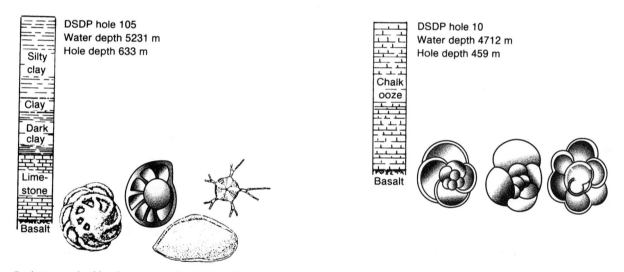

Geologic epoch of fossil association for DSDP drill cores 105 and 10.

_____

_____

Age, in m.y., of geologic epoch for DSDP cores 105 and 10.

_____

_____

Age, in m.y., of basalt, based on magnetic reversal scale for DSDP sites 105 and 10.

_____

_____

Figure 6.7 Continued.

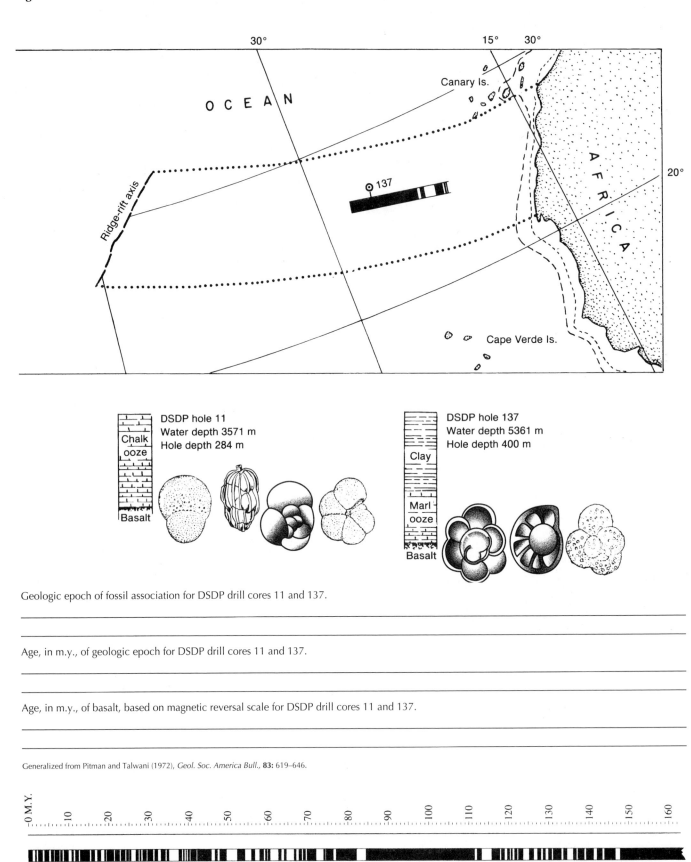

Geologic epoch of fossil association for DSDP drill cores 11 and 137.

Age, in m.y., of geologic epoch for DSDP drill cores 11 and 137.

Age, in m.y., of basalt, based on magnetic reversal scale for DSDP drill cores 11 and 137.

Generalized from Pitman and Talwani (1972), *Geol. Soc. America Bull.,* **83:** 619–646.

Sea-Floor Spreading and Plate Tectonics

not everybody is convinced. Suppose that the drilling at one of the DSDP sites was continued into the basalt for several hundred feet, then passed through it and encountered sediments of Paleozoic age. Would this invalidate the hypothesis? Explain your reasoning.

_____

_____

_____

_____

TERMS

active margin   Plate margin often associated with the convergence of lithospheric plates. Commonly characterized by frequent earthquakes, faulting, volcanic activity, and mountain building.

asthenosphere   A zone within the earth, occurring in the upper mantle between 70 and 250 km, below which seismic waves travel at reduced speed and in which convective flow may occur. The lithosphere rests upon the asthenosphere.

continental drift   A precursor theory to the hypothesis of sea-floor spreading that implied the continents had moved in relation to one another. The mechanism of this theory was so implausible that it was ignored until the more acceptable mechanisms of sea-floor spreading were developed.

convection   The movement of material caused by density differences due to heating. Convection can occur in materials such as mantle rock, and this mechanism is called upon by plate tectonic theory to supply the motive force for the movement of the lithospheric plates.

convergent plate margin   The boundary between two lithospheric plates that are moving toward one another. At this boundary, crust produced at divergent plate boundaries is consumed as the material is pushed/pulled back into the asthenosphere. Also called a **subduction** boundary.

core   The central zone of the earth's interior, constituting about 16% of the total volume of the earth. It is divided into a liquid outer core and a solid inner core and is composed mainly of the metals iron and nickel. The earth's magnetic field originates there. Depth below earth surface: 2900 to 6370 km.

crust   The outermost layer of the earth, defined according to the location of the Mohorovicic discontinuity. This region makes up about 1% of the total volume of the earth. It is approximately 40 to 50 km thick under the continents and about 10 km thick under the oceans.

Curie point   The temperature above which any magnetism is eliminated. Cooling through this temperature retains the magnetic field in existence at that time, providing the material has a magnetic susceptibility.

divergent plate boundary   A boundary between two lithospheric plates that are moving apart, with new oceanic lithosphere being created at the boundary.

isochrons   Lines drawn on a map that connect points at which rocks have approximately the same age.

lithosphere   The outer, rigid portion of the earth. It includes the **crust** and part of the upper **mantle.** Considered to extend to about 70 km depth.

mantle   The zone of the earth below the crust and above the outer core. It is divided into an upper and a lower region and makes up 83% of the total volume of the earth. Depth: 50 km to 2900 km.

ooze   A fine-grained deposit of the deep ocean floor that contains more than 30% skeletal remains of marine organisms.

paleomagnetism   The natural remanent magnetism found in rocks (and certain minerals), which can be quantified in order to determine the intensity and direction of the earth's magnetic field in the geologic past.

passive margin   The boundary between two plates that are not moving toward (or sliding past) one another.

plate   A rigid, thin segment of the lithosphere, which may be assumed to move horizontally, and which adjoins other plates along zones of frequent earthquake activity.

plate tectonics   A theory of *global* tectonics, where the lithosphere is divided into a number of rigid plates that interact with one another at their boundaries, causing seismic and tectonic activity.

polar wander   Long-period displacement of the earth's poles, which may have occurred during the passage of geologic time.

sea-floor spreading   The process by which new oceanic crust is produced along divergent junctions and conveyed away from those junctions.

subduction   The process of one lithospheric plate descending beneath another.

transform plate boundary   A boundary between two lithospheric plates that are moving side-by-side, relative to one another. At this type of plate boundary, the two plates grind past one another. Earthquakes are very common at these boundaries. The San Andreas fault system in California is an example of a transform plate boundary.

7

Age Relations and Unconformity

CRITERIA FOR AGE RELATIONS

In order to interpret earth history from a package of sedimentary rock beds (or other bodies of rock), it is necessary to know which is older and which is younger. In geology, two types of age determinations can be made. One is a **quantitative age measurement** (often called an **absolute age**). This method gives the age of formation of the rock (or the deposition of the sediment) in terms of years before the present. The *absolute age* of a rock can be determined by the technique of radiometric age dating. This method, discussed in more detail in Chapter 8, is used to determine the age of many igneous and metamorphic rocks, but is not used as often for sedimentary rocks.

For sedimentary rocks (and in some cases certain packages of igneous or metamorphic rocks) a **qualitative age measurement** (often called a **relative age**) is inferred from the spatial relations between rock bodies. A *relative age* does not provide a precise measurement in terms of years before the present but indicates whether the rock (or sediment) is older or younger with respect to another spatially related rock body.

The idea of relative age relationships for sediment and rock was developed by Niels Stens (who was also known as Nicholas Steno; 1638–1687), a Danish physician who was employed by the Grand Duke of Tuscany in Italy. Steno observed and described how sediment was deposited and developed in what are now called *Steno's Laws*. The most important of his observations was the *Principle of Superposition*. This basically states that in any undisturbed process of sediment deposition, the oldest material is on the bottom and the youngest on the top.

Other observations of how rocks are emplaced on the earth led Charles Lyell (1797–1875) to develop another set of important relative age relationship "laws." These two observations relate to situations where the sediment (or rock) has been disturbed by erosion, faulting, or igneous intrusions. The first observation is called the *Principle of Cross-cutting Relationships*. This means that if a package of sedimentary strata is cross-cut by another rock (such as an igneous intrusion), the rock that is cross-cut is older. The second observation deals with the presence of inclusions within the rock. The *"Principle of Inclusions"* indicates that the inclusions must be older than the rock in which they are included.

Table 7.1 reviews these three relative age relationship observations and should be used in your interpretation of Figures 7.1 through 7.8.

Table 7.1 Basic observational criteria to interpret relative ages in rock.

> ***Principle of superposition.*** Rock X is inferred to be younger than rock Y if it rests on top of rock Y. This relation is shown by any outcrop of horizontal beds, one above the other, provided that they have not been disturbed by folding (turned upside down) or faulting (where older beds have been emplaced on top of younger beds). If there is reason to suspect that some form of disturbance has occurred, other criteria must be used.
>
> ***Principle of Cross-cutting Relationships.*** Rock X is inferred to be younger than rock Y if it cuts across or displaces rock Y (see Fig. 7.1).
>
> ***Principle of Inclusions.*** Rock X is inferred to be younger than rock Y if it contains inclusions or fragments of rock Y (see Fig. 7.2).

Figure 7.1 Relative age determination using cross-cutting relationships.

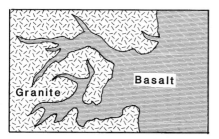 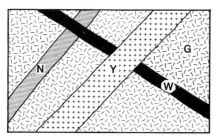 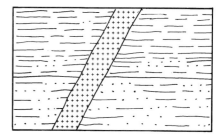

1. **Left** Which is older, the granite or the basalt? Explain your reasoning.

_____

_____

2. **Center** G is granite, and N, Y, and W are **dikes.** Write the letters in correct succession of age, from oldest to youngest. Write a short description of the geologic history of this block from the oldest to the youngest rock.

_____

_____

_____

_____

3. **Right** Shale cut by dike. If the radiometric (absolute) age of the dike is 50 m.y., how old is the shale?

_____

_____

Figure 7.2 Relative age relations from inclusions.

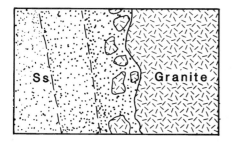 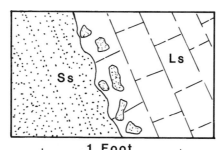 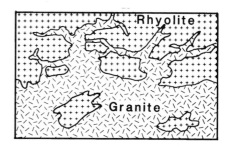

1. **Left** Which is older, the granite or the sandstone? Explain your reasoning.

_____

2. **Center** Which is older, the sandstone or the limestone? Explain your reasoning.

_____

3. **Right** Which is older, the **rhyolite** or the granite? Explain your reasoning.

_____

_____

Figure 7.3 Normal faulting in red and white banded Jurassic siltstone and claystone, near Zion National Park, Utah.

1. What is the probable sedimentary environment of the beds? Explain your reasoning.

    _____

    _____

    _____

    _____

2. Note the small fault, labeled *B,* that is inclined toward the right. Is this fault older or younger than the nearly vertical fault, labeled *A,* that extends from the top to the bottom of the outcrop? Explain your reasoning (refer to Figs. 7.1 and 7.2).

    _____

    _____

    _____

    _____

3. Examine Figure 14.10. Based on that figure, are faults A and B normal or reverse?

    _____

    _____

    _____

    _____

4. Label additional faults observed in the photograph. Are any of the faults related to A or B? Explain your reasoning.

    _____

    _____

    _____

    _____

Figure 7.4 Thick graded bed of immature sandstone, *left,* in contact with interbedded shale and sandstone, *right.* Note inclusions of dark shale in the sandstone bed. Foot rule in photo for scale.

Immature sandstone

Shale inclusions

1. Do the beds get younger toward the left or the right? Explain your reasoning.

    _____

    _____

2. Suggest a probable means by which the shale inclusions got into the sandstone. What would you suggest for the probable tectonic setting and environment (Chapter 5)?

    _____

    _____

Figure 7.5 Chert nodules in clastic limestone of the Mississippian Burlington Fm. near St. Louis, Missouri. Black rule is 6 inches.

1. What is the probable sedimentary environment for deposition (Table 4.1: Figure 4.1)?

    Marine

2. Chert is not an igneous rock; thus, it cannot intrude limestone. Either it formed as clumps of silica gel on the sea floor (same age as limestone), or it has replaced the limestone and is therefore younger. Discuss which of these hypotheses is supported by the spatial relations.

Figure 7.6 Outcrop of **schist** (dark) and granite (light), near Frisco, Colorado.

1. Which is older, the granite or the schist?

2. What evidence can you use to support your conclusion?

3. Based upon your answers, suggest a tectonic process by which these two dissimilar rock types were placed in contact.

Figure 7.7 Vertical airphoto of an area of Archaean to Proterozoic rocks in the Northwest Territories, Canada. Width of area in photograph is about 4 miles. Granite (G) is light, while the **metamorphic** rocks (M) are dark. Straight lines are dikes, the thickest of which (D) is at extreme right.

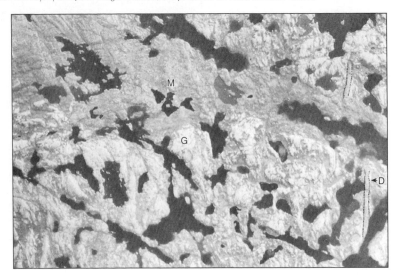

1. The granite has intruded the metamorphic rocks. Which is older, the dikes or the metamorphic rocks?

_____

_____

2. Note the position of the dikes in the airphoto. Based upon your previous answer, what does the displacement of the dikes suggest? Explain your reasoning.

_____

_____

Figure 7.8 Photomicrograph of mature quartz sandstone from the Cambrian Potsdam Fm., Iron Mountain, Michigan. Width of area in photograph is about 2 mm. Crystal faces of quartz, which appear as straight lines, have developed by overgrowth on rounded quartz grains.

1. What is the probable tectonic setting in which the sandstone was deposited (Chapter 5)?

_____

_____

2. When do you think that the quartz overgrowths formed (before or after the transport and deposition of the sand grains)? Explain your reasoning.

_____

_____

THE METHOD OF MULTIPLE WORKING HYPOTHESES

Infer means "arrive at through reasoning from evidence or premises." Earth history can be interpreted using probable inferences from evidence such as the properties of sedimentary rocks and from premises such as the "law" of superposition. Inferences about the past, like other scientific inferences, always have some probability of error. The scientific investigator decreases his or her probability of error by proposing more than one hypothesis to explain the observed facts. This is sometimes called the *method of multiple working hypotheses.* Each hypothesis has certain logical consequences, which lead to the search for additional evidence to confirm or deny it. For example, two alternative hypotheses might be proposed for the origin of a particular bed of sandstone: (1) it was deposited by the wind, or (2) it was deposited by a stream. A logical consequence of the wind deposition hypothesis is that the sandstone should not contain particles too large to be transported by the wind (see Chapter 2). If, on further search in the field for evidence to evaluate the two hypotheses, the sandstone bed is found to contain gravel, the wind hypothesis is discarded.

Although the method of multiple working hypotheses is best suited for complicated situations, its use can be illustrated with a simple example. Let us consider the contact between the diatomite and the brown siltstone, as shown in Figures 7.9 and 7.10. (**Diatomite** is a rock composed of the siliceous coverings of microscopic plants.) In detail, the contact is irregular, and the siltstone contains inclusions of diatomite for a depth of half a meter or more below the contact (Fig. 7.10).

The following working hypotheses can be proposed to account for the nature of the contact:

Hypothesis A   The strata are upside down. The siltstone is actually younger than the diatomite and contains fragments eroded from the diatomite.

Hypothesis B   After the first thin layer of diatomite was deposited on the siltstone, storm waves churned up both the siltstone and the diatomite, incorporating some fragments of diatomite in the siltstone.

Hypothesis C   Diatomite and siltstone were deposited simultaneously for a time, preceding the deposition of pure diatomite.

Hypothesis D   Shortly after deposition of the diatomite began, marine clams burrowed into the siltstone, and their burrows became filled with soft diatomite.

Figure 7.9   Sea cliff at San Gregorio beach, about 40 km south of San Francisco, California. The white stratified layer is diatomite, which is overlain by gravel and underlain by light brown siltstone.

Figure 7.10   Closer view of contact between the diatomite and the underlying siltstone of Figure 7.9.

Study Questions

1. Can any of the hypotheses be rejected because they are not logical? Explain your reasoning. Use Table 7.1.

    _____
    _____
    _____
    _____
    _____

2. What are some of the logical consequences of Hypothesis A? What evidence might be searched for at the outcrop to either confirm or deny the hypothesis?

    _____
    _____
    _____
    _____
    _____
    _____

3. What are some of the logical consequences of Hypothesis D? What evidence might be searched for at the outcrop to either confirm or deny the hypothesis?

    _____
    _____
    _____
    _____
    _____

4. From evidence visible in the photographs, which hypothesis do you think is most probably correct? State your reasons.

    _____
    _____
    _____
    _____
    _____

5. Propose two hypotheses to account for the irregular contact between the diatomite and the overlying conglomerate.

    _____
    _____
    _____
    _____
    _____

UNCONFORMITY

A buried erosion surface (or a surface of nondeposition) that represents a break of significant duration in the geologic record is called a surface of **unconformity,** and the strata that bury the surface are unconformably related to the rocks beneath. The spatial relation at an unconformity is merely one of superposition. However, this break in the rock record, which may amount to tens or even hundreds of millions of years, has obvious significance in the interpretation of the geologic history of the region. The types of unconformity, as illustrated in Figure 7.11, differ as to the arrangement of rocks beneath the erosion surface. It is important to recognize that buried surfaces are not necessarily erosional and even if erosional, they are not necessarily unconformities. For example, a bedding plane is a buried surface that probably represents nondeposition, rather than erosion. Similarly, a buried river channel found in a deposit of fluvial sedimentary rocks represents an erosional surface but probably does not represent a significant break in the geologic record. This is because a river channel can be formed (by erosion) and then filled in a single flood event.

Some useful criteria for recognizing unconformities are as follows:

a. A surface between sedimentary strata and crystalline (either igneous or metamorphic) rocks is an erosion surface if the sedimentary strata show no effects of metamorphism and contain fragments of the underlying crystalline rocks. From this evidence, a break of significant duration (often called a **hiatus**) is inferred, on the premise that intrusive igneous and metamorphic rocks are formed deep in the earth and exposed at the surface only after an episode of tectonic uplift and erosion. This type of unconformity is called a **nonconformity** (Fig. 7.11A).

b. If the underlying strata intersect a surface at an angle, the surface is inferred to be erosional, on the premise that the strata were originally horizontal and must therefore have been first tilted and then beveled by erosion. This type of unconformity is called an **angular unconformity** (Fig. 7.11B). However, if the underlying beds owe their angle to foreset deposition (e.g., cross-bedding; see Fig. 4.3), the inference does not apply, nor is the break likely to be of significant duration.

c. If the sedimentary strata above and below a surface in question are parallel, a break of significant duration is inferred if (1) there is evidence of weathering or soil formation (such ancient soils are called **paleosols**) on the underlying strata; (2) the surface has a noticeable relief, like that of a buried

Figure 7.11 Types of unconformity: (A) **Nonconformity** (strata overlying crystalline rocks). (B) **Angular unconformity** (horizontal strata over tilted strata). (C) **Disconformity** (parallel sedimentary strata separated by a surface of erosion).

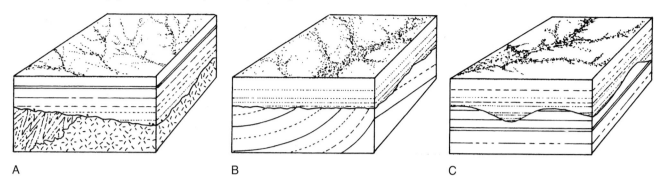

A

B

C

landscape (called **paleotopography**); (3) the association of beds below the unconformity is distinctly different in rock type or origin from the association above it; or (4) fossils below the unconformity are distinctly older than those above

it. These types of unconformities are called **disconformities** (Fig. 7.11C).

Use Figures 7.12 through 7.16 to explore the various types of unconformities.

Figure 7.12 Possible unconformities.

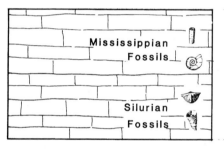
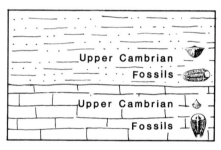
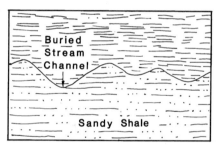

1. **Left** What approximate length of geologic time, in millions of years, is represented by the unconformity?

_____

_____

According to one hypothesis, the Devonian beds are missing because they were deposited and then removed by erosion. What other hypothesis would account for their absence?

_____

_____

2. **Center** Is there any reason to call this an unconformity? Explain your reasoning.

_____

_____

3. **Right** If the beds above and below this unconformity are marine, what events are implied by the unconformity?

_____

_____

If the beds above and below are nonmarine fluvial, what events are implied?

_____

_____

Figure 7.13 Occurrence of events in relation to unconformities.

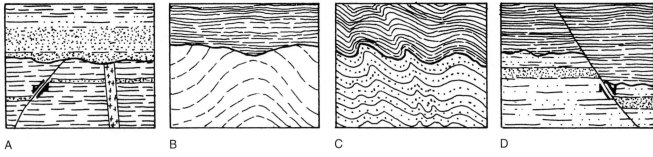

A   B   C   D

1. **Block A** Did the faulting and dike intrusion occur before or after the unconformity?

_____   _____

2. **Block B** What is the type of unconformity? Did the folding occur before or after the unconformity?

_____

3. **Block C** Did the folding occur before or after the unconformity?

_____

4. **Block D** Did the faulting occur before or after the unconformity?

_____

_____

Figure 7.14 Unconformity of Siccar Point, Scotland, where the historical significance of an unconformity was first realized by James Hutton in 1788.

1. What type of unconformity is represented at this location? What are your reasons for thinking this?

_____

_____

2. The vertical beds below the unconformity are graded immature sandstone and shale of Silurian age. What is their probable tectonic setting and environment (Chapter 5)?

_____

_____

3. The beds above the unconformity are coarse, red, immature cross-bedded sandstone of Devonian age. What is their probable tectonic setting and environment (Chapter 5)?

_____

_____

4. Why is an episode of mountain building inferred from this outcrop?

_____

_____

Figure 7.15 Mature quartz sandstone of the Ordovician St. Peter Fm. (white) overlying thin-bedded Ordovician-age dolostone (gray, center), along Interstate 44 near St. Louis, Missouri.

1. What is the probable tectonic setting and environment (Chapter 5)?

_____

_____

_____

2. What is the type of unconformity? Explain your reasoning.

_____

_____

_____

Figure 7.16 Unconformity near Ouray, Colorado. Tilted beds are Proterozoic quartzites (metamorphosed sandstone), while the horizontal beds are Devonian sedimentary rocks.

1. What is the type of unconformity? Explain your reasoning.

_____

_____

_____

2. What is the minimum length, in millions of years, of the break in the geologic record? (See Fig. 8.4 or geologic time scale inside front cover.)

_____

_____

_____

EPISODES OF PRECAMBRIAN HISTORY

While referring to Figure 7.17, answer each of the following statements as probable (P) or not probable (NP). In addition, discuss what evidence or line of reasoning you used to decide. (See Fig. 5.3 for a summary of tectonic settings.)

1. Rock units **Asc** and **As** were deposited in a cratonic setting.

_____

_____

_____

2. Rock units **Co** and **Br** were deposited in an orogenic setting.

_____

_____

_____

3. Granite gneiss **Gg** is older than sediments **Br.**

_____

_____

4. Granite **Gr** is younger than granite **Gg,** and perhaps younger than sediments **Br.**

_____

_____

_____

5. The last period of folding in the region was after the deposition of **Co.**

_____

_____

_____

_____

6. **Asc** and **As** are not necessarily the same age, but both are younger than **Br.**

_____

_____

_____

7. A very long period of erosion intervened between the deposition of **Asc** and **Br.**

_____

_____

_____

_____

8. Sediments **Br** were deposited on the eroded roots of an ancient mountain belt.

_____

_____

_____

_____

Figure 7.17 Geologic cross section of the Algoma Mining District, Canadian Shield. This district is just north of the international boundary, about 480 km directly north of Detroit, Michigan.

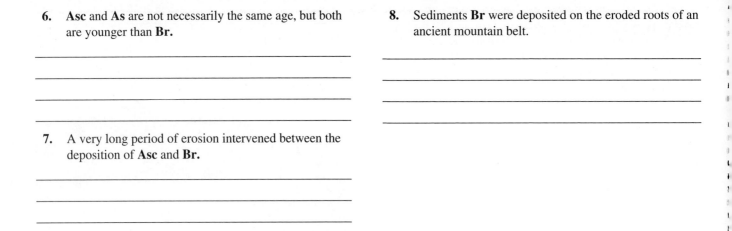

Co   (Cobalt) Quartzite, quartz conglomerate, limestone, siltstone.

Br   (Bruce) Quartzite, conglomerate, limestone, siltstone, arkose sandstone.

Asc Schist derived from basalt; locally contains iron-rich chert.

As   Graywacke sandstone, locally altered to mica schist.

Gg   Granite **gneiss.**

Gr   Granite.

All of the rocks are Proterozoic in age.
Redrawn from the Blind River Map, Geologic Survey of Canada, 1925.

TERMS

absolute (geologic) age   The actual age, expressed in years, of a geologic material or event. Also called a **quantitative age measurement.**

angular unconformity   An unconformity in which older strata are inclined at a different angle than younger strata.

cross-cutting   Presence of a geologic feature such as a vein, igneous intrusion, or fault that cuts across a preexisting feature or rock mass.

diatomite   A sediment composed of the siliceous covering of microscopic plants called diatoms.

dike   A tabular body of igneous rock that cuts across preexisting stratification or structures.

disconformity   An unconformity in which the bedding planes above and below the break are essentially parallel. Usually marked by a visible erosion surface. A *paraconformity* is similar to a disconformity in the parallelism of the bedding planes, but the erosion surface is undetectable.

gneiss   A coarse-grained metamorphic rock with compositional banding and parallel alignment of minerals. Commonly formed by the metamorphism of granite.

granite   A phaneritic igneous intrusive rock composed primarily of orthoclase feldspar, sodic plagioclase, quartz, and minor amounts of muscovite, biotite, and hornblende.

hiatus   A lapse or break in continuity.

inclusion   A mineral or rock particle enclosed within a body of igneous rock, or a mineral particle within a crystal. Gas- or water-filled cavities in crystals are also termed inclusions.

metamorphism   Process by which rock is changed in both mineral composition and form by heat and/or pressure.

nonconformity   An unconformity developed between sedimentary rocks and older rocks (plutonic igneous or metamorphic rocks generally called *crystalline basement*) that had been eroded before the sedimentary rocks were deposited.

paleosol   A buried soil horizon of the geologic past.

paleotopography   The natural or physical surface features of an area at a particular time in the geologic past. This is one aspect of the study of *paleogeography.*

relative (geologic) age   The chronologic placement of a rock body or event in the geologic timescale without reference to its absolute (geologic) age. A relative age expresses whether the object or event is older or younger than another object or event. Also called a **qualitative age measurement.**

rhyolite   An aphanitic or porphyritic igneous extrusive rock having a composition similar to that of granite.

schist   A metamorphic rock having subparallel orientation of flaky minerals (like biotite and muscovite), or needlelike minerals that impart a foliated appearance to the rock.

superposition   (Principle of) The principle that in any sequence of undisturbed strata, the oldest layer is at the bottom, and successively higher layers are successively younger.

unconformity   A buried surface, often produced by erosion, that represents a gap or break in the geologic record of considerable duration.

8

Rock Units and Time-Rock Units

1. A display copy of the standard geologic column and geologic time scale (available from Wards Scientific or Geological Society of America). Students should use the geologic time scale on the inner cover of this laboratory manual for reference to the text and the questions of this and later chapters.

2. Geologic map (with geologic cross section) for the Grand Canyon National Park (available from U.S. Geological Survey; instructor-provided).

3. Geologic Map of North America (instructor-provided) and Plate 1 and Plate 2 (following Chapter 20).

HISTORY OF A LAKE

INTRODUCTION

Episodes in the geologic history of a locality can be inferred from the outcrop. But how can the history at one locality be related to that of other localities, either nearby or far away? The method of doing this is illustrated here by an imaginary situation that, although small in the scale of time and space, will enable you to understand how earth history is reconstructed on a much larger scale. The history of a lake, originally formed by a landslide dam in a river valley, is interpreted by a geologist from the sedimentary beds that accumulated in it (Fig. 8.1).

Recognition of Rock Units, or Formations

A rock unit does not have to be confined to an episode in time or consist of a single **bed,** but geologists usually group the beds of a vertical sequence into rock units called **formations.** The beds in a formation consist either of a single rock type or of related rock types, and the formation is bounded at the top and bottom by a change in rock type, color, texture, or some other recognizable or distinctive feature. Also, a formation should be sufficiently thick (several tens of feet) so that its out-cropping edges will be wide enough to appear on a geologic map. Ideally, it represents a single sedimentary

environment, and its top and bottom boundaries represent a change in environment. In the interpretation of sedimentary rocks, formations may be either combined into **groups** or subdivided into what are called **members.** A member may consist of a single bed or a combination of beds.

Our geologist decided that one exposure, labeled Section 1 in Figure 8.2, contained the most complete succession of rock units and selected this as the **type section** for the units exposed there. From this, the geologist prepared the columnar sections of Figure 8.2.

Fossils

For recognizing formations, the **fossils** contained in the rocks can be regarded merely as distinctive features, like bedding or rock type. If fossils are to be used as indicators of age, they must be accurately identified, and their ranges (time spans of existence) must be obtained from reference works in paleontology. The rock units deposited in the lake contained the following artifacts and fossils:

1. ***Bottleus breweri*** An elongate brown bottle, probably intended to contain beer. Carried by fishermen of the species *Homo sapiens,* who distributed them widely in and around lakes. At Section 1, restricted to Rock Unit Y. (Range: A.D. 1900 to present.)

2. ***Jugus firewateri*** Crockery jug carried by fishermen and widely distributed in and around lakes. At Section 1, restricted to lower part of Rock Unit Y. (Range: A.D. 1850–1910.)

3. ***Valvata sp.*** Species of snail that inhabits the muddy bottoms of lakes. At Section 1, restricted to Rock Unit Y. (Range: Pliocene—Recent.)

4. ***Unio sp.*** Species of clam that inhabits the sandy shores of lakes. At Section 5, restricted to Rock Unit S. (Range: Miocene—Recent.)

5. ***Bison bison*** The western buffalo. Most easily recognized remains are skulls with horns, which are restricted to Rock Unit Q. (Range: known only from the Recent.)

6. ***Parelephas columbi*** A mammoth, relative of the elephant. Tusk is most distinctive fossil remain. Restricted to Rock Unit G. (Range: late Pleistocene.)

Figure 8.1 (A) Bottom sediments of a lake basin in 1920, exposed by draining of the lake. (B) The same sediments in 1975, after being deeply entrenched by streams. Numbers give locations of stratigraphic sections in Figure 8.2.

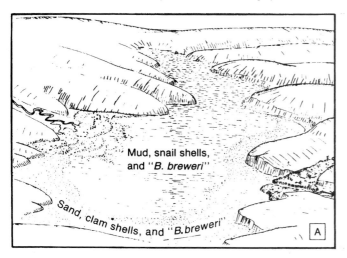

Figure 8.2 Stratigraphic sections at localities numbered in Figure 8.1B.

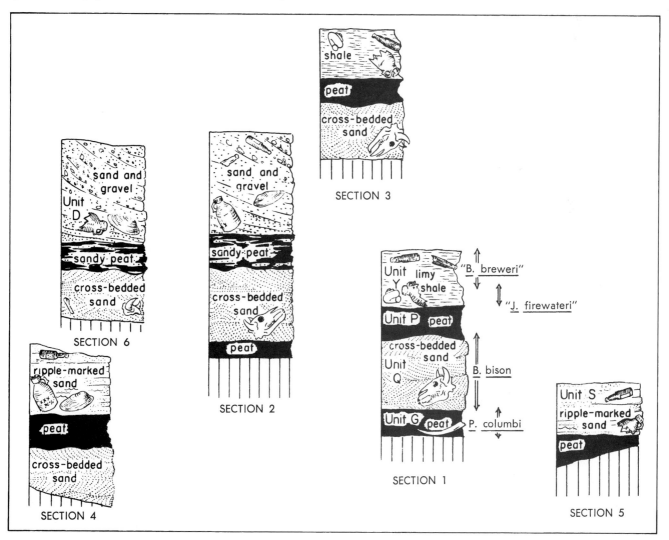

Correlation of Rock Units

Where possible, a **rock unit** is correlated from one outcrop to the next by *tracing it continuously* over the intervening ground. If continuous tracing is not possible, the presence of the unit at the second outcrop can usually be discerned by some *characteristic feature,* such as rock type, color, bedding, or fossil content. Correlation is aided if the unit in question has a distinctive *position in a sequence* of rock units. For example, if a rock unit at one outcrop is underlain by black shale and overlain by red sandstone, it is likely to correlate with a unit having this same position at a second outcrop.

The following questions on correlation refer to the stratigraphic sections in Figure 8.2.

Study Questions

1. Rock Unit Y is labeled at Section 1. At which of the other sections is it present? Which of the methods described in the paragraph above did you use to make the correlation?

_____

_____

_____

2. Rock Unit Q is labeled at Section 1. At which of the other sections is it present? Which of the methods described above did you use to make the correlation?

_____

_____

_____

3. Rock Unit S is identified at Section 5. At which of the other sections is it present?

_____

_____

_____

4. Is the peat at Section 5 most probably correlative with Unit P, or Unit G?

_____

_____

_____

Correlation of Time-Rock Units

A **time-rock unit** consists of an assemblage of strata deposited during a particular interval of time. An ideal example of a time plane, of the sort that bounds time-rock units, is the bottom of a body of water such as a lake. As shown in Figure 8.1A, several different kinds of sediment were being deposited on the bottom of the lake when it was drained, and the top of these is a stratigraphic time plane for the year A.D. 1920. Buried time planes are likely to correspond with bedding planes, but time planes usually cannot be traced very far. A time-rock unit is usually correlated from place to place by its contained fossils, in partic- ular those whose range, or known time span of existence, is restricted to a relatively short span of geologic time. In our example, the ranges of "fossils" characteristic of each time-rock unit are shown by arrows to the right of Section 1 in Figure 8.2.

Study Questions

1. Which rock unit(s) were accumulating at Sections 2 and 6 while Rock Unit Y was accumulating at Section 1?

_____

_____

_____

_____

2. Which rock unit(s) were accumulating at Sections 4 and 5 while Rock Unit Y was accumulating at Section 1?

_____

_____

_____

3. Why is it probable that Rock Unit P was deposited at the same time at all the sections, even though it contains no guide fossils? (P is an example of a stratum that is both a rock unit and a time-rock unit.)

_____

_____

_____

4. Draw dashed lines between Sections 6, 2, 1, and 5 to connect the time-equivalent rock units. These lines are the boundaries of time-rock units.

_____

_____

Inferred History of the Basin

For purposes of historical inference, observational data and correlations are organized in the form of stratigraphic sections (Fig. 8.2) and stratigraphic diagrams (Fig. 8.3). Another useful device is the lithofacies map, which shows the areal distribution of rock types in either a rock unit or a time-rock unit. The distribution of bottom sediments in Figure 8.1A is a **lithofacies** map for the top of a time-rock unit. It is also a **biofacies** map because it shows the areal distribution of organisms.

Study Questions

1. Why does Unit D contain only *J. firewateri* at Section 6 although it contains both *J. firewateri* and *B. breweri* at Section 2 (see Figures 8.2 and 8.3)?

_____

_____

_____

_____

Figure 8.3 Stratigraphic diagram along the line of Sections 6, 2, 1, and 5 from Figure 8.1B. Rock units D, Y, and S constitute a time-rock unit. Units P, Q, and G are both rock units and time-rock units.

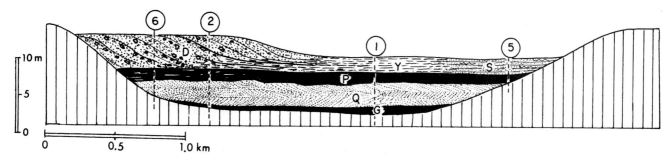

2. We have demonstrated that Rock Units Y and S are the age equivalents of D. Why was sand deposited in the area of Unit S while silt and marl were being deposited in the area of Unit Y?

_____

_____

_____

_____

3. Because of their relatively long ranges, the fossils *Valvata sp.* and *Unio sp.* are not useful for age determination of these strata. But what other usefulness do they have for historical interpretation?

_____

_____

_____

_____

4. Why is Unit D much thicker than Unit Y, even though it was deposited during the same interval of time?

_____

_____

_____

5. Was the lake present when Rock Unit Q was deposited? If not, how was it deposited?

_____

_____

_____

THE STANDARD GEOLOGIC COLUMN

If a single region were known in which sediments had been deposited continuously throughout all of geologic time, the **geologic column** of this region would serve as a standard with which the incomplete histories of other regions could be compared. Unfortunately, no such region exists, and there-fore, a standard geologic column must be constructed as a composite of many local columns. As the early geologists had no idea of how many time units would be represented, the geologic column grew piecemeal. The *standard geologic column* (Fig. 8.4) consists of units that were named and dis-covered in various parts of the world. These relatively well-exposed and complete sequences of strata, located in time by the principle of superposition and correlated by their fossil assemblages, are called **type sections.** They are utilized worldwide as a standard to identify a stratigraphic unit or to give a relative age to a fossil assemblage.

In order to compile the complete standard geologic column, the type sections for all recorded parts of geologic time had to be piled one on top of the next in the proper order, from oldest on the bottom to youngest at the top. As noted above, their **relative ages** had first to be determined by the principle of superposition and by fossil correlation. In the beginning, however, no assumptions were made as to the rel-ative ages of different fossil assemblages; the fossils were simply treated as part of the lithology of the type section. The rocks that contain the most primitive fossils were not placed near the bottom of the geologic column because the fossils were primitive, but because the principle of superposition required that they be placed there. The exercise of Figure 8.5 will give insight into the methods used in constructing the standard geologic column.

All of geologic time is divided into three components (called **eons**): The *Archaean,* the *Proterozoic,* and the *Phanerozoic* ("evident life"). The period of time that in-cludes the Archaean and the Proterozoic is informally termed the *Precambrian* and comprises 87% of all geologic time. The Archaean began about 4.6 billion years ago and ended at 2.5 Ga (billion years ago). During the Archaean, there was no life and the atmosphere and the hydrosphere of the earth was much different from today. The Proterozoic extends from 2.5 Ga to 542 Ma (million years ago). In the Protero-zoic, the atmosphere and hydrosphere of the earth began to have free oxygen and life originated first as simple bacteria and single–celled organisms. In the Late Proterozoic (around 700 to 600 Ma), life became more complex, with some of the first soft- and hard-bodied organisms. The Proterozoic-Phanerozoic boundary at 542 Ma is based on abundant (and diverse) hard-shelled animal fossils in the rock record.

Figure 8.4 Standard geologic column. (After International Stratigraphic Chart, 2004, International Commission on Stratigraphy)

Reprinted by permission of National Research Council of Canada.

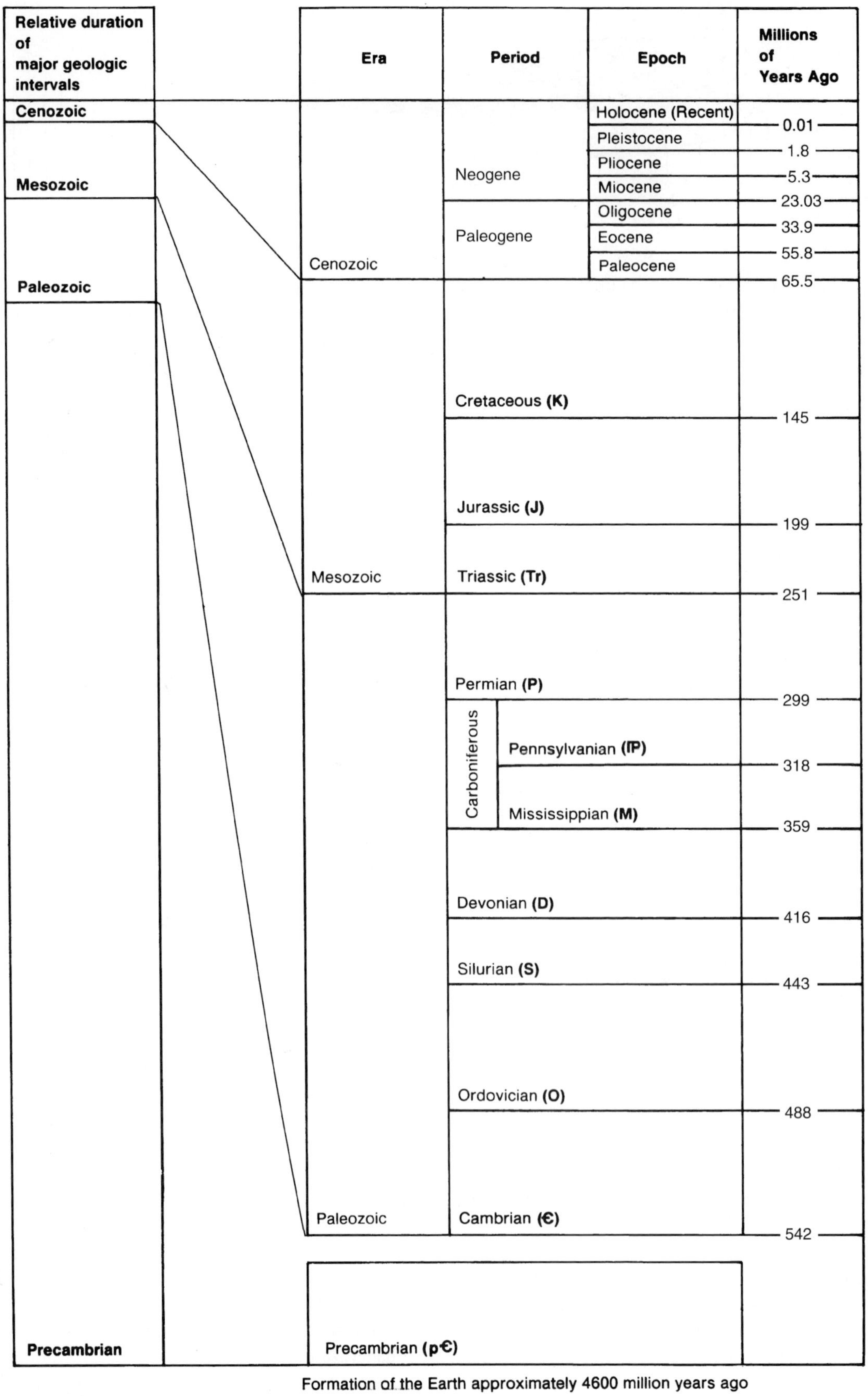

Era	Period	Epoch	Millions of Years Ago
Cenozoic	Neogene	Holocene (Recent)	0.01
		Pleistocene	1.8
		Pliocene	5.3
		Miocene	23.03
	Paleogene	Oligocene	33.9
		Eocene	55.8
		Paleocene	65.5
Mesozoic	Cretaceous (K)		145
	Jurassic (J)		199
	Triassic (Tr)		251
Paleozoic	Permian (P)		299
	Carboniferous — Pennsylvanian (℗)		318
	Carboniferous — Mississippian (M)		359
	Devonian (D)		416
	Silurian (S)		443
	Ordovician (O)		488
	Cambrian (€)		542
	Precambrian (p€)		

Relative duration of major geologic intervals: Cenozoic, Mesozoic, Paleozoic, Precambrian

Formation of the Earth approximately 4600 million years ago

Exercise: The Length of Geologic Time

Long intervals of time, such as hundreds of millions of years, are difficult for the mind to grasp (much like a trillion dollars or a million stars in the night sky). In order to understand the "true" length of time that has passed since the formation of the earth, one must put geologic time into perspective.

Materials

This exercise can be done either indoors or, weather permitting, outdoors. For the outdoor exercise the instructor should find a long (~300 ft) sidewalk and provide chalk and a tape measure (in feet and inches or meters and centimeters). For the indoor exercise, a roll of register tape (for a calculating machine or cash register), a long hallway (or laboratory), and a scale (with the smallest increments of 0.1 inch should be provided).

Procedure

Break into groups of two and determine the total distance of either the sidewalk or register tape. From the geologic time scale on the inside of the front cover, compute the appropriate distance at which each eon, era, period, and epoch (for the Cenozoic) occurred, using the total length of the sidewalk or register tape. Include the following events in addition to those from the geologic time scale. Mark on sidewalk (using chalk) or register tape (using pencil or pen) the events that occurred.

(BYA = billion years ago; MYA = million years ago)

Date	Event	Computed Distance
4.6 BYA	Formation of the Earth	
4.0 BYA	Oldest rocks found on Earth	
2.8 BYA	Oldest stromatolites in the fossil record	
1.8 BYA	Banded iron formations	
1.0 BYA	Grenville Orogeny	
630 MYA	Ediacaran faunas	
525 MYA	Trilobite fossils in the record	
520 MYA	Formation of Gondwanaland supercontinent	
512 MYA	Burgess Shale fossils	
490 MYA	Taconic Orogeny	
470 MYA	First land plants	
410 MYA	First jawed fishes	
375 MYA	First insects	
300 MYA	Alleghany Orogeny	
250 MYA	Formation of Pangaea and Laurasia continents with the breakup of Gondwanaland	
230 MYA	First dinosaurs	
225 MYA	First mammals	
160 MYA	First birds	
120 MYA	First flowering plants	
65 MYA	Extinction of the dinosaurs	
45 MYA	First uplift of the Himalaya Mountain Range	
8 MYA	First humans (early hominids)	
10,000 years	End of the last glacial period (Ice Age)	

The Phanerozoic has been divided into three major subdivisions, termed *eras.* The oldest era is the *Paleozoic,* from 542 to 251 Ma, followed by the *Mesozoic,* from 251 to 65 Ma, and lastly, the *Cenozoic,* the era in which we live.

These eras (Fig. 8.4) are divided into shorter time units called **periods.** The rocks formed during these periods constitute a time-rock unit called a **system.** Thus, the distinctive fossil assemblages of a particular system allow identification of the given period within the geologic column in which the organism lived. This we will consider in more detail in Chapters 10 and 11.

The use of time-rock units to specify relative time as based on their fossil assemblages sometimes leads to confusion in nomenclature. Table 8.1 shows the relation between time, rock, and time-rock units. Note that although Table 8.1 indicates the relative scale for each time, time-rock, or rock unit, the *only agreement is between individual subunits in time and time-rock units.* Thus, the temporal equivalent of the time-rock unit stage is age.

Furthermore, as a general rule, when talking about a time unit, you may indicate the age relation such as *Early* Cambrian versus *Late* Cambrian, but with a stratigraphic unit, it is always *Lower* Cambrian versus *Upper* Cambrian.

Figure 8.5 Columnar sections of hypothetical regions. The sections represented have been divided into the time-rock units called **systems.** The systems have been correlated from one region to the next on the basis of similarities in lithology, including fossil content. Using the principle of superposition, correlate each group and compose a geologic column by writing the letter designations of the systems in proper order, from the oldest on the bottom to the youngest, in the blank geologic columns at left.

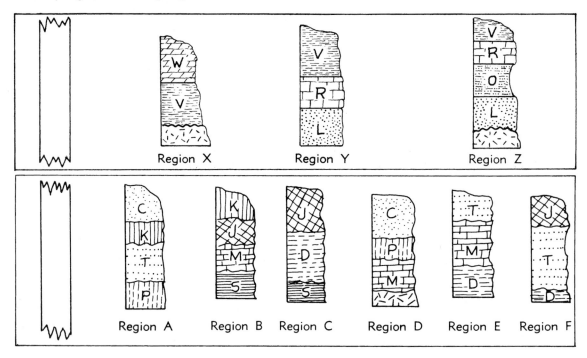

Table 8.1 Nonmenclature for Time, Time-Rock, and Rock Units. Descending Order Implies Subdivision into Successively Smaller Units

Time	Time-Rock	Rock
Era	Erathem	Group
Period	System	Formation
Epoch	Series	Member
Age	Stage	Bed

 ## ABSOLUTE AGE DETERMINATION

In our discussion of the geologic time scale and the rock column that it encompasses, we have been using relative ages. Only since the discovery of radioactivity have scientists been able to assign absolute values to the geologic time scale and to determine how many years ago the various eras, periods, and epochs began and ended.

Absolute dating by the use of radioactive decay provides a natural "clock" that starts when certain types of rocks are formed. **Radioactive decay** is the spontaneous nuclear disintegration of isotopes of certain chemical elements into their stable daughter isotopes. As disintegration occurs, energy is released and can be detected by a Geiger counter or similar device. The decay rate for isotopes is stated in terms of **half-life:** the amount of time it takes for a given amount of a radioactive parent isotope to be reduced (converted to its daughter isotope) by one-half (Table 8.2). By comparing the amount of a radioactive (parent) isotope in a particular rock with its decay (daughter) product, the geologist can determine how much time has elapsed since the rock formed.

Some caveats about absolute ages must be mentioned. Although the *rate* of decay (called the **decay constant** and expressed by the symbol λ) for a particular radioactive isotope is considered to be a constant and reasonably well known, the amounts of daughter product and/or the radioactive parent isotope may have been altered since the rock formed. These variations may cause the age measured to be younger (by removal of daughter product or by addition of the parent isotope) or older (by addition of daughter product or removal of parent isotope) than what the age would be if no alteration had occurred (Fig. 8.6).

Study Questions

1. In Figure 8.6A, the general relation of a half-life is portrayed. The decay constant (λ) is related to the half-life by this mathematical relation:

$$T_{half\text{-}life} = \frac{0.693}{\lambda}$$

Table 8.2 Radioactive Isotopes Commonly Used for Absolute Age Determinations

Parent Isotope	Daughter Isotope	Half-life (M.Y.)	Materials
^{238}U	^{206}Pb	4498	zircon, uranite
^{235}U	^{207}Pb	713	zircon, uranite
^{232}Th	^{208}Pb	13,890	zircon, uranite
^{87}Rb	^{87}Sr	48,800	micas, igneous, and metamorphic rocks
^{40}K	^{40}Ar (gas)	1310	micas, hornblende, igneous, and metamorphic rocks
^{147}Sm	^{143}Nd	106,000	igneous rocks
^{14}C	^{14}N (gas)	5730 *years*	wood, charcoal, bone, cloth, water, animal tissue

Note that ^{14}C has a very short half-life in comparison to the other radioactive isotopes and that its daughter product, ^{14}N, is a gas. Both this isotope pair and the $^{40}K - ^{40}Ar$ pair can give erroneous results because the daughter product can be easily lost. In all cases, the best approach to evaluating an absolute age is to use several isotope pairs and bracket the age of the sample.

For each parent-daughter isotope pair in Table 8.2, determine the decay constant (λ). What are the units of the decay constant?

^{238}U decay constant (λ) _____

^{235}U decay constant (λ) _____

^{232}Th decay constant (λ) _____

^{87}Rb decay constant (λ) _____

^{40}K decay constant (λ) _____

^{147}Sm decay constant (λ) _____

^{14}C decay constant (λ) _____

2. To answer the following question, we will first outline a simplified explanation of the derivation of the general age equation. This derivation is in general terms so that it is applicable to all radioactive dating techniques (except fission tracks).

Although the decay constant (λ) differs for each parent-daughter pair, all radioactive decay follows the exponential curve of Figure 8.6A. This constancy allows for the relation of the parent to daughter isotope by the general age equation:

$$N = N_O \left(\exp\right)^{-(\lambda)t}$$

where N is the number of parent atoms measured today, N_o is the number of parent atoms at time t in the past when the rock formed, λ is the decay constant for the particular radioactive parent, and *(exp)* is the log e. This equation is not very useful for calculating ages because it is not possible to determine N_o without first knowing t. This difficulty can be circumvented by recognizing that if neither parent nor daughter atoms are lost or gained except by radioactive decay, the number of daughter atoms plus the number of parent atoms (measured today) equals the number of original parent atoms. Therefore, if $N_o = N + D$, where D is the number of daughter atoms measured today,

$$D = N \left[\left(\exp\right)^{(\lambda)t} - 1\right]$$

As an example, suppose the analysis of a rock gives a daughter ratio (in this case for Sr): $(^{87}Sr/^{86}Sr) = 0.7468$ and a parent ratio: $(^{87}Rb/^{86}Sr) = 7.72$. We use the isotope ^{86}Sr for the denominator because it does not decay spontaneously and we can assume that its abundance has been relatively constant since the time the rock formed. In order to use the general age equation, we must know the initial daughter ratio, in this case $(^{87}Sr/^{86}Sr)$initial. For this example, we can assume its value is 0.7100 (a value that can be determined by the isochron method shown in Fig. 8.6). By rearranging the general age equation and solving for $t_{formation}$, the resulting equation is

$$t_{formation} = \frac{1}{\lambda} ln\left[\frac{D - D_O}{N} + 1\right]$$

Substituting in the values, we obtain: $t_{formation} = 3.35 \times 10^8$ years (335 Ma).

A final substitution, involving an assumption that the total number of daughter atoms is equal to the number of daughter atoms the system started out with at the formation of the rock plus those formed by radioactive decay, gives

$$D = D_O + N\left[\left(\exp\right)^{(\lambda)t} - 1\right]$$

where D_o is the number of daughter atoms that were present when the rock formed. Generally, it is easier to measure ratios of the daughter and parent atoms than to count them individually. Thus, the general form of the age equation is

$$\frac{D}{K} = \frac{D_O}{K} + \frac{N}{K}\left[\left(\exp\right)^{(\lambda)t} - 1\right]$$

where K is an isotope of the same element that is stable (i.e., does not radioactively decay).

This equation has the form of a straight line (i.e., y = mx + b). If the samples chosen to evaluate an absolute age formed at the same time, the measured ratios will form a linear array as shown in Figure 8.6C. The best-fit straight line through the data points is

Figure 8.6 Radioactive decay (A and B). Decay of the parent atoms (A) follows an exponential curve. At time zero, no daughter atoms are present (B). After one half-life, 50% of the parent atoms remain, and 50% have been converted to the daughter product (compare a given point on Fig. 8.6A with the corresponding point on Fig. 8.6B to see the proportionality).

Isochron diagrams (C and D). (C) Four samples taken from igneous rock formed at the same time and analyzed for their Rb and Sr isotopic ratios form a straight line (an **isochron**). The intercept of the isochron with the ordinate yields the initial isotope ratio *at the time of formation of the igneous rock.* The slope of the isochron gives the time that has elapsed since the rock crystallized. (D) At time t_0, the igneous rock has crystallized; there is no slope to the isochron. At time t_1, the ^{87}Rb (parent) has decayed to ^{87}Sr (daughter), and each sample has evolved in its ^{87}Sr/^{86}Sr ratio, moving from t_0 to t_1. The slope of the evolution line is controlled by the particular decay constant of the radioactive species. At time t_2, more ^{87}Rb has decayed to ^{87}Sr, increasing the slope of the isochron.

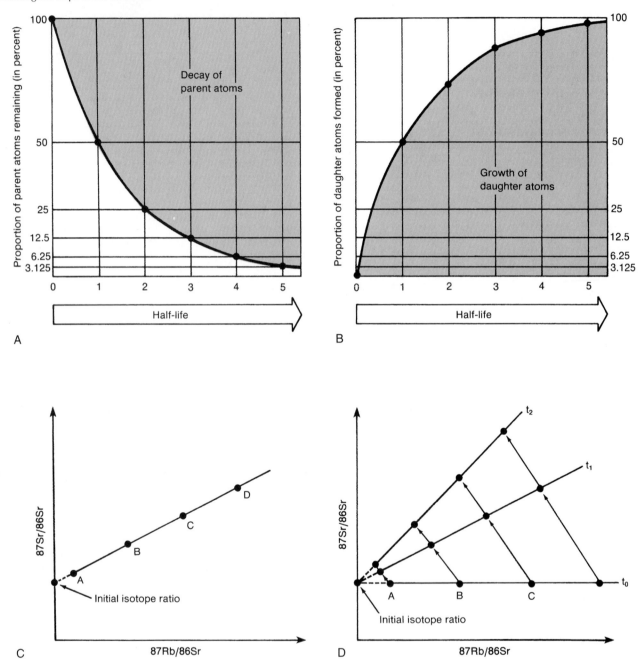

called an **isochron** ("equal time"). The slope of the line is a measure of the age of the samples. The steeper the line, the older the rock. The intercept of the isochron is the initial ratio of the daughter isotope at the time of formation of the rock.

3. Using the data in Table 8.3, plot the sets of data on a parent-daughter diagram (Fig. 8.6C) and determine graphically the age of each set. Then use the general form of the age equation derived in the preceding discussion and recalculate the ages. Explain what errors could arise

Table 8.3 Rubidium—Strontium Mass Spectrometry Data

A. Salisbury pluton, Rowan County, North Carolina (from Fullagar, P.D., 1971, *Geol. Soc. Am. Bull.,* v. 82, 409–416).

Sample	$^{87}Sr/^{86}Sr$	$^{87}Rb/^{86}Sr$
A-2	1.1970	86.7
C-2	0.7468	7.72
D-3	0.8341	22.8
F-3	0.9979	50.7
G-6	0.7111	1.26
I-1	0.7132	1.78
I-2	0.7239	3.90

B. Whitewater sediments (graywacke) surrounding the Sudbury lopolith, Sudbury, Ontario (from Fairbairn, H. W., 1968, *Can. J. Earth Sci.,* 5, 707–714).

Sample	$^{87}Sr/^{86}Sr$	$^{87}Rb/^{86}Sr$
R5945A	0.8749	7.47
R5946	0.7328	1.11
R5945	0.9379	9.11
R5945B	0.8783	7.13
R5948B	0.8191	4.35
R5949A	0.8809	7.65
R5943A	0.7545	2.06
R5939A	0.7563	1.95
R5942B	0.8147	4.77
R5947A	0.8160	4.45
R5948A	0.8162	4.62
R5946A	0.8408	5.63
R5949B	0.8505	6.40

by using only an age determination derived from the graphical approach. Comment on what assumptions you must use to assign an age using the age equation.

_____

_____

_____

_____

_____

_____

FORMATIONS OF THE COLORADO PLATEAU

The Grand Canyon of the Colorado River

1. Compare Figure 8.8 with the lower part of the **columnar section** of Figure 8.7, and draw on Figure 8.8 the unconformable contact between the Vishnu Schist (dark, not stratified) and the basal Cambrian sandstone (T).

2. The Redwall Limestone (R) forms the prominent cliff in the foreground and, in the left side of the photo (but not on the right), the next high cliff above the Cambrian sandstone. Draw in the contact at the top of the Redwall Limestone.

3. The next prominent cliff, near the top of the canyon, is formed by the white Coconino Sandstone (C). Draw in the contact at its base.

4. At the right side of the photo, a prominent cliff (G)—not present at the left—occurs below the Redwall cliff. This cliff is formed by a quartzite member of the Grand Canyon Series of Precambrian age (see Fig. 8.7). Why doesn't this cliff appear on the left side of the photo?

_____

_____

_____

5. What is the type of unconformity between the Vishnu Schist and the Grand Canyon Series?

_____

_____

Between the Grand Canyon Series and the Cambrian?

_____

_____

If each of these indicates an episode of mountain building, which is likely to have been of longest duration?

_____

_____

6. Ordovician and Silurian rocks are missing from the Grand Canyon section, and the Devonian is represented only by an eroded and discontinuous layer of limestone (see columnar section, Fig. 8.7). What is the type of unconformity at the base of the Devonian?

_____

_____

_____

About what length of geologic time does this unconformity represent (in million years)? (See geologic time scale inside front cover.)

_____

_____

Figure 8.7 *Upper right,* location map for Colorado Plateau. *Left,* generalized columnar section for western part of Colorado Plateau. *Bottom,* geologic structure section from Pausaugunt Plateau south to the Grand Canyon (see N-S cross-section line on inset location map).

Modified from Erwin Raisz, *Landforms of the United States.*

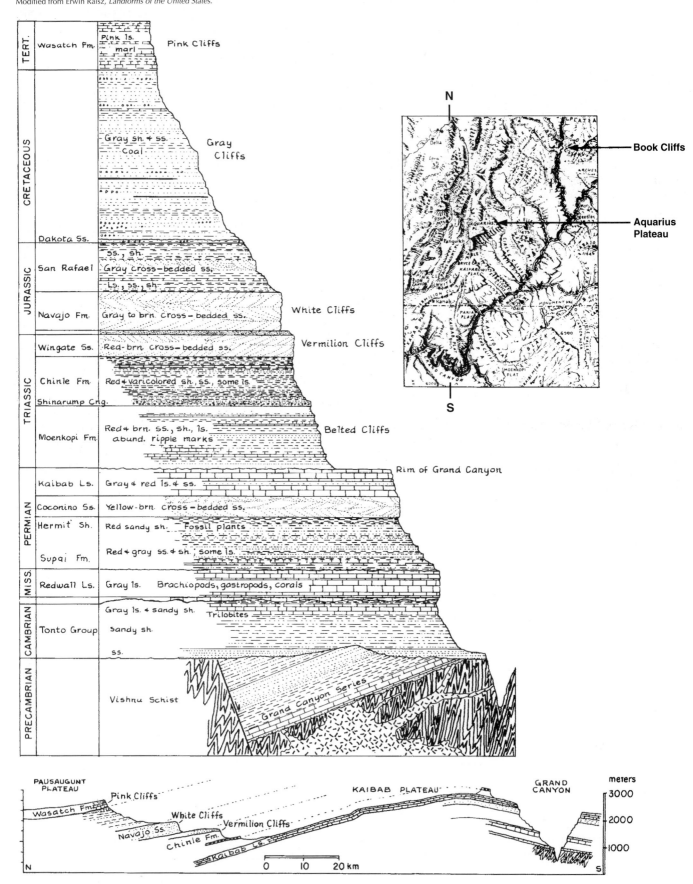

Figure 8.8 View of Grand Canyon from Hopi Point.

Courtesy of R. G. Leudke, U.S. Geological Survey.

7. What was the probable tectonic setting (refer to Fig. 5.3) and sedimentary environment (Table 4.1) of the Colorado Plateau in the Cambrian?

_____

_____

In Mississippian time?

_____

_____

In Permian time?

_____

_____

Capitol Reef National Monument

Capitol Reef is located in south-central Utah about 150 miles northeast of the Grand Canyon, and just south of the Aquarius Plateau on the location map of Figure 8.7. Although the formations of the Colorado Plateau are generally horizontal (the dip is much exaggerated in the structure section of Fig. 8.7), there are some broad folds, and the beds at Capitol Reef dip because they are on the flanks of the Waterpocket Fold.

Study Questions

1. In Figure 8.9, the main cliff at the upper right (the Wingate Sandstone) is about 350 ft (107 m) high. Draw in the contact at the base of the Wingate.

2. The Moenkopi Fm. forms a line of lower cliffs ("Belted Cliffs") just above the road. It consists here of red-brown shale, red sandstone, and limestone; crossbedding and ripple marks are common. The top of the Moenkopi is usually easy to find because it is capped by the resistant Shinarump Conglomerate. In this cliff, however, the Shinarump is a very thin layer. Draw in the top of the Moenkopi at the uppermost

Figure 8.9   West side of Capitol Reef, view looking northeast.

Courtesy of J. R. Stacy, U.S. Geological Survey.

resistant bed, beneath a lighter colored shale. Notice that determining the contact relation between formations is not always easy.

3.   What was the sedimentary depositional environment of the Moenkopi Fm.?

4.   To what formation does the Shinarump conglomerate belong?

5.   What formation would be expected below the Moenkopi?

6.   Using the structure section of Figure 8.7, list which rock types form more resistant cliffs and which form less resistant slopes.

7.   Note, at the upper center of Figure 8.9, the white, domelike erosional forms, from which Capitol Reef takes its name. In what formation would these erosional forms develop? If the climate here was humid, instead of semi-arid, would this formation show a significantly different weathering pattern? Why or why not?

TERMS

bed   The smallest lithostratigraphic unit of sedimentary rocks. A bed is a well-defined, easily identifiable stratum or layer that has distinctive characteristics such as lithology or fossil content.

biofacies   The biological aspect of a stratigraphic unit, as reflected by its fossil content.

columnar section   A graphic illustration of the vertical sequence of rock units in an area. Lithology on the columnar section is shown by standard symbols, and the thicknesses of rock units are drawn to scale.

decay constant (Symbol: λ)   A constant, characteristic of a particular nuclear species, which expresses the probability that an atom of the species will decay in a given time interval.

formation   A mappable, lithologically distinct body of rock having recognizable contacts with adjacent rock units. Formations may be combined into **groups** or subdivided into **members.**

eon   The largest time unit in the geologic time scale, measured usually in billions of years. For example, the Precambrian is composed of the Archaean and Proterozoic eons.

fossils   The remains or traces of organisms that lived in the geologic past.

geologic column   A composite diagram depicting in columnar form the sequence of rock units for a given locality or region.

half-life   The time necessary for a radioactive substance to lose one-half of its radioactivity.

isochron   On a plot of parent radioactive isotope versus daughter product, the straight line drawn to fit the data points. This line of equal age is only valid if the rocks that are being dated formed at the same time. The slope of the isochron is a measure of the age of the samples; the intercept is the initial value of the daughter isotope in the rocks at their formation.

isotope   One of two or more species of the same chemical element, having the same number of protons in the nucleus, but differing from one another by having a different number of neutrons. The isotopes of an element have slightly different physical and chemical properties, owing to their mass differences, by which they can be separated.

lithofacies   A mappable subdivision of a stratigraphic unit based upon mineral and paleontological characteristics. Generally, this term signifies a distinguishable rock unit that was formed under common environmental conditions of deposition, without regard to age or geological setting.

period (geologic)   A major geologic time interval and subdivision of a geologic era. For example, the Mesozoic Era is subdivided into the Triassic, Jurassic, and Cretaceous geologic periods.

radioactive decay   The spontaneous disintegration of the atoms of certain **nuclides** (isotopes) into new nuclides, which may be stable (i.e., not decay spontaneously) or undergo further decay until a stable nuclide is formed. Radioactive decay involves the emission of energetic particles (which can be detected by a Geiger counter or other instruments) and the generation of heat.

relative (geologic) age   The chronological placement of a rock body or event in the geologic time scale without reference to its absolute (geologic) age. A relative age expresses whether the object or event is older than or younger than another object or event.

rock unit   A body of rock defined and identified by its composition, texture, color, or structural features without regard to fossils or time boundaries.

system (geologic)   A major time-rock unit representing the rocks that were formed during a geologic period. Thus, the rocks of the Cambrian System were deposited during the Cambrian Period.

time-rock unit   Term used for rock bodies formed during a particular segment of geologic time, and thus having synchronous lower and upper boundaries. They are also called **chronostratigraphic units.**

type section   An originally described sequence of strata that forms a stratigraphic unit. It is used as a comparison worldwide and ideally is located in an area where it has the maximum thickness and gives total exposure of the section.

9

The Advance and Retreat of Ancient Shorelines

PALEOGEOGRAPHY

What did our planet look like millions or even hundreds of millions of years ago? We know its present geography, but what was its ancient **geography** or **paleogeography?** Answering that question is one of the most fascinating objectives of historical geology.

In order to reconstruct the geologic past, it is necessary to know what kinds of sediments were deposited, what areas of the earth did they cover, how did the deposits vary in thickness, and what kinds of fossils, if any, did they contain. The task usually involves preparation of **facies** and **isopach** maps. A facies map depicts lateral changes in a time-rock unit. For example, a unit may change from nearshore sandstones to offshore limestones. Facies maps may be excellent indicators of ancient environments.

An isopach (EYE-so-pack) map uses lines drawn through points of equal thickness to show how a rock or time-rock unit changes across a given area. The lines of equal thickness are called *isopach lines*. Isopach maps are useful in paleogeographic reconstructions, especially in determining basin shapes and the position of ancient shorelines.

Study Questions

1. Figure 9.1 is a simple illustration of how a very generalized **paleogeographic map** is constructed. On a base map of a region, the locations of marine rocks and nonmarine or terrestrial rocks of a particular age are plotted (m = marine, t = terrestrial).

   Draw a line on the map that would indicate the approximate edge of the sea that covered much of western North America during the Late Cretaceous.

2. The zero line on the map marks the maximum lateral extent of the sedimentary rock unit used as a basis for the paleogeographic map. Beyond the zero line, the unit does not exist. The zero line may be the result of erosion (the unit once extended farther, but that portion

Figure 9.1 Paleogeographic map of North America in the Late Cretaceous.

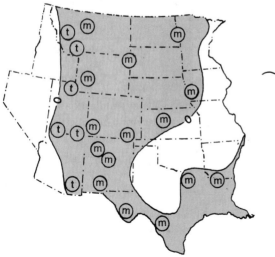

~ 0 — Edge of Upper Cretaceous time-rock unit (Series)

(m) Fine-grained shales and chalky limestones that grade westward to thicker shales and sandstones. Fossils of marine organisms locally abundant.

(t) Sandstones and shales with interbedded coal beds. Some localities yield fossils of land plants and dinosaurs. Unit thickens to over 3000 m in some areas near the western zero line. Thick conglomerates near western margin.

was destroyed by erosion). It may also, however, represent the edge of the basin of deposition and thus would truly indicate the original extent of the unit. What lithologic characteristics of the rocks adjacent to the zero line might one look for as an indication that the zero line represents the true margin of the basin of deposition?

_____

_____

_____

_____

_____

3. Based on the nature of rocks near the western zero line, what inference can be made about the probable topography of western Idaho, Nevada, and western Arizona during the Late Cretaceous?

_____

_____

_____

4. Figure 9.2 is a combined facies and isopach map of Lower Silurian sedimentary rocks in the eatern United States. What facies would be encountered on a journey from Virginia to Missouri?

_____

_____

5. How would you describe the probable source region for the conglomerates, sandstones, and other clastic sedimentary rocks?

_____

_____

_____

6. How do you account for the absence of Lower Silurian rocks in the elongate area trending across Tennessee and Kentucky?

_____

_____

_____

7. Does the western zero line in Figure 9.2 represent a shoreline or the border of a region in which Lower Silurian rocks may have been deposited but have been lost to erosion? What is the reason for your answer?

_____

_____

_____

_____

TRANSGRESSION AND REGRESSION OF THE SEA

Seas have advanced across (**transgressed**) and receded (**regressed**) from the continental craton many times during geologic history, as has been established from mapping the distribution of the distinctive rock types or **lithofacies** that were deposited along ancient shorelines. These fluctuations of sea level are often caused by changes in the dimensions of the ocean basins brought about by processes of plate tectonics (i.e., divergent or convergent plate motions) or by changes in the volume of ocean water resulting from accumulation and melting of continental glaciers. But changes in *apparent* sea level may also occur by either subsidence or uplift of landmasses relative to the ocean.

At Time A in Figure 9.3, the following lithofacies were being deposited: (A) nonmarine silt, clay, and coal; (B) marine sandstone; (C) marine shales; and (D) marine limestone. Notice how the advance of the sea from Time A to Time C is recorded by the positions of the lithofacies. That lithofacies associated in lateral succession may also be associated in vertical succession is an example of **Walther's Law** (Chapter 5).

Study Questions

1. In Figure 9.3, might each of the lithofacies present at Time C be properly regarded as a separate formation? Explain your reasoning.

_____

_____

_____

2. If so, would all parts of a formation be the same age? How could you tell?

_____

_____

_____

3. How many time-rock units might be distinguished? What would be the basis for separation?

_____

_____

_____

4. Assume that the shoreline regresses from Time C to a position that it occupied at Time A. What would be the succession of lithofacies, from bottom to top, that would appear at the position of the arrow?

_____

_____

_____

Figure 9.2 Lithofacies and isopach map of Lower Silurian rocks in the eastern United States. The isopach lines represent thickness, which reaches 400 ft over a large area in Pennsylvania and about 1000 ft at a few localities inside the 400-ft contour.

After T. W. Amsden (1955), *Bull. Amer. Assoc. Petroleum Geol.,* **39**:60–74.

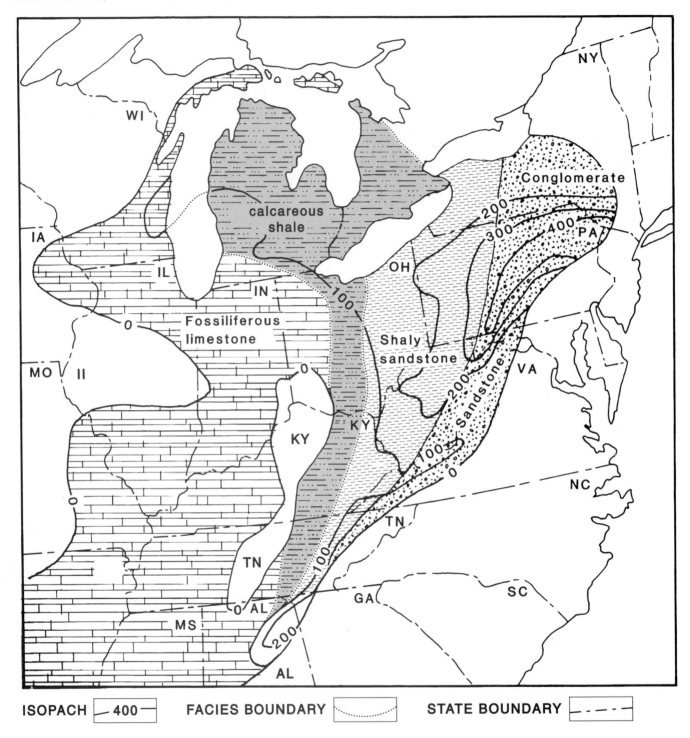

ISOPACH 400   **FACIES BOUNDARY**   **STATE BOUNDARY**

Figure 9.3 Sections W, X, and Y depict a transgressing shoreline as recorded by lithofacies. Transgression produces an onlap relation, in which finer offshore facies overlie coarser nearshore facies, as emphasized in the inset block. The cross section labeled Z depicts the lithofacies relation during a marine regression, where coarser nearshore facies overlie finer offshore facies.

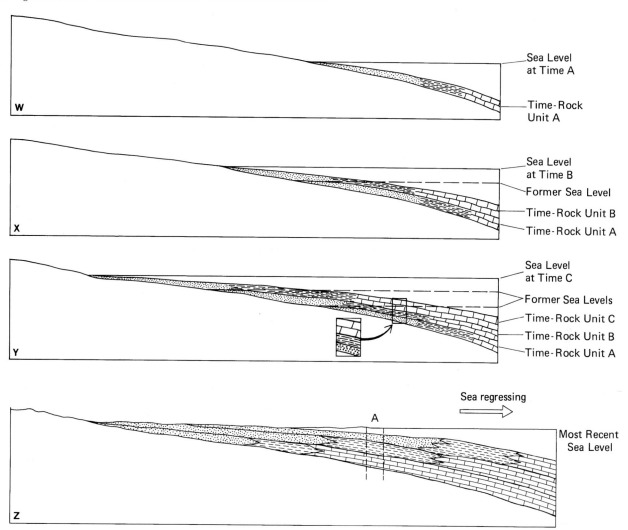

SHORELINES OF CAMBRIAN TIME

Study Questions

1. In Figure 9.4, what kind of sediment, in terms of textural and compositional maturity, was deposited along the shore of the Cambrian Sea?

   _____

   _____

   _____

   What kind of sediment was deposited farther out to sea?

   _____

   _____

   _____

2. During the interval of time between the Middle and Late Cambrian, did the sea transgress or regress? Explain your reasoning.

   _____

   _____

   _____

   When did the sea cover western Colorado: Middle or Late Cambrian time?

   _____

   _____

3. What evidence depicted in Figure 9.4B indicates the presence of islands in the Late Cambrian Sea?

   _____

   _____

   _____

Figure 9.4 **Lithofacies maps** of the Cambrian. (A) Lithofacies at a time in the Middle Cambrian. (B) At a time in the Late Cambrian. (Star indicates location of Glenwood Springs, Colorado.)

After C. Lochman (1957), *Geol. Soc. Amer. Mem.* **67**, vol. 2.

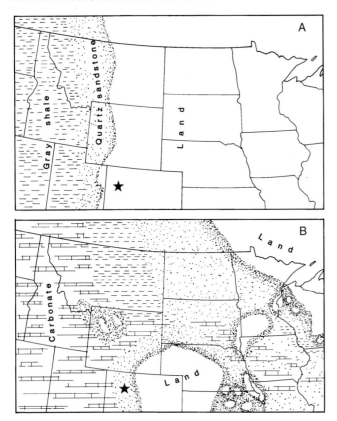

4. Based on the lithofacies maps, is the sandstone in Figure 9.5 of Middle or Late Cambrian age?

_____

_____

5. The Sawatch Sandstone shown in Figure 9.5 overlies crystalline igneous rocks that are cut by a light-colored pegmatite dike. On Figure 9.5, draw the contact between the Sawatch Sandstone on the crystalline igneous rocks. What kind of unconformity separates the Sawatch Sandstone from the underlying crystalline rocks?

_____

_____

_____

_____

ANCIENT SHORELINES IN THE BOOK CLIFFS OF UTAH

The Mancos Shale and the Mesaverde Group are clearly exposed in the Book Cliffs of northern Utah for a distance of

Figure 9.5 Cambrian sandstone of the Sawatch Fm. overlying Proterozoic basement (P) near Glenwood Springs, Colorado.

almost 322 kilometers. The Mancos Shale is a **formation** that consists dominantly of one rock type: shale. If a formation consists of several rock types, the term *formation* is used with its geographical name. The geographical part of a formation name (such as "Mancos") usually comes from some locality where the formation is prominently exposed. The term **group** is applied to two or more successive formations that belong to the same rock association.

In the study region for the following questions, the Mesaverde Group consists mainly of sandstone, but it includes numerous beds of carbonaceous shale and coal.

Study Questions

1. Draw a horizontal line (a time line) in Figure 9.6 that crosses all of the lithofacies, including coal. What is the lateral succession of facies, from left to right?

_____

_____

_____

Figure 9.6 Diagram of Mancos Shale and Mesaverde Group as exposed along the Book Cliffs between the towns of Castlegate and Green River, Utah.

From *Central Utah Coals* (1966), Bull. **80,** Utah Geological Survey. Reprinted by permission of Utah Geological Survey.

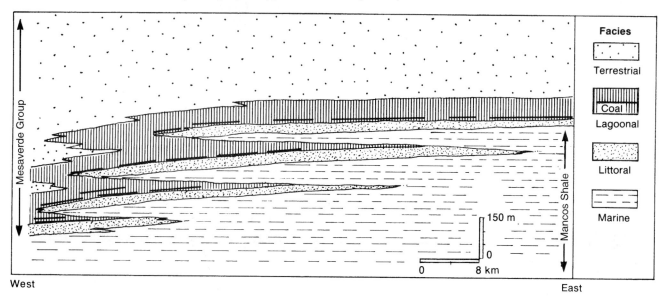

Figure 9.7 Coal beds in sandstone tongue of the Mesaverde Group. Road cut at Castlegate, Utah.

2. The tongues of Mesaverde sandstone that penetrate the Mancos have been attributed (from fossil oysters and sedimentary structures) to the littoral or shore (coastal) environment, probably similar to modern barrier beaches. Were these tongues of sand (and their overlying coal beds) deposited as the sea transgressed or only as it regressed? Explain your reasoning (see Figure 9.7).

If the coal represents a nonmarine swamp, and the sandstone a beach or offshore island, did the shoreline advance or retreat between deposition of the lowermost sandstone bed and that of the overlying coal bed? Explain your reasoning.

3. As Cretaceous time progressed, toward which direction (Fig. 9.6) did the shoreline move?

4. Above the coal beds (dark lines and labeled) in the cliff face in Figure 9.8 are two tongues of Mancos Shale separated by Mesaverde Sandstone. Label these formations directly on the photograph. On hillsides (rather than roadcuts) how do thick exposures of shale differ from thick, well-cemented sandstone exposures in their topographic appearance?

Figure 9.8 Book Cliffs north of Helper, Utah. Mancos Shale underlies slope at base of lowermost cliff of Mesaverde sandstone, near top of which, coal beds appear as dark lines.

See location map, Figure 8.7.

Coal

TERMS

craton   The portion of a continent that has been tectonically stable for several hundred million years, which is underlain by deformed Precambrian crystalline igneous and metamorphic rocks (often called *crystalline basement*). Most cratons have a central core region (called the *shield*) where the crystalline basement rocks are exposed at the surface. Surrounding the shield is a region of flat-lying or gently tilted strata, mainly sedimentary, called the *platform,* which is underlain at varying depths by crystalline basement.

facies   General appearance or aspect of a rock body, as distinct from other parts of the same unit or adjacent units. For example, a time-rock unit might have a quartz sandstone facies at one locality, and a shale facies at another.

facies map   A map that shows the distribution of different rock types occurring within a designated stratigraphic unit, without regard to the position or thickness of individual beds.

formation   A mappable, lithologically distinct body of rock having recognizable contacts with adjacent rock units. Formations may be combined into **groups** or subdivided into *members.*

isopach   A line drawn on a map through points of equal thickness.

isopach map   A map that shows the thickness of a bed, formation, or other body throughout a geographic area.

lithofacies   A mappable subdivision of a stratigraphic unit based upon mineral and paleontological characteristics. Generally, this term signifies a distinguishable rock unit that was formed under common environmental conditions of deposition, without regard to age or geologic setting.

lithofacies map   A map that shows the areal variation in the lithologic characteristics of a stratigraphic unit.

paleogeographic map   A map that shows the reconstructed physical geography of part of the earth at a particular time in the geologic past.

paleogeography   The study and description of the physical geography of the geologic past. Attempts to reconstruct the surface features at a given time in the past or the successive changes of surface relief over a given interval of time.

regression   The retreat of the sea from land areas with the deposition of terrestrial sediments over marine sediments.

rock unit   A body of rock defined and identified by its composition, texture, color, or structural features without regard to fossils or time boundaries.

time-rock unit   Term used for rock bodies formed during a particular segment or geologic time, and thus having synchronous lower and upper boundaries. They are also called *chronostratigraphic units*.

transgression   The spread of the sea over land areas with the deposition of marine sediments over terrestrial sediments.

Walther's Law   Observation that within a given sedimentary cycle, the same succession of facies that occurs laterally is also present in a vertical succession that has no evidence of erosional or depositional unconformity.

10

Fossils and Their Living Relatives: Protists, Sponges, Corals, Bryozoans, and Brachiopods

Note to Instructors

Because it is difficult to adequately describe representatives of each of the eight major phyla of frequently fossilized invertebrates in a single laboratory period, we have arbitrarily divided the fossil exercises into two chapters. In this chapter, we consider the unicellular invertebrates, as well as the Porifera, Cnidaria, Bryozoa, and Brachiopoda. In Chapter 11, mollusks, arthropods, echinoderms, and plant fossils are considered. The use of fossils (including trace fossils) in stratigraphic and paleoenvironmental studies is examined in Chapter 12.

INTRODUCTION

WHAT IS A FOSSIL?

If we inquire into the Latin origin of the word **"fossil,"** we find that the term refers to anything dug out of the earth. Modern usage, however, restricts the meaning to remains or traces of life of the geologic past. It is difficult to assign absolute limits in years to the expression "geologic past." Most paleontologists agree that in order for the remains of an organism to be regarded as a fossil, the organism must have lived prior to the beginning of recorded human history. Remains of life from Pleistocene (Ice Age) or older rocks are definitely fossil. In addition to its rather intangible quality of age, a fossil must also provide some evidence of at least part of the anatomy of the organism that produced it. Even tracks, trails, borings, burrows, and preserved excrement are considered fossils *(trace fossils)* because they provide clues to an animal's size, shape, and mode of life.

Although there are rare examples of fossils in lava flows or occurring as distorted forms in metamorphic rocks, most invertebrae fossils are found in shallow marine sedimentary rocks. Life flourishes on the floors of shallow seas, and burial by falling sediment is a relatively continuous and often rapid process. At certain localities, however, abundant fossils of nonmarine animals and plants can be found. For example, remains of dinosaurs are often found in fluvial (river), lacustrine (lake), deltaic, or eolian sedimentary rocks.

PRESERVATION OF FOSSILS

When one considers the many factors that tend to destroy an organism after its death, it seems remarkable that fossils are as common as they are. Chemical decomposition, erosion, attack by scavengers, and a multitude of geologic processes make the odds against fossilization formidable. The possession of hard skeletal parts and rapid burial in sediment increase the chances that the animal or plant will leave a fossilized relic of itself.

To many, the term fossil implies **petrifaction**—literally, a transformation into stone. This may occur by gradual addition of a chemically precipitated substance into pore spaces **(permineralization),** or by a molecular exchange of substances that were once part of the organism with other substances carried in by percolating water solutions **(replacement).** Fossils replaced or permineralized by calcium carbonate, silica, or iron sulfide are common. A process of fossilization known as **carbonization** results when soft tissues are preserved as films of carbon. Soft-bodied organisms such as jellyfish, worms, and graptolites, as well as leaves of trees, may be preserved in this way. However, not all blacks and shiny fossil films are the result of carbonization. For example, many of the soft-bodied animals of the Burgess Shale fauna from a famous locality in British Columbia are preserved as black, reflective films composed of calcium and magnesium aluminosilicates (Fig. 10.1). The remarkably diverse and exquisitely preserved Burgess Shale organisms are evidence of the veritable explosion of life during the Cambrian Period.

Another form of preservation, common for the shell-bearing invertebrates, develops when the shell material is progressively removed by leaching, so as to leave a void in the rock bearing the surficial features of the original shell. Such fossils are called external **molds.** If the cavity on the

Figure 10.1 *Hyolithus carinatus.* The photograph illustrates the nature of preservation in fossils of the Middle Cambrian Burgess Shale. *Hyolithus carinatus* is a hyolith. The shell is subtriangular in cross section and had a small operculum that served to close the opening of the shell. Two curved appendages probably helped to support the animal on the ocean floor.

Courtesy of Chip Clark, Museum of Natural History, Smithsonian Institution.

inside of the shell is filled with sediment and that filling preserved, it becomes an *internal mold*. A **cast** is formed when the void between the internal and external mold is filled with mineral matter. Thus the cast becomes a replica or model of the original shell (Fig. 10.2).

Replacements, permineralizations, carbonizations, molds, casts, tracks, trails, and imprints are all examples of preservation. In geologically young rocks, bone or shell material may be preserved without any significant alteration. Unaltered soft tissue requires extraordinary conditions for preservation. Insects trapped in amber and mammals found frozen in Arctic soils are dramatic, but very rare, examples of such preservation.

In addition to their interest in the discovery and identification of fossils, paleontologists examine fossils and the rock enclosing them in order to determine what occurred to the original organisms from the time they died to the time they reached their fossilized state. The term *taphonomy* refers to the study of events affecting an organism from death to fossilization. Taphonomic study may reveal if a lineage of organisms became extinct as a result of environmental changes, or whether the absence of fossils attributed to extinction was merely caused by dissolution, or other modes of destruction of fossils, after their burial. It may inform us if the organisms actually lived at the site where they were discovered or were carried to that place from a distant, and possibly very different, habitat.

CLASSIFICATION AND NOMENCLATURE

Because of the large numbers of organisms that have lived or are now alive, random naming would result in much confusion. Realizing this, the Swedish naturalist Carolus Linnaeus (1707–1778) established a system of naming animals

Figure 10.2 Diagram to explain molds and casts. (A) Cross section through the valves of a clam that has been buried and filled in with sediment. In (B), the original shell matter is dissolved, leaving a cavity or mold in the sediment or rock. In (C), mineral matter has been precipitated within the mold to form a cast that resembles the original shell. (D) The impression of the exterior of one valve of the clam in the enclosing sediment forms an external mold. (E) The lithified sediment that once filled the original shell or the cast constitutes an internal mold.

From Levin H. L., *Ancient Invertebrates and Their Living Relatives,* Upper Saddle River, NJ: Prentice Hall, 1999.

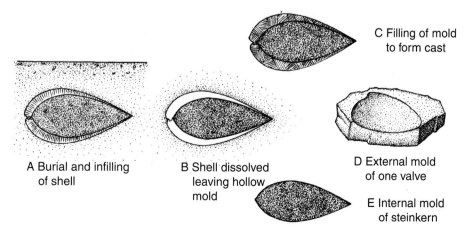

C Filling of mold to form cast

A Burial and infilling of shell

B Shell dissolved leaving hollow mold

D External mold of one valve

E Internal mold of steinkern

Table 10.1 Hierarchy of Taxonomic Groups

Kingdom	Animalia	Animalia
Phylum	Arthropoda	Chordata
Class	Crustacea	Mammalia
Order	Decapoda	Carnivora
Family	Nephrosidae	Canidae
Genus	*Homarus*	*Canis*
Species	*Homarus americanus*	*Canis familiaris*
Individual	a common lobster	a pet dog

and plants in 1758. The Linnaean scheme recognized structure as the basis for classification and established a system that includes binomial nomenclature at the species level. The name of a species consists of two parts, the generic name and the trivial name. For example, the scientific name for a dog is *Canis familiaris.* Both names are italicized or underlined, and the generic name is capitalized.

The basic category in taxonomy is the **species.** Species may be defined as groups whose members will interbreed and (except for sex differences) are more like one another than like any member of another similar group. The individuals, which are ultimate items of classification, can thus be grouped into species, the species combined into genera (singular, *genus*), the genera into families, and so on to successively larger divisions (Table 10.1). In this scheme, the smaller the category is, the more similar are its members.

The use of genus and species provides an efficient and precise nomenclature. The student may ask his professor to see a specimen of *Mytilus edulis,* and both the student and professor know that a certain species of common clam must be produced. It is difficult to guess the professor's reaction if the student asked to see a "whaddayacallit made of two lopsided, partly joined discs."

The definition of species must be modified when one is studying the fossil remains of extinct organisms. Obviously, the paleontologist cannot directly determine the ability or inability to interbreed. He therefore considers a species as a group of organisms whose individual differences both are small in comparison with their differences from other groups and are of the kind and magnitude to be expected in interbreeding populations.

When Linnaeus formulated his binomial system of nomenclature, he assumed that all life could be divided into two large categories, namely, the *Kingdom Animalia* and the *Kingdom Plantae.* In Linnaeus's time, the distinction between plants and animals seemed readily apparent. Shortly thereafter, however, unicellular creatures were studied that were simultaneously photosynthetic, like plants, and capable of ingesting food and moving about like animals. For such unicellular organisms having intergrading combinations of plant and animal characters, the biologist Ernst Haeckel proposed a third kingdom, the *Kingdom Protista,* in 1886. (The term Protoctista has recently been offered as a replacement for

Figure 10.3 Modern stromatolites formed within the intertidal zone at Shark Bay, Australia.

Courtesy of J. W. Schopf.

Protista.) Protists include a vast array of yellow-green and golden-brown algae and single-celled creatures once simply called "protozoans." Eventually, even the three-kingdom classification proved to be inadequate.

The *Kingdom Monera* was established for asexually reproducing, unicellular microorganisms so primitive that they lack a cell nucleus and special cell organelles ("little organs" located within the larger cell that carry out certain cellular functions). Various bacteria, as well as the cyanobacteria (formerly called blue-green algae), are members of the Kingdom Monera. For geologists, the most familiar fossil formed by the activities of monerans like cyanobacteria are the stromatolites. *Stromatolites* are thinly laminated accumulations of calcium carbonate having rounded, cabbagelike, branching, or columnar shapes. Living cyanobacteria can be observed to form similar structures today (Fig. 10.3). Fine particles of calcium carbonate settle between the minute filaments and strands of the sticky algal layer and are bound within the gelatinous sheath. Successive additional layers result in fine laminations. The remains of filaments and spheres of cyanobacteria with fossil stromatolites (Fig. 10.4) indicate that ancient stromatolites formed in the same way. The cyanobacteria are photosynthetic organisms

Figure 10.4 2.3 billion-year-old fossil stromatolites from southern Africa.

Courtesy of J. W. Schopf.

and thus liberate oxygen. The presence of stromatolites over 3 billion years ago indicates that these organisms played a significant role in the production of atmospheric oxygen in the earth's early atmosphere.

Stromatolites range in age from the Archean to the present. Modern forms prefer the intertidal zone where their tops are at the high-water mark. Today's stromatolites are relatively short (rarely more than a meter tall), whereas some Proterozoic stromatolites exceeded 6 m in height.

The final kingdom of the five-kingdom classification consists of the Fungi. Fungi require such special placement because they depend on a supply of organic molecules for nutrition, as do animals, yet they absorb nutrients as do plants. Fossils of Fungi are rare, but fungal spores are known to occur in ancient Precambrian cherts.

The five-kingdom classification is based on the observable traits of organisms. These traits are important indicators of evolutionary relationships. Another indicator of relationships is the structure of large molecules in the cell such as ribonucleic acid (RNA). Comparison of these molecules in different kinds of organisms indicates that such superficially dissimilar groups as plants, animals, and fungi are actually related, and that the kingdom Monera of the five-kingdom system actually contains two distinct kinds of microorganisms. To account for the evolutionary relationships revealed by molecular biology, it has been proposed that all life be divided into three great groups termed *domains*. They are named *Bacteria, Archaea,* and *Eukaryia.* The domain Bacteria includes the cyanobacteria, purple sulfur bacteria, and several non-photosynthetic groups of microbes. Heat-tolerant microbes called thermophiles and methane-producing bacteria are placed in the Archaea. Both the *Bacteria* and *Archaea* consist of organisms that would be placed in the Monera of the older, traditional classification. All other forms of life have cells with a discrete nucleus. They constitute the domain Eukarya.

Study Questions

1. In life, nearly all snails (Class Gastropoda) build their shells out of calcium carbonate. What type of fossilization has probably occurred in a fossil snail that is composed of silica?

   _____

   _____

2. A stratum exposed at the base of a vertical cliff contains the fossil shells of clams resting in their "living position" within the rock. Several meters above this stratum, another bed contains the same species of clams, but the shells are abraided, broken, and jumbled within the rock. What taphonomic inference can be made from these observations?

   _____

   _____

   _____

3. Why is the older biological classification of all organisms into either the Kingdom Plantae or the Kingdom Animalia no longer followed?

   _____

   _____

   _____

4. Considering that sandstone is more permeable than shale, would one be more likely to find the *unaltered* remains of a clam in sandstone or shale? Why?

   _____

   _____

   _____

5. Why are there more fossils of organisms that lived in the ocean than there are of organisms that lived on the continents?

   _____

   _____

   _____

6. What is the basis for the inference that Precambrian stromatolites lived in relatively shallow, well-lighted parts of the sea?

   _____

   _____

   _____

7. What is the relation between the spread of Precambrian stromatolites and the evolution of the earth's atmosphere?

_____

_____

_____

_____

8. Describe a hypothetical scenario in which the body of a dinosaur would eventually become a fossil in floodplain deposits of the Cretaceous Period.

_____

_____

_____

_____

_____

FOSSIL AND LIVING PROTISTS

Protists include a variety of microorganisms, some clearly "animal" in their characteristics and some that combine some of the traits of both plants and animals. Because they have hard parts that are preserved as fossils, coccolithophorids, diatoms, foraminiferida, and radiolaria are protists used by geologists in stratigraphic studies. In addition to their abundance in the fossil record, these protists are so small that they can be recovered intact even in broken chips of rock brought to the surface in the course of drilling for oil. The geologic ranges of some frequently fossilized protists are depicted in Figure 10.5.

Coccolithophores

Coccolithophores are unicellular, planktonic, golden-brown **algae.** The usually spherical cell of the coccolithophore is called a **coccosphere** (Figure 10.6), and is covered by intricate calcium carbonate skeletal structures called **coccoliths** and **discoasters** (Fig. 10.7). Although usually less than 30 micrometers in diameter (1 micrometer equals 0.001 mm), these tiny structures are marvels of geometrically precise construction.

   Most coccoliths take the form of two elliptical discs, stacked one above the other and joined at the center by a hollow stud. The discs are concave on one side, so that they are able to fit closely around the outside of the spherical coccolithophorid cell. Each disc of the coccolith is itself composed of still smaller elements uniformly arranged in overlapping circular or spiral patterns. With astonishing precision, the living cell exerts an exact control on the size, shape, and arrangement of the smaller elements that compose the coccolith.

   Discoasters are simpler in construction than coccoliths. They are recognized by their radiate or stellate symmetry. The common genus _Discoaster,_ for example, is a star-shaped

Figure 10.5   Geologic ranges of some frequently fossilized Protists.

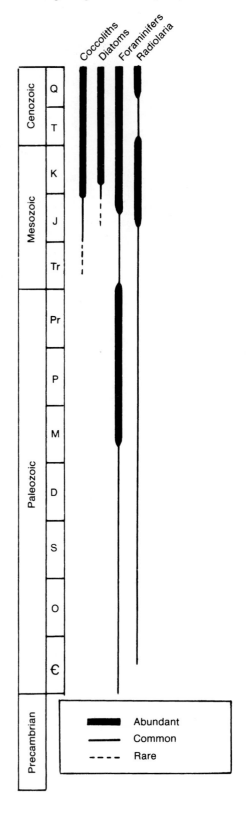

disc composed of variously shaped radiating arms. *Braaru-dosphaera* has five plates arranged in pentaradial fashion. Each five-plate unit fits together with its neighbors to form a structure rather like a geodesic dome.

Coccolithophores began to occur abundantly during the Jurassic. Thereafter, they were particularly numerous and diverse during the Cretaceous, Eocene, Miocene, and Recent (Fig. 10.5).

Study Questions

1. Most coccoliths are recovered from soft, calcareous sediments like chalk. Why are they rarely obtained from well-indurated rocks or recrystallized rocks?

_____

_____

_____

_____

2. Label two discoasters and two coccoliths directly on Figure 10.7.

3. In samples of the ocean floor, coccoliths are most abundant in pelagic calcareous oozes. They are rarely encountered at great depth or in the coldest parts of the ocean. How do you account for this distribution?

_____

_____

_____

_____

4. What would be the maximum diameter (in micrometers) of a coccolith whose greatest diameter was 0.0025 mm?

_____

_____

_____

_____

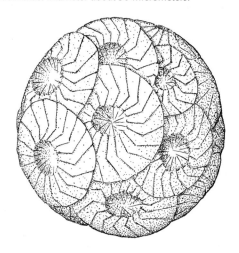

Figure 10.6 A complete coccosphere of the coccolithophorid **Cyclococcolithus.** Diameter about 30 micrometers.

Figure 10.7 Coccoliths and discoasters of Tertiary age.

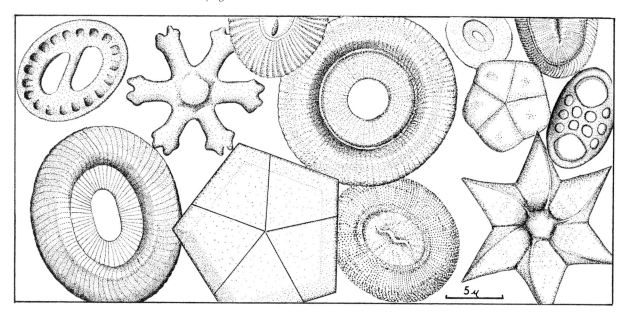

5 μ

5. What would be the maximum (oldest) geologic age for a sedimentary rock containing coccoliths (see Fig. 10.5)?

_____

_____

_____

_____

Diatoms

Diatoms (Fig. 10.8) are microscopic, one-celled aquatic algae, having beautifully constructed siliceous cell walls that are in two parts and fit together like the halves of a tiny pill box (Fig. 10.9). Myriads of diatoms live in the surface waters of oceans and lakes. The delicate siliceous coverings (frustules) of dead diatoms rain down continuously onto the floors of the seas, and they may accumulate to form a lightweight porous rock called diatomite.

Study Questions

1. Diatomite is quarried for use as an abrasive (e.g., in toothpaste) and as a chemical filter. On what property does its use as an abrasive depend? Why might the remains of diatoms make good chemical filters?

_____

_____

_____

_____

_____

_____

Figure 10.8 Living and fossil diatoms (magnified about 300×).

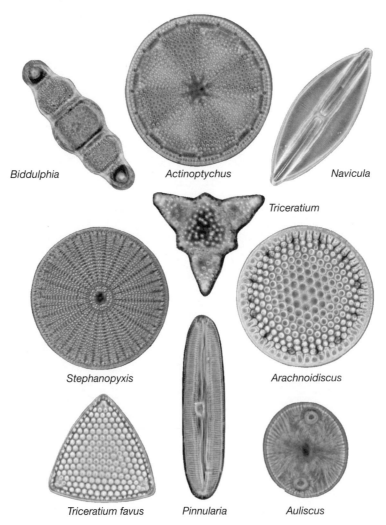

Biddulphia   *Actinoptychus*   *Navicula*

Triceratium

Stephanopyxis   *Arachnoidiscus*

Triceratium favus   *Pinnularia*   *Auliscus*

Figure 10.9 A freshwater diatom. The siliceous covering of the cell consists of two overlapping parts called valves. The inner or lower part is the hypotheca, and the upper part is the epitheca (magnified 500 times).

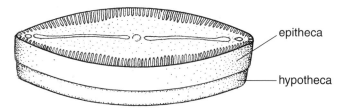

epitheca

hypotheca

2. Examine the slide of diatoms with the aid of the microscope and substage lamp. Prepare a simple sketch of three different forms in the space below.

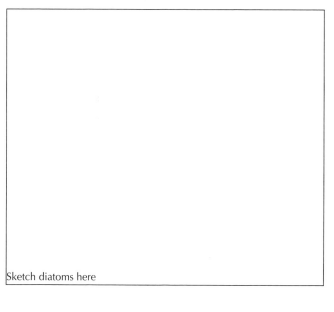

Sketch diatoms here

Receptaculitids

Receptaculitids are stratigraphically useful early Paleozoic fossils that remind one of the seed-bearing central area of a sunflower. As a result, amateur collectors have dubbed them the "sunflower corals." They are neither flowers nor corals, however. Most paleontologists believe they are lime-secreting green algae. The receptaculitid named *Fisherites* (Fig. 10.10) is a guide fossil used in correlating Ordovician to Devonian strata.

Foraminiferida

The members of the Order **Foraminiferida** (Fig. 10.11) are nearly all marine organisms that build their shells, called **tests,** by adding chambers in one or more rows, in coils, or in spirals. Foraminiferida means pore bearing, and refers to the large number of small holes *(foramina)* that perforate the tests of some species, through which stream thin pseudopodia of protoplasm. A larger opening, the **aperture,** is also usually present. Although the majority of tests are composed of calcium carbonate, some construct their shells of sediment

Figure 10.10 *Fisherites* (formerly known as *Receptaculites*).

Courtesy of Ward's Natural Science Establishment, Rochester, New York.

particles, some form them of a chitinous substance, and others secrete siliceous shell material. The chambers are separated from one another by **septa,** which are indicated on the exterior of the shell by **sutures.**

Most species of foraminiferida are bottom dwellers *(benthic).* Fewer species, but large populations, are free floating *(planktonic).* The empty shells of these planktonic foraminiferida (Fig. 10.12), along with cocoliths accumulate in large numbers in the deep-sea sediment known as calcareous ooze. In the past, many calcareous oozes have been lithified to form a variety of limestone known as chalk.

The rarity of foraminiferida remains in rocks older than Silurian may imply a lack of hard skeletal parts in older forms. Early foraminiferida constructed spherical, tubular, or chambered shells composed of cemented particles of fine sediment. By Late Paleozoic time, a group known as fusulinids became abundant and widespread. The fusulinids developed marvelously complex internal shell structures (Fig. 10.12) but take their name from their fusiform shape, which resembles a grain of wheat (see *Fusulina* and *Schwagerina* on Fig. 10.11). A milestone in foraminiferal history came in the Cretaceous, when planktonic forms (Fig. 10.13) made their appearance and expanded rapidly. Since the Cretaceous, the entire order has flourished and provided many useful guide fossils for stratigraphic correlation.

Study Questions

1. Foraminiferida index or guide species *Uvigerinella sparsicostata, Uvigerina gallowayi,* and *Nonion affinis* are found to be characteristic of a stratum penetrated in an oil well located at A in Figure 10.14. The index species were recovered at a depth of 1950 m. The same well encountered an oil-producing sandstone stratum 30 m thick at a depth of 2650 m. A second well is begun

Figure 10.11 Foraminiferida.

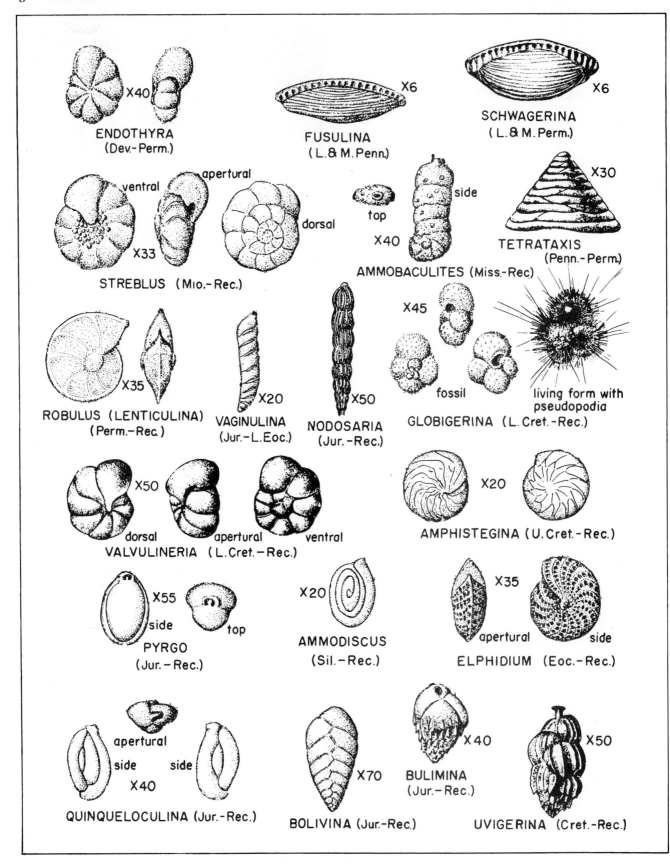

ENDOTHYRA
(Dev.-Perm.)
X40

FUSULINA
(L.& M. Penn.)
X6

SCHWAGERINA
(L.& M.Perm.)
X6

STREBLUS (Mio.-Rec.)
ventral   apertural   dorsal
X33

AMMOBACULITES (Miss.-Rec)
top   side
X40

TETRATAXIS
(Penn.-Perm.)
X30

ROBULUS (LENTICULINA)
(Perm.-Rec.)
X35

VAGINULINA
(Jur.-L.Eoc.)
X20

NODOSARIA
(Jur.-Rec.)
X50

GLOBIGERINA (L.Cret.-Rec.)
X45   fossil   living form with pseudopodia

VALVULINERIA (L.Cret.-Rec.)
dorsal   apertural   ventral
X50

AMPHISTEGINA (U.Cret.-Rec.)
X20

PYRGO
(Jur.-Rec.)
side   top
X55

AMMODISCUS
(Sil.-Rec.)
X20

ELPHIDIUM (Eoc.-Rec.)
apertural   side
X35

QUINQUELOCULINA (Jur.-Rec.)
apertural   side   side
X40

BOLIVINA (Jur.-Rec.)
X70

BULIMINA
(Jur.-Rec.)
X40

UVIGERINA (Cret.-Rec.)
X50

Figure 10.12 Thin section of a fusulinid showing complex internal structure. Length of specimen is 6 mm.

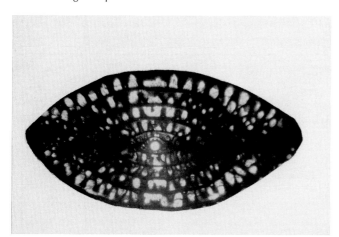

Figure 10.13 The planktonic foraminiferida *Globigerinoides*.

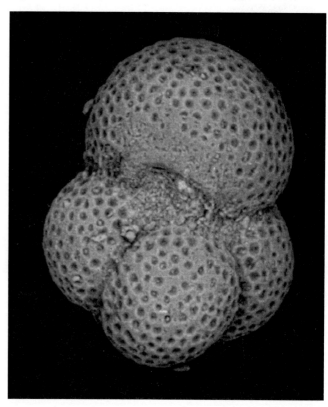

at location B, 2 km from the first. In the second well, the index species are encountered at a depth of 1350 m.

a.   At what depth will the oil zone be reached at the second location? State the assumptions you used to reach your conclusion.

_____

_____

_____

_____

_____

_____

b.   Assume that the shale and sandstone have not changed in thickness, and complete the rock column for well B.

2.   In Figure 10.15 a distinctive foraminiferida assemblage is encountered at a depth of 410 m in well #34. In well #62, an identical assemblage is encountered at 400 m and again at 1200 m. In well #71, the assemblage occurs at 1200 m.

a.   On the diagram, draw in a fault to indicate how faulting may have caused the repeated assemblage in well #62.

Figure 10.14 Use of foraminiferida in predicting drilling depth.

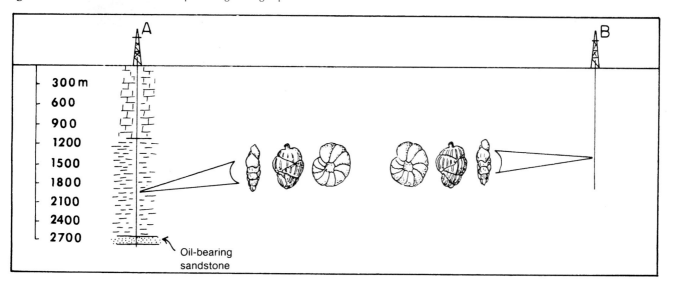

Figure 10.15 Recognizing subsurface structures with index fossils.

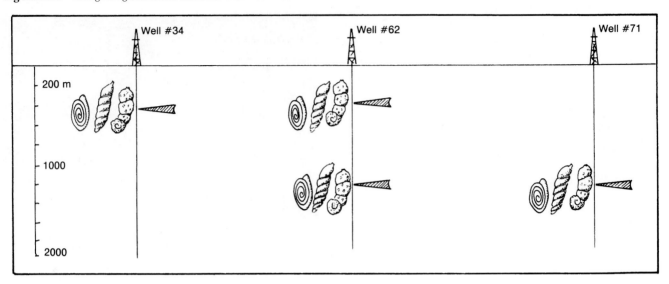

b. What is the name given to this type of fault? (See Chapter 14 for a summary of fault terminology.)

_____

_____

_____

3. Would **benthonic** or **planktonic** foraminiferida be the most useful for indicating the depth of water in which a stratum was deposited? Why might an oil geologist be interested in this type of information?

_____

_____

_____

4. Why are planktonic foraminiferida generally considered more useful in intercontinental stratigraphic correlation than benthonic foraminiferida?

_____

_____

_____

5. In some areas of the present sea floor, foraminiferida tests are accumulating at the rate of 1 cm per 1000 years. At this rate, how thick a deposit would have accumulated during the 65-million-year duration of the Cenozoic era, if no allowance is made for compaction?

_____

_____

_____

6. Examine the prepared slide of fossil foraminiferida. Since these slides are not transparent, light should be reflected onto the slide surface from a position above and slightly to the left of the microscope stage. Your instructor will describe the manner in which these fossils have been prepared for study.

a. Determine the generic name (genus) of as many of the specimens as you can by comparing them with the illustrations in Figure 10.11 and any additional aids provided.

b. Sketch any three of the genera, and label _aperture_, _suture_, and _chamber_.

Figure 10.16 Range chart.

Stratigraphic distribution of some genera of the foraminiferida														
Periods ⬇ Genera														
Quaternary														
Tertiary														
Cretaceous														
Jurassic														
Triassic														
Permian														
Pennsylvanian														
Mississippian														
Devonian														

7. On Figure 10.16, show by means of vertical bars the geologic ranges for *Ammodiscus, Endothyra, Fusulina, Ammobaculites, Globigerina, Bolivina,* and *Quinqueloculina.*

   a. What is the maximum possible geologic range for rock containing only *Endothyra?*

   _____

   _____

   b. What is the age of a formation containing both *Robulus* and *Endothyra?*

   _____

   _____

   _____

   c. What is the advantage, if any, of having more than one index fossil when making age determinations?

   _____

   _____

   _____

   d. Which is the better index fossil, *Fusulina* or *Ammodiscus?* Why?

   _____

   _____

   _____

Figure 10.17   Radiolaria (all magnified 120×). Identification Key: **1.** *Podocyrtus;* **2.** *Theocorys;* **3.** *Dictyostrum;* **4.** *Eucyrtidium;* **5.** *Panartus;* **6.** *Heliodiscus;* **7.** *Lamprocyclus;* **8.** *Dendrocircus;* **9.** *Hexaconthium;* **10.** *Anthocyrtium;* **11.** *Astractura.*

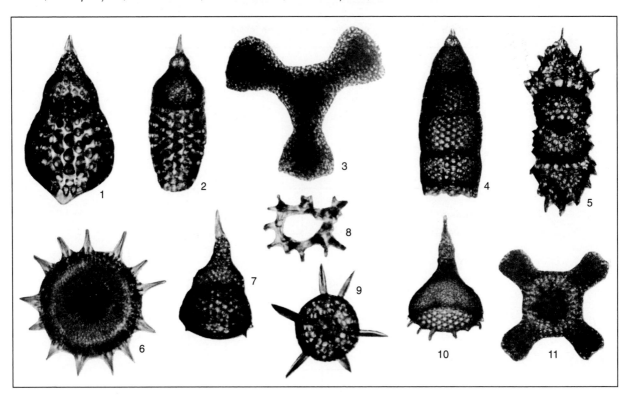

Radiolaria

Radiolaria are planktonic marine microorganisms in which the protoplasm is surrounded by a delicate, often beautifully fili-greed skeleton (Fig. 10.17). The skeleton is almost always composed of opaline silica ($SiO_2 \cdot H_2O$). In some regions of the sea, radiolarian tests fall like a microscopic rain upon the sea floor to accumulate as a mucklike deposit called "radiolarian ooze." Because of the increased concentration of carbon dioxide at depths below 4500 meters, the calcareous shells of foraminiferida dissolve, but the siliceous radiolarian skeletons are relatively unaffected and tend to accumulate. The tests of radiolaria are generally modified helmet-shaped or spherical. Some forms, especially those with spherical tests, have long spines projecting from their rims, which may have contributed to buoyancy.

Study Questions

1. How might one interpret the depositional environment of a stratum containing numerous radiolarian fossils but few preserved foraminiferida?

   _____

   _____

   _____

   _____

2. Why might radiolarian tests be less likely to be recovered intact from well cuttings than tests of foraminiferida?

   _____

   _____

   _____

   _____

3. Examine the slide of radiolaria with the aid of a microscope and substage lamp. Sketch two different genera in the space below.

SPONGES

Sponges are primitive multicellular animals belonging to the Phylum Porifera (literally, *pore bearers*). They make up a relatively conservative evolutionary sideline, and have been in existence since the Late Proterozoic (Fig. 10.18). All but one group of sponges are marine, and they inhabit all the seas at all depths from the strand line to the deepest abyss. If they are living in calm water, sponges tend to grow into tall, symmetrical, and bushy forms, but they develop low, encrusting shapes when growing in areas where currents exist. Sponges with skeletal elements composed of calcium carbonate require some sunlight and thus grow at depths of less than 200 m. As might be expected, the siliceous spicule-producing "glass sponges" are tolerant of cold temperatures and great depths. Cold, deep waters may not, however, have been the habitat of ancient glass sponges. In the Devonian of New York and Pennsylvania, for example, glass sponges occur in strata apparently deposited on the sandy bottom of a shallow sea.

Although sponges vary widely in form and size, the basic plan is that of a much-perforated vase with walls modified by folds and canals. Most modifications appear to be a response toward attaining greater food-gathering surface and protecting the feeding cells. The body is attached at the bottom to the sea floor, or it may be connected at the base to lateral tubes leading to other members of the colony (Fig. 10.19). In a simple sponge, there is a central cavity that has an opening (**osculum**) at the top.

The body wall is composed of two layers of cells with a layer of mesenchyme between them. The outer layer consists of hexagonal protective cells, whereas the inner layer is composed of collared flagellated feeding cells (**choanocytes**). Pore cells (*porocytes*) provide openings for water to enter the sponge.

The paleontologically important **spicules** (Fig. 10.20) are secreted by amoeboid cells located in the mesenchyme. The spicules support the body. They also provide criteria for identification and greatly improve the chances for preservation. Spicules may be calcareous or siliceous, single or multiple rayed, or variously modified into pitchfork, tuning fork, or tripod shapes. Some sponges do not secrete mineral spicules but secrete a leathery substance called spongin.

There are five taxonomic classes of Porifera. These are the *Desmospongea*, *Hexactinellida* (formerly *Hyalospongea*), *Calcarea* (formerly *Calcispongea*), *Sclerospongea* (coralline sponges), and a problematic but important group known by the name *Stromatopora*. Desmospongids have spicules composed of silica, spongin, or both. Hexactinellids have siliceous spicules, whereas the spicules in the calcarids are composed of calcium carbonate.

Figure 10.18 Geologic range of the phylum Porifera. The fossil record for sponges is not an abundant one, but it is of long duration.

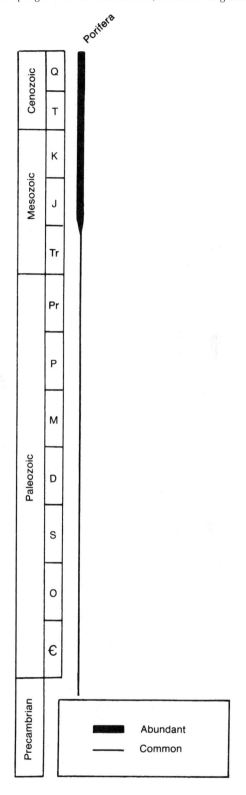

Figure 10.19 Features of a simple sponge.

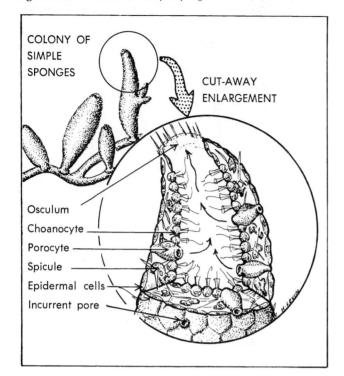

COLONY OF SIMPLE SPONGES

CUT-AWAY ENLARGEMENT

Osculum
Choanocyte
Porocyte
Spicule
Epidermal cells
Incurrent pore

The stromatoporates (Fig. 10.21) are problematic because they bear a strong resemblance to both sponges and corals. Although they exhibit certain canal and skeletal patterns seen in sclerosponges, stromatoporates (also referred to as stromatoporoids) lack spicules. For this reason, some paleontologists prefer to place them in the Cnidaria along with the corals.

Regardless of their taxonomic placement, stromatoporates are important fossils that grew prolifically in Early Paleozoic seas. Their skeletons consist of fine, closely spaced pillars and partitions. The skeletal structures are so small and intricate that microscopic examination is essential for identification.

The fossil sponges illustrated in Figures 10.20 will help you to identify some of the sponges in your study set. The drawings include one representative of problematic spongelike fossils called *archaeocyathids*. Archaeocyathids built extensive reefs in many tropical areas of the world during the Cambrian. Their cone-shaped skeletons, with perfo-

rated double walls and septalike partitions, suggest affinities to both corals and sponges. Archaeocyathids, however, are regarded as sufficiently distinct to warrant their placement in a separate phylum, the Archaeocyatha.

Study Questions

1. How many rays can you count on the spicules of *Astraeospongium?* Might there be additional rays perpendicular to the ones you have counted?

_____

_____

_____

_____

2. In Figure 10.22, what is the age of the stratum above the layer of bentonite (altered volcanic ash)? What is the age of the stratum below the bentonite layer?

_____

_____

_____

_____

_____

3. Was the ash (now bentonite) deposited in a marine or continental environment?

_____

_____

_____

_____

4. Name a characteristic of archaeocyathids that suggested to early investigators that they might be members of the Phylum Porifera.

_____

_____

_____

_____

Figure 10.20 Sponges and spongelike fossils.

Partly from Collinson, C. W., 1959, Illinois State Geological Survey, Educational Series 4.

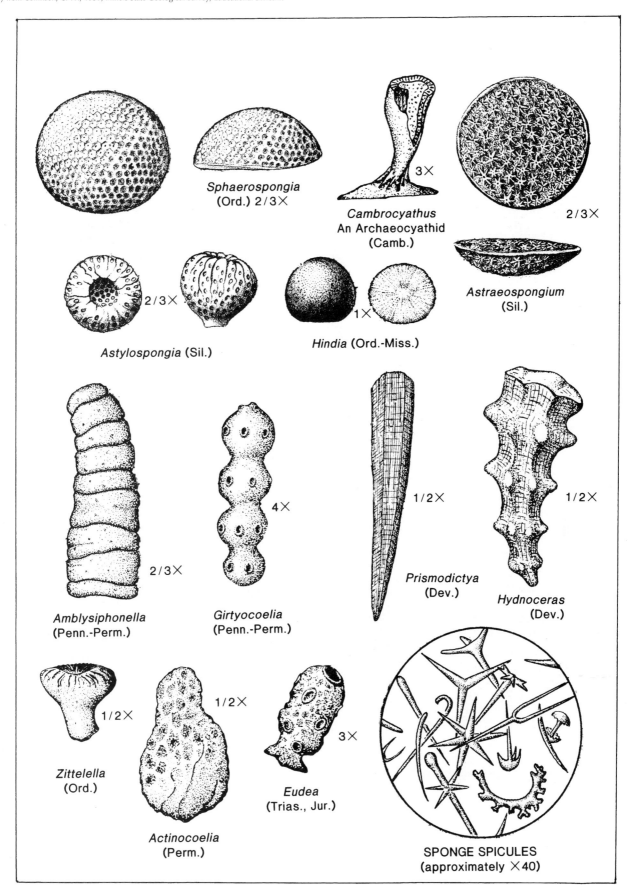

Sphaerospongia
(Ord.) 2/3×

Cambrocyathus
An Archaeocyathid
(Camb.) 3×

2/3×

Astylospongia (Sil.) 2/3×

Hindia (Ord.-Miss.) 1×

Astraeospongium
(Sil.)

Amblysiphonella
(Penn.-Perm.) 2/3×

Girtyocoelia
(Penn.-Perm.) 4×

Prismodictya
(Dev.) 1/2×

Hydnoceras
(Dev.) 1/2×

Zittelella
(Ord.) 1/2×

Actinocoelia
(Perm.) 1/2×

Eudea
(Trias., Jur.) 3×

SPONGE SPICULES
(approximately ×40)

Figure 10.21 Fragment of the stromatoporoid *Actinostroma*. The small raised areas on the top surface are called mamelons, and the rootlike grooves on the summits of the mamelons are termed astrorhizae. Mamelon astrorhizae are probably the skeletal traces of a water-conducting system. Vertical tubes shown on the front and sides of the specimen are astrorhizal canals. (X6)

Figure 10.22 Two beds of limestone separated by an easily weathered layer of bentonite (a clay formed from the alteration of volcanic ash). The geology hammer marks the location of the bentonite layer. The bed above the bentonite layer contains the fossil sponge *Astraeospongium*. The bed below the bentonite layer contains *Zittelella*.

CORALS AND RELATED CNIDARIANS

The Phylum **Cnidaria** (formerly Coelenterata) comprises a large and diverse group of Proterozoic to Recent animals, which include corals, sea anemones, sea fans, the *Hydra,* and the jellyfish (Figs. 10.23 and 10.24). All cnidarians are aquatic, and most are marine. They are characterized by a saclike body, with tentacles around the mouth of the sac. Special stinging cells assist the coelenterate in capturing live food. These cells, called *cnidoblasts,* are a distinctive feature of the phylum.

The body plan assumed by cnidarians may be that of either the polyp or the medusa. In the polyp, the sac is attached to some object at one end, and the mouth and tentacles are at the other. Corals and sea anemones have this form. The medusa or jellyfish form is similar although it is inverted so that the mouth and tentacles lie below the sac (see Fig. 10.24). In some species, both body forms alternate in the life cycle.

Jellyfish are the oldest known fossilized representatives of the Phylum Cnidaria. Earliest finds are Late Proterozoic, and jellyfish have persisted since that time without significant change. Because they have no hard parts, fossil jellyfish are found only rarely, usually as impressions in what was once the soft mud of the sea floor.

Paleontologically, the corals are the most important of the Cnidaria. Corals have been important members of the marine biosphere since the Ordovician (Fig. 10.25). The living coral polyp secretes a calcareous cup **(theca)** in which it resides (Fig. 10.26) and from which it grows upward and outward. If the coral is solitary, the cup often takes a horn shape. Most corals, however, live in colonies composed of great numbers of individual theca bundled together, sometimes saving space in prismatic arrangement. The theca of some corals may be divided by vertical plates **(septa)** that mark the position of deep folds in the body wall of the animal. Septa lend support to the polyp and separate layers of tissue, thus increasing the digestive area. As the animal grows, it partitions off its lower part with horizontal plates called *tabulae.*

Corals are classified on the basis of the nature and arrangement of their septa and other skeletal features. For example, the common Paleozoic "horn corals," Order Rugosa, inserted their septa at only four locations during adult growth, and are therefore called *tetracorals* (see Fig. 10.26). In other Paleozoic corals, the tabulae are the most obvious morphologic feature of the theca, and septa are absent or represented by vertical rows of short spines or low ridges. These are the tabulate corals of the Order Tabulata. They include many colonial forms, such as the so-called "honeycomb" and "chain" corals. (See *Favosites* and *Halysites,* Fig. 10.27.) Both the Tabulata and Rugosa became extinct at the end of the Paleozoic, and most Cenozoic corals are members of the Order Scleractina. They are commonly called *hexacorals* as a reminder of their six-fold septal symmetry.

Figure 10.23 Diversity of living cnidarians.

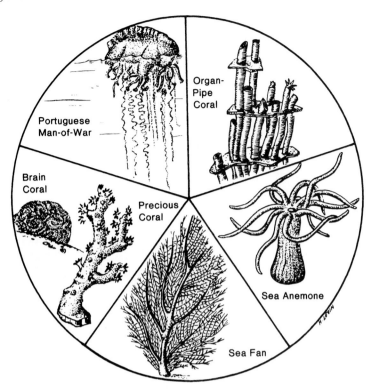

Figure 10.24 (A) Drawing of *Hydra* with section cut away. This tiny, tentacled cnidarian (3 to 10 mm in height) lives in freshwater bodies. As with many other cnidarians it is radially symmetrical. (B) Enlargement of the body wall of *Hydra* to show types of cells. (C) The polyp (left) and medusa (right) body plans of cnidarians. (D) A cnidocyte containing an undischarged nematocyst.

From Levin, H. L., *Ancient Invertebrates and Their Living Relatives,* Upper Saddle River, NJ: Prentice Hall, 1999.

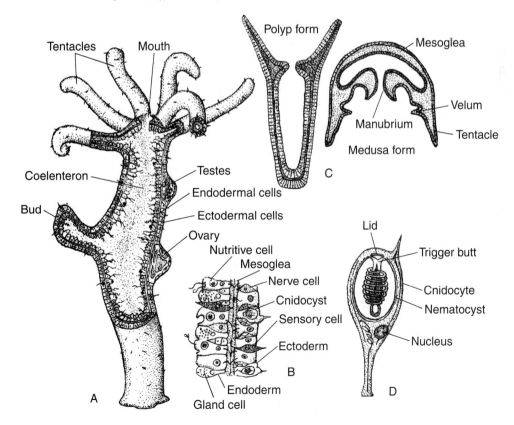

Figure 10.25 Geologic range of some important groups of corals.

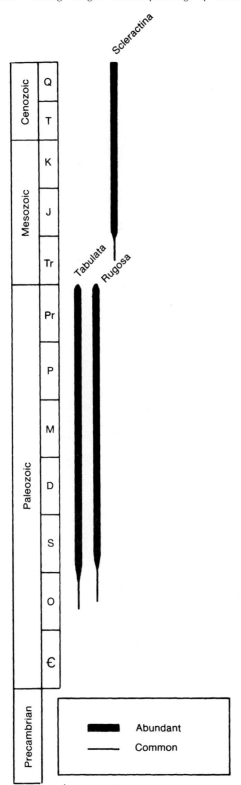

Figure 10.26 Presumed polyp-theca relationship in a Paleozoic horn coral.

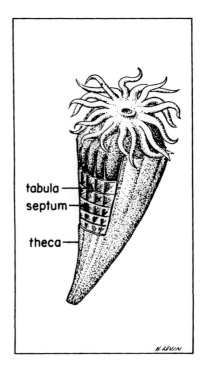

Reef corals imply clear (low terrigenous sediment influx), warm, and shallow marine waters. Today, the majority of reef builders live at depths of about 15 m or less. Their optimum temperature preference seems to be between 25°C and 29°C, and this tends to restrict them to latitudes between 30°N and 30°S. Reef corals have a mutually beneficial relationship with algae, called **symbiosis.** This relationship partly accounts for the fact that reefs do not grow at water depths greater than about 30 m because the algae require adequate light for photosynthesis.

Study Questions

1. What function do septa serve in a coral? What is the function of tabulae?

   _____

   _____

   _____

2. If fossil corals are found in a limestone ledge exposed in the Rocky Mountains at an elevation of 1500 m, what geographic changes must have occurred since the time the corals were living?

   _____

   _____

   _____

Figure 10.27 Cnidaria.

Partly from Collinson, C. W., 1959, Illinois State Geological Survey, Educational Series 4.

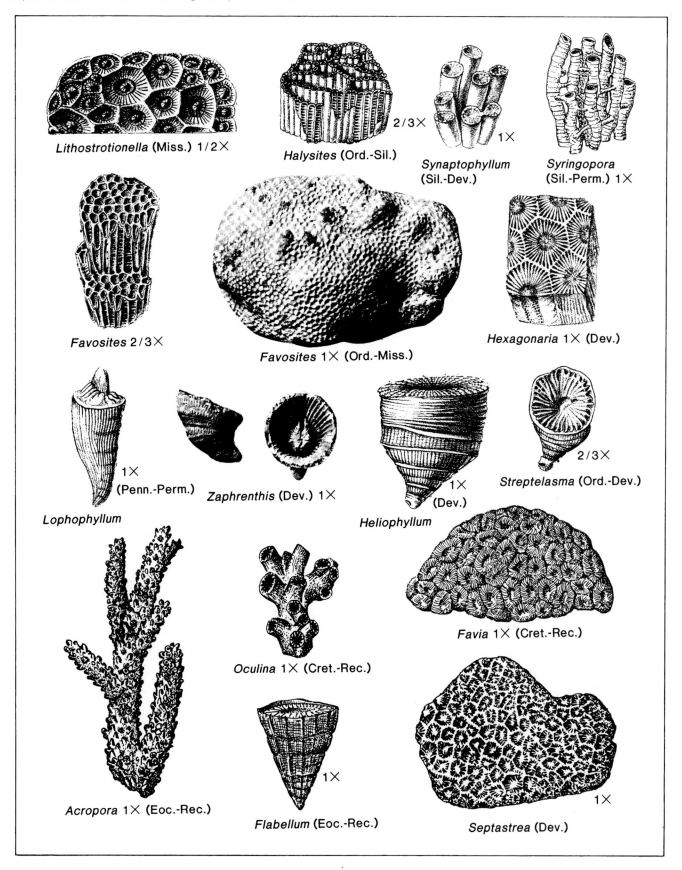

Lithostrotionella (Miss.) 1/2✕

Halysites (Ord.-Sil.)

Synaptophyllum (Sil.-Dev.) 2/3✕ 1✕

Syringopora (Sil.-Perm.) 1✕

Favosites 2/3✕

Favosites 1✕ (Ord.-Miss.)

Hexagonaria 1✕ (Dev.)

Lophophyllum 1✕ (Penn.-Perm.)

Zaphrenthis (Dev.) 1✕

Heliophyllum 1✕ (Dev.)

Streptelasma (Ord.-Dev.) 2/3✕

Acropora 1✕ (Eoc.-Rec.)

Oculina 1✕ (Cret.-Rec.)

Favia 1✕ (Cret.-Rec.)

Flabellum (Eoc.-Rec.) 1✕

Septastrea (Dev.) 1✕

3. Certain Pennsylvanian strata in the Arctic island of Spitzbergen (Lat. 78°N) contain extensive deposits of reef-building corals. From this occurrence, what do you infer about the climate of this region in Pennsylvanian time? On what assumption is this inference based? Is more than one interpretation possible?

_____

_____

_____

_____

4. Examine the study specimens of recent and fossil corals. In the space below, sketch a tabulate coral, a rugose coral, and a hexacoral. Label tabulae, septa, and theca, and lightly outline a restored profile of a polyp for each coral type.

Tabulata

Rugosa

Hexacoralla

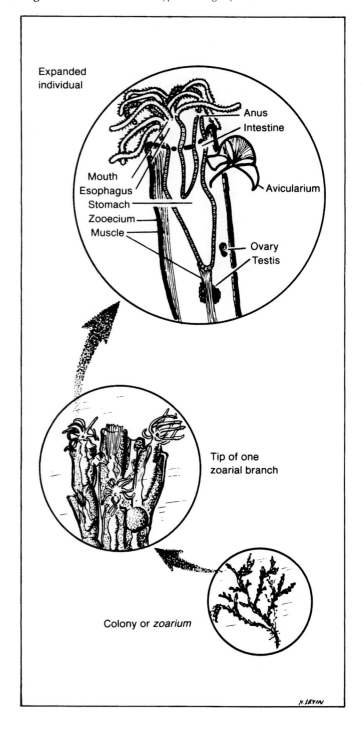

Figure 10.28 Structures of a typical living bryozoan.

THE BRYOZOA

Bryozoans, now classified as Phylum Ectoprocta, include a large group of animals that grow in colonies, appear mosslike to the naked eye, and bear a superficial resemblance to some Cnidarians. However, the similarity cannot be pressed too far, for a bryozoan has both a mouth and an anus, and a complete V-shaped digestive tract. Most fossil bryozoans secreted a calcareous twiglike, matlike, or frondlike colony called a **zoarium.** Each tiny animal **(polypide)** was housed in a cuplike cavity or **zooecium** (Fig. 10.28). Because of their small size, the zooecia appear as tiny pinpoint depressions on the outside of the zoaria. In most living species, the mouth is surrounded by a horseshoe-shaped structure bearing ciliated tentacles called the **lophophore.**

Fossil bryozoans are abundant in Paleozoic carbonate strata (Fig. 10.29). Because of their small size, thin sections of zoaria must be prepared and examined under the microscope. However, a few of the common forms can be classified into the following artificial categories for recognition in hand specimens.

Figure 10.29 Geologic range of bryozoans.

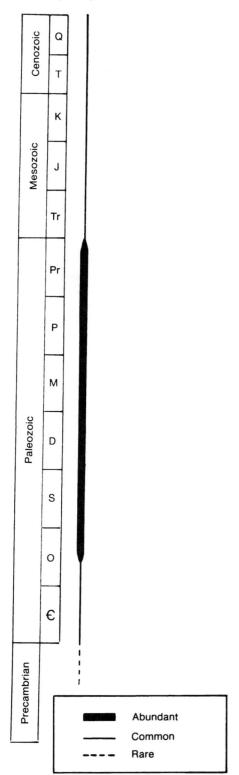

▬▬	Abundant
—	Common
- - - -	Rare

1. The "branching twig" or ramose forms are especially characteristic of rocks of the Ordovician. At first glance, they resemble erect-branching shrubbery with stems about the thickness of pencils. Low mounds, called monticules, and flat areas, called maculae, may be spaced regularly over the surface. The maculae in the genus *Constellaria* have a distinctive star shape (Fig. 10.30). *Hallopora* (see Fig. 10.30) is characterized by abundant nodose monticules.

2. Lacy or *fenestellate* bryozoans appeared in the Silurian and became especially abundant during the Mississippian. A typical genus, *Fenestella* (see Fig. 10.30), rather resembles petrified lace, and is best seen along the weathered surfaces of limestone beds. In these delicate bryozoans, the zooecial openings occur in rows along the vertical branches, and the branches are connected by crossbars (dissepiments). The tiny windows framed by branches and dissepiments are called **fenestrules** (L. *fenestra,* window).

3. Perhaps the most unusual fossil bryozoan is *Archimedes.* Sometimes called the "corkscrew fossil," *Archimedes* consists of a calcareous helicoid spiral that served as the axial support for fronds of lacy bryozoa (see Fig. 10.30). Normally, the fragile fronds are broken away, and only the axis can be collected. The generic name is derived from the axial spiral, which resembles the "Archimedes Screw," by which that famous Greek is presumed to have lifted water from wells. Although most prevalent in Mississippian strata of eastern and central North America, *Archimedes* migrated slowly westward, reaching Russia by the Permian.

4. Another unusual shape in bryozoans is seen in the star-shaped *Evactinopora* (see Fig. 10.30). It is a common fossil in the Mississippian rocks of the central United States.

Study Questions

1. Study the specimens of fossil and living (preserved) bryozoans. Sketch the specimens designated by your instructor, and note the position of a zooecium on each drawing by an arrow.

2. Compare the size of a zooecium in a "branching twig" bryozoan with the theca of a colonial tabulate coral. Is size a useful means for distinguishing bryozoans from corals?

_____

_____

_____

_____

_____

_____

Figure 10.30 Fossil bryozoa. Individual zooecia are represented by the tiny dots on *Tabulipora* and *Evactinopora,* and by the small pits adjacent to large fenestrules in *Fenestella.*

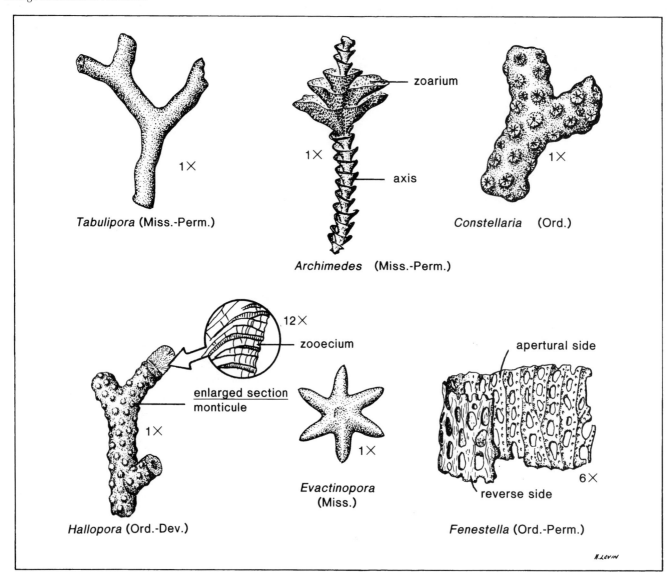

3. What advantage, if any, do fenestra provide to the polypides of a bryozoan like *Fenestella?*

_____

_____

_____

_____

_____

_____

_____

_____

4. *Archimedes* and most other bryozoans live sedentary lives on the sea floor. How then did they spread from one place to another?

_____

_____

_____

5. What organs present in the coelum of a bryozoan are **not** present in the coelenteron of a coral polyp?

_____

_____

_____

THE BRACHIOPODA

Brachiopods have been inhabitants of the earth from the Late Proterozoic and Early Cambrian to the present (Fig. 10.31). During the Paleozoic, however, they were far more abundant and diverse than today. On first seeing a fossil brachiopod, one is reminded of the two parts or valves of a clam shell. However, closer examination shows that the two valves of a brachiopod are not identical, as is the case with most common clams. The valves of a brachiopod (Fig. 10.32) are designated **pedicle** (ventral) and **brachial** (dorsal), whereas those of a clam are referred to as right and left. In the articulate brachiopods, the valves are hinged along the posterior edge by teeth and sockets. Inarticulate brachiopods lack a definite hingement, the valves are held together only by muscles, and the shell is composed of a mixture of calcium carbonate and an organic horny substance called chitin.

Most brachiopods attach themselves to the sea floor by a fleshy stalk or **pedicle** (see Fig. 10.33). In the inarticulate group, the pedicle simply emerges between the two valves, whereas in articulate brachiopods, the pedicle emerges from an opening (the **pedicle opening**). In most articulates, the valves are composed of calcium carbonate and are variously ornamented by concentric growth lines and radial ridges **(costae)** or threadlike ridges called **costellae**. Some forms developed a midradial fold on one valve, with a trough **(sulcus)** in a corresponding position on the other valve. These radial and concentric features probably served to strengthen the shell and aided in directing water movement into and out of the shell.

Among the more conspicuous soft parts within the valves is the **lophophore**. The lophophore (Fig. 10.33) consists of two coiled ciliated tentacles whose function is to keep the water between the valves in circulation, so as to distribute oxygen throughout the tissues and remove carbon dioxide. The water currents generated by the cilia on the lophophores also direct suspended food particles in toward the mouth, from which the food passes into the stomach and intestine.

Modern brachiopods are exclusively marine and occur at most water depths in all latitudes. However, they are not common, and tend to occupy a subordinate position in the seas. Even if this were not true, comparative ecological studies between living and fossil brachiopods would be difficult because modern forms are not sufficiently similar to most extinct forms. One exception is found in the inarticulate brachiopod *Lingula,* which has the distinction of being probably the oldest known genus in the animal kingdom. Species of this genus are found today in the Pacific and Indian Oceans. They prefer warm or tropical waters of subnormal salinity, and live at depths less than 183 m. The shells of living *Lingula* (see Fig. 10.33) are similar to those of fossil *Lingula* in Cambrian rocks over 500 m.y. old.

Figure 10.31 Geologic range of the phylum Brachiopoda.

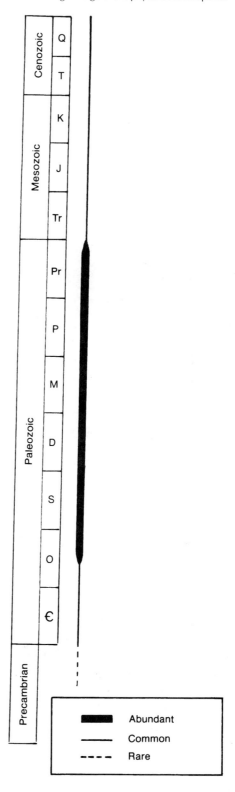

Figure 10.32 Features of brachiopod valves. The brachiopod depicted is *Cererithyris,* a terebratulid of Jurassic age.

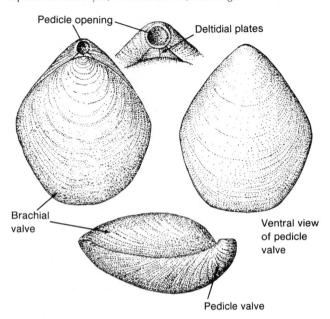

Figure 10.33 Recent brachiopods.

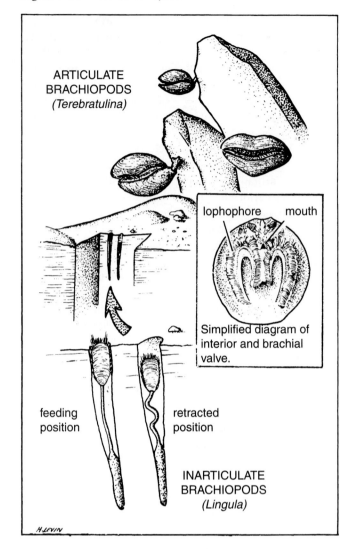

The articulate brachiopods quickly surpassed their linguloid relatives in numbers and diversity. Because of their rapid rate of evolutionary change, they are excellent **index fossils.** The more common articulates can, for convenience, be referred to one of the groups listed below. Each group is taxonomically equivalent to an order or suborder of the Phylum Brachiopoda.

1. **Orthids**   Subcircular, flatly biconvex brachiopods with straight hinge line and radial ornamentation. Geologic range: Cambrian to Permian, abundant in Ordovician. Representative genera are *Platystrophia* (Fig. 10.34), *Enteletes* (Fig. 10.34), and *Hebertella* (Fig. 10.35).

2. **Pentamerids**   Biconvex, often smooth, thickly rounded, elongate forms with short hinge line as in *Pentamerus* (Fig. 10.35). Geologic range: Middle Cambrian to Late Devonian, abundant in Silurian.

3. **Strophomenids**   Compressed forms with straight hinge lines and usually fine radial ornamentation (costellae). Geologic range: Ordovician to Late Jurassic, abundant in Ordovician and Devonian. Typical genera are *Strophomena, Sowerbyella, Rafinesquina* (Fig. 10.34), and *Leptaena* (Fig. 10.35).

4. **Spiriferids**   The interior of the brachial valve of the spiriferid contains a calcareous **helicoid** coil, which in life surrounded the lophophore (as in *Mucrospirifer,* Fig. 10.34). Because the coil extends laterally, the enclosing valves are elongated laterally and give many of the spiriferids a winglike appearance. Geologic range: Middle Ordovician to Late Jurassic, abundant in Devonian. Other representative genera are *Platyrachella* (Fig. 10.35) and *Cyrtina* (Fig. 10.34).

5. **Atrypids**   Atrypids are also spire-bearing but differ from spiriferids in the direction of coiling of the spiral. In atrypids, the axis of coiling is in the dorsal-ventral (up and down) direction and not parallel to the hinge line. As a result, atrypids develop hemispherical brachial valves to accommodate the coil. Geologic range: Middle Ordovician to Early Mississippian, abundant in Devonian. *Atrypa* is a typical and abundant representative of this group (Fig. 10.35).

Figure 10.34 Fossil brachiopods.

From Collinson, C. W., 1959, Illinois State Geological Survey, Educational Series 4.

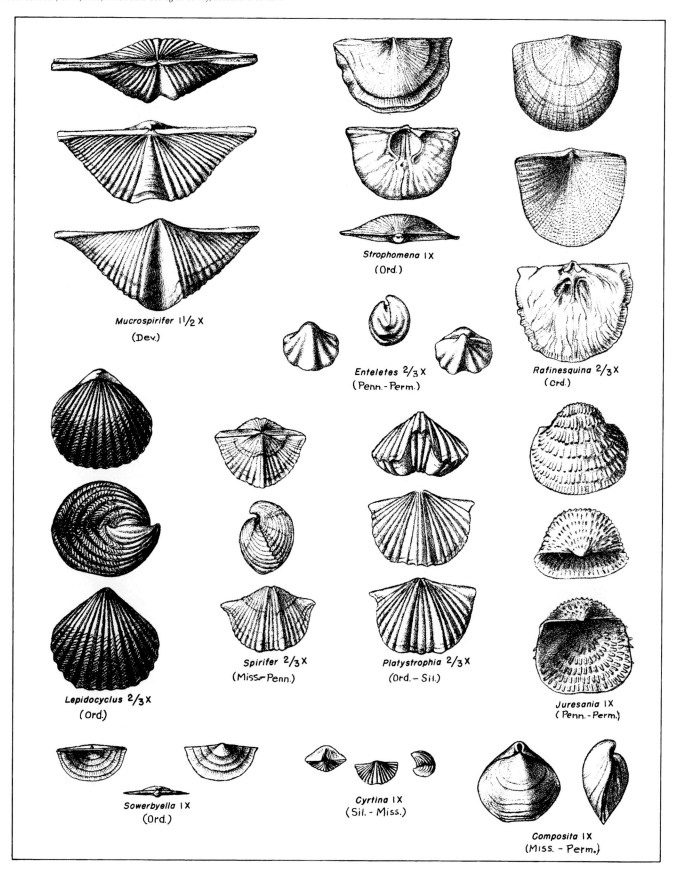

Mucrospirifer 1 1/2 X
(Dev.)

Strophomena 1X
(Ord.)

Enteletes 2/3 X
(Penn.-Perm.)

Rafinesquina 2/3 X
(Ord.)

Lepidocyclus 2/3 X
(Ord.)

Spirifer 2/3 X
(Miss.–Penn.)

Platystrophia 2/3 X
(Ord.–Sil.)

Juresania 1X
(Penn.-Perm.)

Sowerbyella 1X
(Ord.)

Cyrtina 1X
(Sil.-Miss.)

Composita 1X
(Miss.-Perm.)

Figure 10.35 Fossil brachiopods.

Partly from Collinson, C. W., 1959, Illinois State Geological Survey, Educational Series 4.

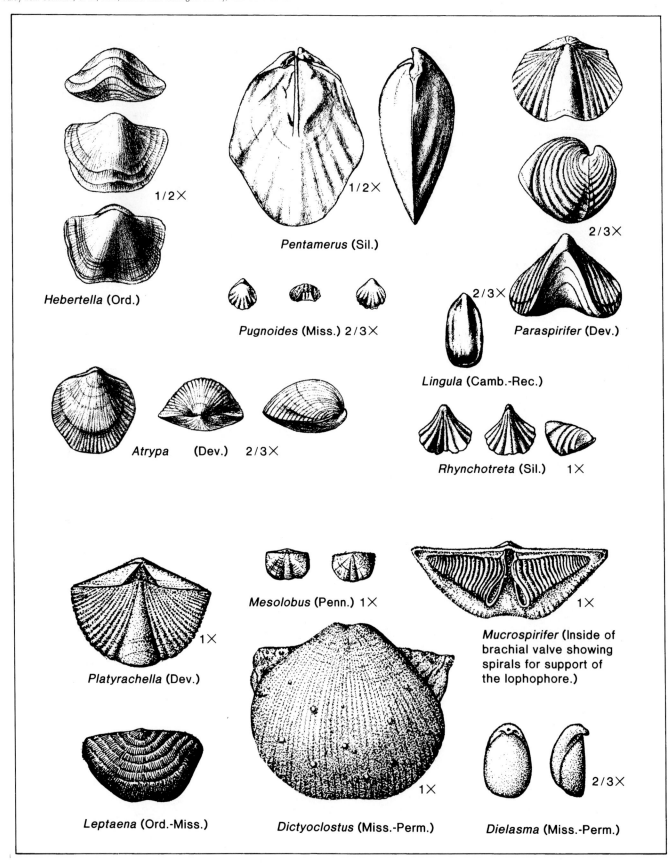

Hebertella (Ord.)

Pentamerus (Sil.)

Pugnoides (Miss.) 2/3×

Paraspirifer (Dev.)

Lingula (Camb.-Rec.)

Atrypa (Dev.) 2/3×

Rhynchotreta (Sil.) 1×

Platyrachella (Dev.)

Mesolobus (Penn.) 1×

Mucrospirifer (Inside of brachial valve showing spirals for support of the lophophore.)

Leptaena (Ord.-Miss.)

Dictyoclostus (Miss.-Perm.)

Dielasma (Miss.-Perm.)

6. **Athyridids**   These forms are characterized by an elongate beak and an internal spiral of the atrypid type. The shell is mostly biconvex and lacking in strong radial ornamentation. Geologic range: Late Ordovician to Jurassic, abundant in Mississippian and Pennsylvanian. A typical and common genus is *Composita* (Fig. 10.34).

7. **Productids**   The most characteristic features of productids are the conspicuous spines and broken spine bases located over the general exterior surface of the valves or along the hinge line. The larger forms possess flattened or slightly concave brachial valves and strongly convex pedicle valves. Geologic range: Late Ordovician to Late Permian, abundant in Mississippian through Permian. *Juresania* (Fig. 10.34) and *Dictyoclostus* (Fig. 10.35) are representative genera.

8. **Rhynchonellids**   Rhynchonellids are strongly beaked forms, usually with strong, accordionlike plications extending radially from the beak. The shells are triangular to rounded in outline, and the hinge line is short. Geologic range: Middle Silurian to Recent, abundant Ordovician through Mississippian and in Jurassic. *Lepidocyclus* (Fig. 10.34), *Pugnoides,* and *Rhynchotreta* (Fig. 10.35) are representative of this group.

9. **Terebratulids**   In terebratulids, the lophophore is supported by calcareous loops. Most species have the pedicle opening located within the overhanging beak, and although some are costate, most are smooth and have a streamlined appearance. Geologic range: Devonian to Recent, abundant in Devonian, Jurassic, and Cenozoic. The living *Terebratulina* (Fig. 10.33) and the fossil *Dielasma* (Fig. 10.35) are representative.

Certain of the brachiopod groups characterize rocks of particular geologic periods. For example, a stratum containing abundant orthids and strophomenids is likely to be Ordovician in age. If pentamerids are common, one would suspect the stratum to be Silurian. Spiriferids tended to dominate brachiopod faunas during the Devonian, whereas productids characterized the post-Devonian periods of the Paleozoic. The Late Permian was a time of widespread extinction among brachiopods, and in the Mesozoic, terebratulids and rhynchonellids predominated. Terebratulids and species of *Lingula* persist into the modern world.

Study Questions

1. Examine the valves of a brachiopod and a clam such as *Mercenaria* (see Fig. 11.4). Which of these bivalved invertebrates has identical valves?

    _____

    _____

    In which animal does the plane of symmetry cross the valves and hinge line through the beaks?

    _____

    _____

2. How would you evaluate the genus *Lingula* as an environmental indicator?

    _____

    _____

    _____

    _____

3. By comparing the specimens in the study set with the illustrations, try to identify the brachiopods as to genus. For specimens that do not closely resemble one of the illustrations, assign one of the group names on the basis of the verbal descriptions. Write your identifications on a separate sheet.

4. Sketch a side and top view of one of the brachiopods in the study set and label the brachial valve, pedicle valve, radial markings, concentric markings, fold, and sulcus. On which valve is the sulcus usually located in your study specimens? Which valve is usually larger?

    _____

    _____

    _____

5. A clam shell is opened automatically by the elasticity of a ligament along the hinge line. At rest, the clam is open. Articulate brachiopods open their valves by contracting a pair of diductor muscles, and work must be performed to keep the valves agape. How do these two different valve mechanisms affect the way clams and brachiopods appear as fossils? Are you able to find any evidence in support of your answer in the study set?

    _____

    _____

    _____

    _____

6. What would be the probable age (geologic system) of limestone strata containing the following brachiopods?

    a. *Mucrospirifer, Platyrachella,* and *Paraspirifer?*

    _____

    b. *Strophomena, Sowerbyella, Rafinesquina,* and *Hebertella?*

    _____

    c. *Pentamerus, Rhynchotreta,* and the sponge *Astraeospongia?*

7. Invertebrates of what phylum, other than members of the Brachiopoda, possess a lophophore?

_____

_____

_____

_____

_____

8. During what geologic period or periods were the following brachiopod groups particularly abundant?

Orthida _____

Strophomenida _____

Pentamerida _____

Productidina _____

Spiriferida _____

TERMS

algae   Informal name for a group of photosynthetic mostly aquatic Protista that range in size from microscopic single cells to large unicellular forms like seaweeds. Diatoms and coccolithophores are unicellular algae.

aperture (on **foraminiferida**)   The relatively large opening at the surface of the last formed chamber.

benthonic (or **benthic**)   Marine life that are bottom-dwelling.

brachial valve   Valve of a brachiopod to which the calcareous supports for the lophophore are attached. Typically, the brachial valve is the smaller valve and may have a small beak.

brachiopoda   Marine invertebrates whose shells are composed of two valves that are oriented as dorsal (brachial) and ventral (pedicle).

bryozoa   A phylum of colonial, mostly marine, tiny animals that bear lophophores and build calcareous skeletal structures.

carbonization   A type of fossil preservation in which most of the organic matter is decayed, leaving a thin carbon-film impression of the organism.

cast   A replica of an organism or part of an organism, formed when sediment fills a mold of that organism.

choanocyte   Special cells lining the internal cavity of a sponge, and characterized by a delicate collar of protoplasm through which passes a flagellum.

cnidaria   A phylum of marine invertebrates characterized by a hollow body cavity, radial symmetry, and stinging cells. (Includes jellyfish, corals, and anemones.)

coccoliths   The calcareous plates that form the external covering of a coccolithophore.

coccolithophores   Marine, planktonic, biflagellate, golden-brown algae that typically secrete coverings of discoidal calcareous platelets called **coccoliths.**

coccosphere   The mineralized preservable shell of a coccolithophore.

costae   Radially arranged prominent ridges on the surface of bivalved invertebrates.

costellae   Radially arranged striae or fine ridges on the surface of bivalved invertebrates. Costellae are smaller and less prominent than costae.

diatoms   Microscopic golden-brown algae that secrete a delicate siliceous *frustule* (shell).

discoasters   Calcareous, often star-shaped or radiate, microscopic structures believed to be secreted by certain coccolithophores.

fenestrules   Window-like openings that give a lacy-like appearance to colonies of fenestellate bryozoans.

foraminifers   Informal name for the Order Foraminiferida, consisting of mostly marine, unicellular animals that secrete **tests** (shells) of calcium carbonate.

fossil   The remains or evidence of the existence of organisms that lived in the geologic past.

helicoid   Formed or arranged in a spiral.

index fossil   A fossil that identifies and dates the strata or succession of strata in which it is found. Also called a *guide fossil.*

lophophore   A coiled or lobed ciliated appendage extending from the mouth area in brachiopods and bryozoans. The lophophore's function is primarily feeding.

mold   An impression or imprint of an organism (or part of an organism) that is left in sediment or rock.

osculum   Large opening in a sponge that makes possible the outward flow of water from the internal cavity to the exterior.

pedicle   Fleshy stalk by which a brachiopod attaches itself to a substrate.

pedicle opening   Opening along or adjacent to the brachiopod hinge that serves for emergence of the pedicle.

pedicle valve (of **brachiopod**)   The brachiopod valve, by convention considered ventral, to which the pedicle is attached.

permineralization   A mode of fossilization in which voids in an organic structure, such as bone, are filled with mineral matter.

petrification   The general process of converting organic matter such as bone, shell, or wood into a durable substance such as calcium carbonate or silica.

planktonic (or **planktic**)   Free-floating, mostly microscopic aquatic organisms.

polypide   The individual bryozoan animals (also called a zooid).

radiolaria   Protozoans that secrete delicate, often beautifully filigreed skeletons of opaline silica.

receptaculitids   Paleozoic, discoidal, perforate calcareous fossils having skeletal elements in spiral rows.

replacement   A fossilization process by which the original hard tissue is replaced after burial by inorganically precipitated mineral matter.

septa   Partitions that support the vertical mesenteries in the theca of corals and that separate the chambers in cephalopods and foraminifers.

species   A group of organisms, either plant or animal, that may interbreed and produce fertile offspring having similar structure, habits, and functions.

spicule   A minute calcareous or siliceous structure, having highly varied and often characteristic forms, occurring in and serving to stiffen and support the tissues of various marine invertebrates.

sulcus   A concave longitudinal depression on the surface of either valve of a brachiopod.

sutures   Lines formed by the intersection of septa with the inner wall of cephalopod conchs or foraminiferida chambers.

symbiosis   The relationship that exists between two different organisms that live in close association, with at least one being helped without either being harmed.

test   Term that refers to the skeletons of protozoans and certain other animals.

theca   The cuplike or tubular skeleton structure that surrounds the polyp of corals.

zoarium   The skeleton of an entire bryozoan colony.

zooecium (in **Bryozoa**)   The calcareous or chitinous skeleton of an individual bryozoan.

11

Fossils and Their Living Relatives: Mollusks, Arthropods, Echinoderms, Graptolites, and Plants

WHAT YOU WILL NEED

1. Study set of fossil bivalves (Class Bivalvia), gastropods, cephalopods, trilobites, cystoids, blastoids, and echinoids.

2. Prepared slide of fossil ostracodes.

3. Rock samples exhibiting graptolites and fossil plants.

 ## THE MOLLUSCA

To many people, mollusks are the most familiar of marine invertebrates. Most members of the phylum possess external calcareous shells, which can be readily collected at coastal areas around the world. The soft fleshy parts of some mollusks are widely used as food by humans. Such familiar animals as snails, slugs, clams, oysters, chitons, squids, and octopuses (Fig. 11.1) are included in the Phylum Mollusca.

Typically, mollusks are unsegmented animals with bilateral symmetry. A fold called the **mantle** has the shell-building function, and a muscular portion of the body (the foot) has a primarily locomotor function. In some forms, like the squid and cuttlefish, the foot is modified into tentacles. Respiration in most forms is accomplished by gills. Well-developed circulatory organs (heart, blood vessels), digestive glands (liver, kidney, intestines), and nervous system are evidence of advanced development. Mollusks vary in size from microscopic to the extraordinary dimensions seen in the giant squid, which may grow to lengths of 15 m.

Bivalvia (Formerly Pelecypoda)

Bivalvia are mollusks with a hatchet-shaped foot, layered gills, right and left **valves,** bilobed mantle, and no definite head. The margins of the mantle are modified to form tubes (siphons) for intake and exhaust of water. Incoming water moves over the gills where food particles are trapped and carried into the digestive system.

The valves in bivalves are held together at the dorsal edge by an elastic ligament. Concentric growth lines originate at a dorsal, slightly swelled area called the **umbo.** The terminal end of the umbo is called the **beak,** and in most forms is directed anteriorly. The inside of an empty valve contains markings that represent places of attachment of organs and muscles. In most shells, large, rounded scars of the adductor

muscles can be found dorsally at either end. An exception is seen in oysters and scallops, which have only one adductor muscle. A thin line **(pallial line),** usually paralleling the free margin, marks the place of attachment of the mantle to the inner surface of the shell (Fig. 11.2). A posterior indentation of the pallial line, called the **pallial sinus,** marks the position of the foot retractor muscle and provides information about the size of the siphons. Because siphon size is indicative of habits, the pallial sinus is of particular use in fossil study.

All Bivalvia are aquatic, and the majority are marine. With a few exceptions, they are sedentary organisms that live buried in the mud or sand. Some species can either attach themselves to or burrow into rock or wood. The Bivalvia are typically gregarious and in many localities are enormously abundant over large areas. In general, an abundance of species and individuals of fossil marine bivalves implies shallow-water deposition. The shells of modern deep-water forms are often thin, translucent, and small. Except for mud-dwellers, shallow-water species possess thicker, more robust, ornamented and colored shells.

Bivalves made their first appearance in Cambrian time but did not become notably varied or numerous until the Mesozoic (Fig. 11.3). Many thick-shelled, oysterlike forms like *Gryphaea* (Fig. 11.4) are found in Jurassic and Cretaceous strata. In these bivalves, the lower valve became large and massive, whereas the upper valve was reduced in size to a mere cap. Species of scallops like *Pecten* (Fig. 11.4) are good local guide fossils for Cenozoic rocks.

Study Questions

1. Sketch an external and internal view of the modern clam shell in your study set and label the pallial line, pallial sinus, growth lines, muscle scars, and hinge line.

Figure 11.1 Diversity of living mollusks.

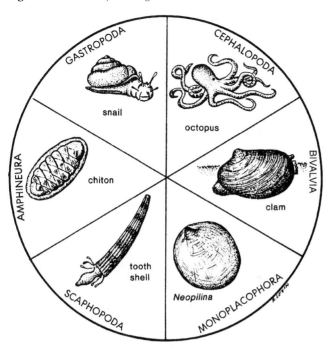

Figure 11.2 Features of the shell of a bivalve.

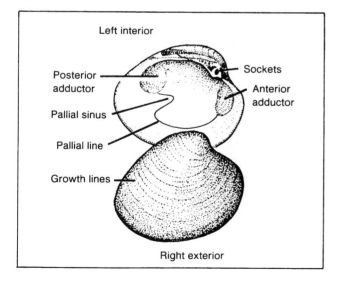

Figure 11.3 Geologic range of gastropods and bivalves.

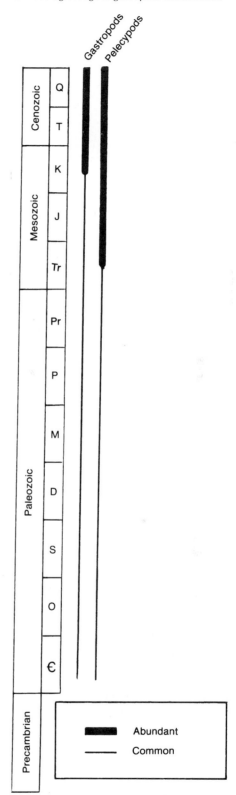

Figure 11.4 Fossil bivalvia.

Partly from Collinson, C. W., 1959, Illinois State Geological Survey, Educational Series 4.

Mercenaria (Jur.-Rec.)

Grammysia (Sil.-Miss.)

Trigonia (Jur.-Rec.)

Cardiopsis (Miss.-Penn.)

Exogyra (Jur.-Cret.)

Allorisma (Miss.-Penn.)

Aviculopecten (Sil.-Perm.)

Ctenodonta (Ord.-Sil.)

Ostrea (Trias.-Rec.)

Pecten (Miss.-Rec.)

Gryphaea (Jur.-Cret.)

Cardium (Trias.-Rec.)

Nuculana (Sil.-Rec.)

Inoceramus (Jur.-Cret.)

2. By comparing the study specimens with the illustrations, try to identify the bivalves as to generic name.

_____

_____

_____

_____

3. At Pozzuoli, Italy, borings of marine bivalves are found about halfway up the vertical columns of a ruined building that once stood in a Roman marketplace. From this, what do you infer about changes in sea level here during the past two thousand years?

_____

_____

_____

_____

_____

_____

_____

4. What explanation may be offered for the observation that heavy-shelled bivalves are mainly nearshore dwellers?

_____

_____

_____

_____

_____

5. In general, bivalves have equal valves, in contrast to brachiopods, which have unequal valves. Which bivalves depicted in Fig. 11.4 are an exception to this generalization?

_____

_____

Which bivalve is most like a brachiopod in having valves that are *nearly* bilaterally symmetrical?

_____

_____

_____

6. Bivalves are *filter feeders*. What does this mean?

_____

_____

_____

7. What differences in symmetry of shells exist between a bivalve like *Mercenaria* (Fig. 11.4) and a brachiopod like *Atrypa* (Fig. 10.35)?

_____

_____

_____

Provide a sketch to illustrate the differences in the space below.

Gastropoda

Over 14,000 species of fossil **gastropods** have been described, and probably an even greater number of species exists today. Members of the class build a one-piece, coiled shell, although some forms do not secrete a shell. The typical gastropod has a distinct head with mouth, eyes, and tentacles, and a ventral flattened foot. Undulatory motion of the foot provides locomotion. The head and foot can be retracted into the shell, and in some forms, the opening (aperture) of the shell can be closed by an **operculum** when the foot is withdrawn. The shell **(conch)** develops as an expanding tube that coils around a hollow axis **(columnella).** That part of the conch representing one revolution is called a **whorl,** and the surface of contact between whorls is expressed by a line called a *suture* (Fig. 11.5). Ornamentation is varied, and may consist of growth lines, ribs, spines, nodes, or combinations of these features.

Since their first appearance in the Cambrian (Fig. 11.3), gastropods have adjusted to a variety of environments. Many live on or burrow into the bottoms of seas or lakes. Others have evolved into air-breathing land dwellers.

In general, an abundance of marine gastropods implies shallow, warm, water. Of living species, the greater number inhabit tropical waters where robust, thick-shelled, highly ornamental species thrive. The number of species decreases in deeper and cooler parts of the sea. Thin-walled conchs with little ornamentation are often characteristic of deeper water forms.

Figure 11.5 Gastropod anatomical features.

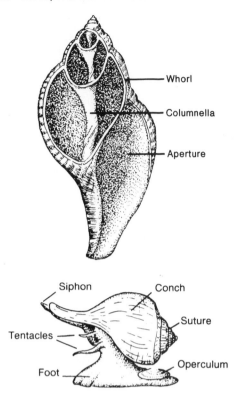

Whorl

Columnella

Aperture

Siphon   Conch

Suture

Tentacles

Operculum

Foot

Study Questions

1. Examine the gastropods in your study set and identify as many as you can by comparison with the illustrations (Fig. 11.6).

2. Sketch one of the fossil snail specimens and label aperture and growth lines.

3. Does the conch coil in a plane or a spiral?

4. Most gastropods coil in a spiral. Which of the gastropods depicted in Figure 11.6 coil in a plane (planispiral coiling)?

Cephalopoda

The **cephalopods** are probably the most advanced and complex of the mollusks. They are oceanic animals and include octopuses, squids, cuttlefish, and nautiluses. The pearly or chambered nautilus is a living form that provides many clues to the habits and soft anatomy of fossil cephalopods. In *Nautilus* (Fig. 11.7 and 11.8), the bilaterally symmetrical body has a prominent head that bears paired, image-forming eyes. Tentacles are present as a modification of the anterior portion of the foot. As in other members of the Mollusca, gills, mantle, and siphon are present. Locomotion is accomplished by expulsion of water from the mantle cavity through a tubular fleshy funnel. In a sense, the animal is jet propelled.

Superficially, conchs of cephalopods resemble conchs of some gastropods. However, most cephalopods coil in a plane, whereas the majority of gastropods are helicoid. An even more important distinction is that the cephalopod conch (Fig. 11.8) is divided into chambers (**camerae**) by transverse partitions called **septa.** The soft parts reside in the final chamber (living chamber), and new septa are added behind the visceral mass as it grows forward in the conch.

A thin, porous, calcareous tube termed the **siphuncle** (Fig. 11.9) extends through the septa. The siphuncle secretes gas through its porous wall into the camerae, thus regulating the animal's buoyancy. The juncture of the septa with the inner surface of the conch is marked by lines called *sutures.* The configuration of a suture is determined by the shape of the periphery of a septum. If the edge of the septum is smooth, like that of a saucer, then the suture will be similarly smooth. If the septum is fluted, like the margin of a pie shell, this will be reflected in the zigzag pattern of the suture.

The folds of the sutures that are convex toward the aperture of the conch are called *saddles.* Lobes are folds that are convex toward the first-formed part of the conch. The pattern of sutures provides the means for identifying genera and species of cephalopods and is also the basis for distinguishing between two great orders of externally shelled cephalopods, the Nautiloidea (Fig. 11.10) and the Ammonoidea (Fig. 11.11). The Nautiloidea have smooth or broadly undulating sutures and central siphuncle location. They were abundant and widely distributed during the Paleozoic. Nautiloids declined after that time, and today the chambered nautilus is the only survivor.

Figure 11.6 Fossil gastropods.

Partly from Collinson, C. W., 1959, Illinois State Geological Survey, Educational Series 4.

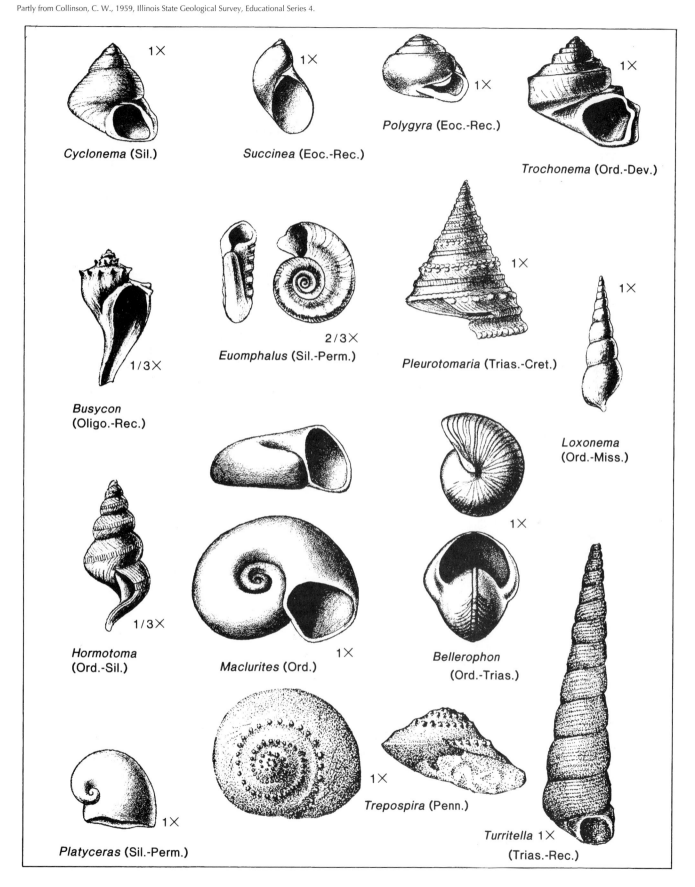

Cyclonema (Sil.)

Succinea (Eoc.-Rec.)

Polygyra (Eoc.-Rec.)

Trochonema (Ord.-Dev.)

Busycon (Oligo.-Rec.)

Euomphalus (Sil.-Perm.)

Pleurotomaria (Trias.-Cret.)

Loxonema (Ord.-Miss.)

Hormotoma (Ord.-Sil.)

Maclurites (Ord.)

Bellerophon (Ord.-Trias.)

Platyceras (Sil.-Perm.)

Trepospira (Penn.)

Turritella (Trias.-Rec.)

Figure 11.7 The living nautiloid cephalopod *Nautilus belauensis*. Note the prominent eye (which has no lens and functions much like a pin-hole camera), the many tentacles, and the protective hood that blocks the opening of the shell when the animal withdraws inside.

Courtesy of W. Bruce Saunders.

Figure 11.8 Conch of *Nautilus*, sawed in half to reveal the large living chamber, septa, and septal necks through which the siphuncle passed. The siphon is soft tissue, and thus not present in this specimen. (Maximum diameter of conch is 19 cm.)

Ammonoids differ from nautiloids in their more complex sutures (Fig. 11.12) and their smaller, more ventrally located siphuncle. Paleozoic ammonoids have sutures with smooth lobes and saddles. Such sutures are called *goniatitic*. Most of the Triassic ammonoids have sutures with smooth saddles and denticulated lobes *(ceratitic)*. A few Triassic ammonoids and most Jurassic and Cretaceous forms have *ammonitic* sutures, with complex wrinkling on both saddles and lobes. The geologic range of each of the major cephalopod groups is depicted on Figure 11.13.

Some members of the Class Cephalopoda have internal rather than external shells. An example is the cuttlefish, whose internal cuttlebone is placed in bird cages as a source of calcium for pet birds. Only one order of internal-shelled cephalopods, the Belemnoidea, is important geologically. (Fig. 11.14). The unchambered guard portion of fossil **belemnite** shells are often bullet-shaped and take on a brown coloration. Belemnites are useful guide fossils in rocks of Jurassic and Cretaceous age.

Figure 11.9 Section of *Nautilus* in plane of coiling.

From Levin, H. L., *Ancient Invertebrates and Their Living Relatives,* Upper Saddle River, NJ: Prentice Hall, 1999.

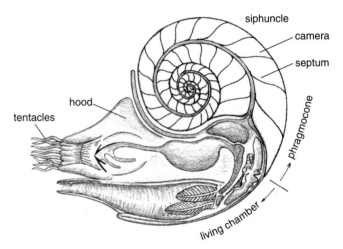

Figure 11.10 Some nautiloid cephalopods. (A) *Plectronoceras,* ×2 (Cambrian). (B) *Ascoceras,* ×0.5 (Silurian). (C) *Phragmoceras,* ×0.5 (Silurian). (D) *Lituites,* ×0.5 (Ordovician). (E) *Ophioceras,* ×2.7. (F) *Trochoceras,* ×0.2 (Devonian). (G) *Michelinoceras,* ×0.7 (Ord.-Trias.). (H) *Endoceras,* ×0.2 (Ordovician). (I) *Bactrites,* ×1 (Sil.-Perm). (J) *Actinoceras,* ×3.0 (Ord.-Sil.) (K) *Aturia,* ×03.0 (Paleocene-Miocene).

From Levin, H. L., *Ancient Invertebrates and Their Living Relatives,* Upper Saddle River, NJ: Prentice Hall, 1999.

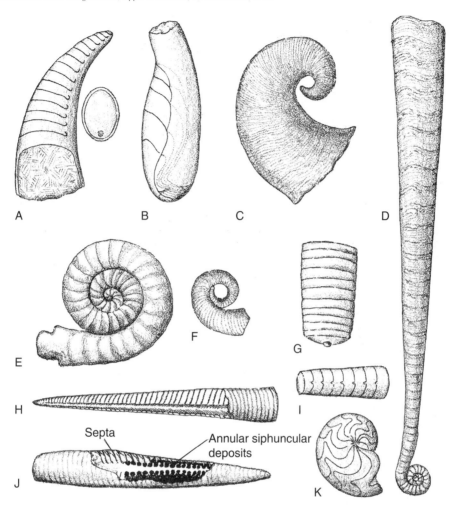

Study Questions

1. By comparison with the illustrations (Figs. 11.10 and 11.11), identify as many of the cephalopods in your study set as you can. Prepare a sketch of one specimen and label suture type.

2. How do cephalopods differ from gastropods with regard to shell structure and portion of conch occupied by soft parts?

_____

_____

_____

_____

_____

3. Examine Figure 11.8 or an actual conch of *Nautilus* that has been sawed in half. Note the holes in the septa that provide for the siphon. Does *Nautilus* have a continuous siphuncle?

_____

_____

_____

Figure 11.11 Ammonoids. (A) *Agoniatites* (with suture pattern), ×0.35 (Devonian). (B) *Clymenia,* ×0.35 (Devonian). (C) *Tornoceras,* ×0.7 (Devonian). (D) *Prolecanites,* ×1.2 (Mississippian). (E) *Meekoceras,* ×0.7 (Triassic). (F) *Phylloceras* (with suture pattern), ×0.7 (Trias.-Cret.). (G) *Scaphites* (with suture pattern), ×0.8 (Cretaceous). (H) *Baculites,* ×0.4 (Cretaceous). (I) *Ancyloceras,* ×0.12 (Cretaceous). (J) *Placenticeras,* ×0.7 (Cretaceous).

From Levin, H. L., *Ancient Invertebrates and Their Living Relatives,* Upper Saddle River, NJ: Prentice Hall, 1999.

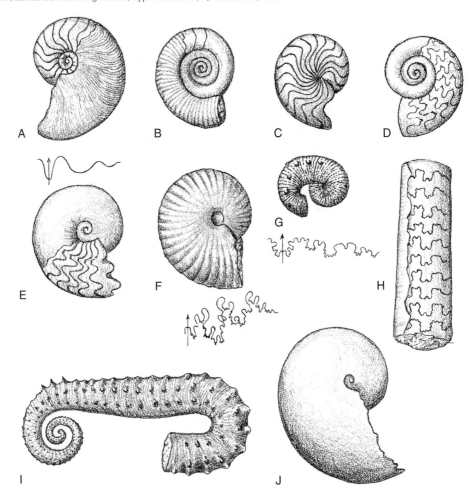

4. Figure 11.15 represents a straight conch of a nautiloid similar to *Michelinoceras* (see Fig. 11.10). The areas shown in black represent soft parts or shell material. If the chambers contain only gas, what would be the orientation of the conch when passively suspended in water? Would this orientation seem advantageous for swimming? If the assumption is made that a horizontal position is preferred, how might this be accomplished in evolving generations of cephalopods? (Hint: Consider adding "balast" or changing the pattern of coiling as ways to achieve horizontality.)

    _____

    _____

    _____

    _____

    _____

5. Figure 11.16 depicts changes in the number of genera of cephalopods and fishes through a segment of geologic time. From the graph, suggest a possible hypothesis for the extinction of ammonites and reduction of nautiloids at the end of the Cretaceous.

    _____

    _____

    _____

    _____

    _____

    _____

6. Cephalopods have been more useful than gastropods in stratigraphic correlation and as index fossils. Suggest a reason (or several reasons) why they are more useful.

_____

_____

_____

_____

_____

_____

_____

_____

Figure 11.12 Cephalopod suture patterns.*

*By convention, illustrations of fossil coiled cephalopods have living chamber placed uppermost, inverted from the normal living position.

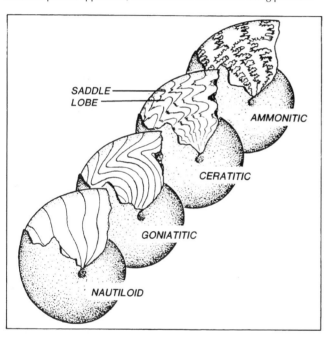

Figure 11.13 Geologic ranges of major groups of cephalopods.

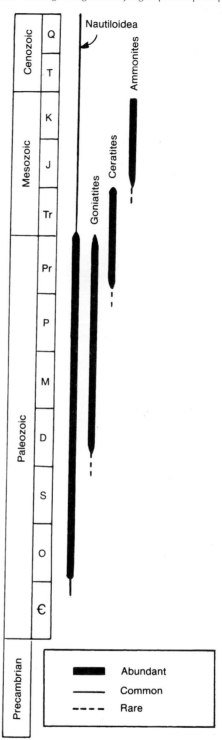

Figure 11.14 Restoration of a belemnite.

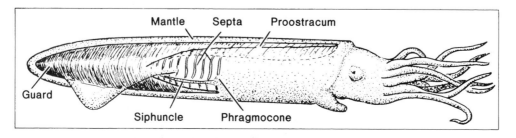

Figure 11.15 Problem of buoyancy in orthocone cephalopods.

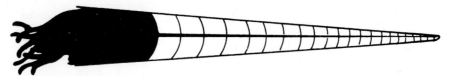

Figure 11.16 Change through time in number of fish genera and number of cephalopod genera.

From Newell (1962), *J. Paleontology,* **36:** 605.

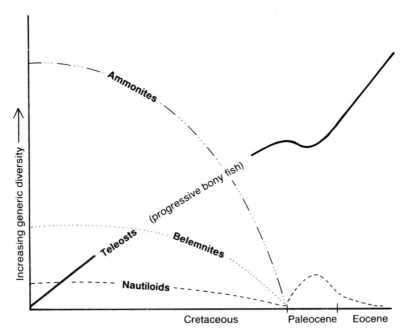

7. The diagram (Fig. 11.17) illustrates an inclined coal seam. Miners working the seam find that it ends abruptly at point A. The abrupt truncation suggests that a fault plane has been encountered. A company geologist examines the hanging wall at A and finds belemnites of the same species as those that had been discovered earlier in a formation exposed west of the milling plant. What are the prospects for additional coal recovery east of the fault? What type of fault is represented? (Refer to Chapter 14 for a summary of fault nomenclature.)

_____

_____

_____

_____

_____

_____

_____

Monoplacophora, Amphineura, and Scaphopoda

Three classes of the Phylum Mollusca are of lesser importance to geologists. These are the Class **Monoplacophora** (Cambrian to Recent), **Amphineura** (Upper Cambrian to Recent), and **Scaphopoda** (Ordovician to Recent). Monoplacophorans (Fig. 11.18) have small, conical shells usually less than one inch in diameter. Amphineurans (Fig. 11.19) include the chitons or "sea mice." Their shell is composed of eight transverse plates that cover and protect the dorsal surface of the animal. Scaphopods or "tooth shells" are small mollusks characterized by a tubular shell open at both ends and shaped like a miniature tusk (Fig. 11.20).

THE ARTHROPODA

The number of species in the arthropod phylum is truly immense. Over 80 % of living animal species are arthropods, and the numbers of individuals represented is astronomical.

Figure 11.17 Coal mine.

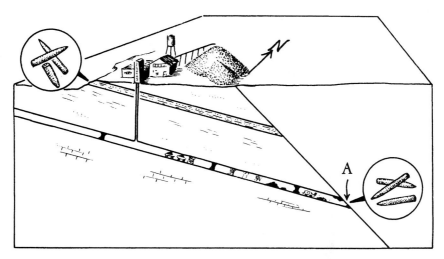

Figure 11.18 Monoplacophorans.

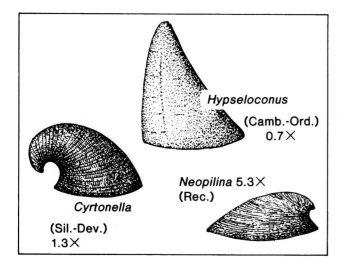

Hypseloconus
(Camb.-Ord.)
0.7×

Neopilina 5.3×
(Rec.)

Cyrtonella
(Sil.-Dev.)
1.3×

Figure 11.20 Scaphopods.

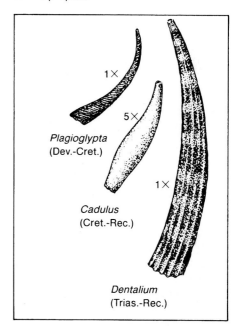

1×

5×

Plagioglypta
(Dev.-Cret.)

1×

Cadulus
(Cret.-Rec.)

Dentalium
(Trias.-Rec.)

Figure 11.19 A modern amphineuran.

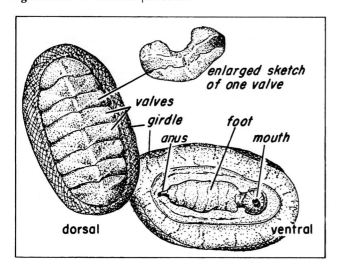

enlarged sketch
of one valve

valves
girdle
foot
anus
mouth

dorsal
ventral

Included here are the lobsters, spiders, insects, centipedes, and a host of other animals too numerous to mention.

The arthropods possess **chitinous** exterior skeletons, segmented bodies, paired and jointed appendages, and highly specialized nerve and sensory organs. Among the arthropods are creatures well adapted for life in and on the surface of the land, for flight, and for existence in all types of natural water bodies. Their success in populating diverse environments is extraordinary, and yet only a relatively few groups have left a significant fossil record. Even the insects, so numerous and diverse today, left a scant fossil record. Preservation is difficult for animals that die in terrestrial environments. Some fossilized insects have been recovered at a few exceptional localities, but the vast majority of sedimentary rocks do not contain insect remains. In terms of their use in stratigraphic studies, the most important arthropods are the trilobites and ostracods. Both of these groups have left an impressive fossil record.

Trilobites (Figs. 11.21 and 11.22)

The skeleton of a trilobite consisted of a dorsal shield, thin ventral membrane, and external covering of the appendages. The **chitin** composing the skeleton was thickened and hardened by calcium carbonate in areas not requiring flexibility. The name *trilobite* means "three lobes" and refers to the division of the dorsal shield into three longitudinal portions, the axial lobe and two lateral or pleural lobes. Three major body regions can be identified as the **cephalon,** the segmented *thorax,* and a posterior portion, the **pygidium.** The pygidium is believed to have evolved by the welding together of posterior thoracic segments. The appendages of the trilobite were located on the ventral surface but are rarely preserved. Sediment would tend to cling readily to this irregular ventral surface, and it is usually impossible to free the fragile appendages from the rock. The smoother dorsal surface can be more easily broken from the sedimentary matrix.

Growth was accomplished in trilobites, as in many arthropods, by molting. Thin places in the exoskeleton called sutures (Fig. 11.21) not only imparted flexibility but also made the molting process easier. Although many trilobites apparently were blind, the majority developed either eyes of the simple, single-lens type or compound eyes composed of a large number of separate visual bodies. In some forms, the separate facets were covered by a continuous protective cornealike layer, whereas in others, each tiny facet had its own equally small cornea. For the latter, the eyes have a reticulated or sievelike appearance.

Trilobites were entirely marine, for their remains are found in association with corals, crinoids, brachiopods, and cephalopods. Most forms appear to have been bottom dwellers, and many evidently burrowed or crawled over the bottom mud or sand of ancient seas. A smaller number show adaptations, suggesting a pelagic existence as either floaters or swimmers.

If biological success can be measured in terms of complexity of structure, variability of form, and numbers of individuals, then trilobites must be considered enormously successful. They first appeared about 600 million years ago, and wherever shallow-water marine Cambrian rocks are found, trilobites are likely to be the most abundant fossils. New species of trilobites appeared at a rapid rate during the Cambrian and Ordovician, but soon thereafter they began to decline, finally becoming extinct in the Permian (Fig. 11.23).

Study Questions

1. By comparison with the illustrations (Figs. 11.21 and 11.22), identify as many of the trilobites in the study set as you can. What types of preservation are represented?

   _____

   _____

   _____

2. Figure 11.24 shows the molted cephalon of the Ordovician trilobite *Calliops.* What kind of eyes did this trilobite possess?

   _____

   _____

3. Sketch the most perfectly preserved trilobite specimen and label the pygidium, thorax, axial lobe, pleural lobes, and cephalon.

4. How do trilobites grow? Is it possible to have several fossil trilobites from a single individual?

   _____

   _____

   _____

5. The eyes of trilobites can be used to make inferences about their living habits. How would you match the eye adaptations with the living habits?

   a. Dwellers on surface of sea floor

   b. Active swimmers

   c. Ooze dwellers

   d. Burrowers

   _____ Eyes on the dorsal surface of cephalon
   _____ Eyes on ventral side of cephalon
   _____ Eyes on end of long stalks on dorsal side of cephalon
   _____ Eyes lacking or functionless

Figure 11.21 Trilobites, plate A.

From Levin, H. L., *Ancient Invertebrates and Their Living Relatives,* Upper Saddle River, NJ: Prentice Hall, 1999.

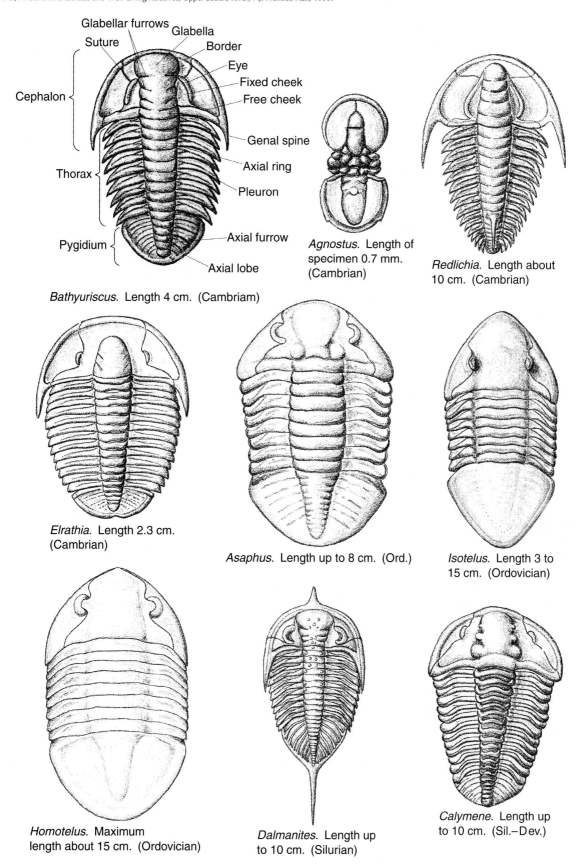

Bathyuriscus. Length 4 cm. (Cambriam)

Agnostus. Length of specimen 0.7 mm. (Cambrian)

Redlichia. Length about 10 cm. (Cambrian)

Elrathia. Length 2.3 cm. (Cambrian)

Asaphus. Length up to 8 cm. (Ord.)

Isotelus. Length 3 to 15 cm. (Ordovician)

Homotelus. Maximum length about 15 cm. (Ordovician)

Dalmanites. Length up to 10 cm. (Silurian)

Calymene. Length up to 10 cm. (Sil.–Dev.)

Figure 11.22 Trilobites, plate B.

From Levin, H. L., *Ancient Invertebrates and Their Living Relatives,* Upper Saddle River, NJ: Prentice Hall, 1999.

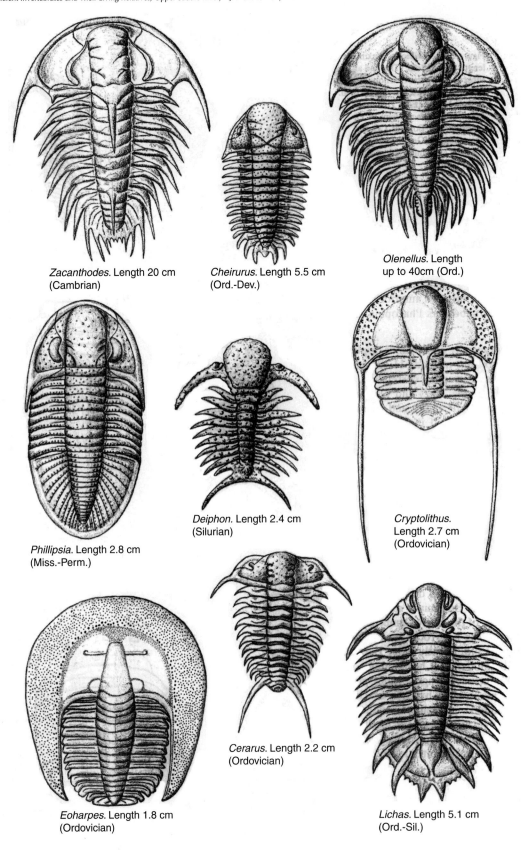

Zacanthodes. Length 20 cm (Cambrian)

Cheirurus. Length 5.5 cm (Ord.-Dev.)

Olenellus. Length up to 40cm (Ord.)

Phillipsia. Length 2.8 cm (Miss.-Perm.)

Deiphon. Length 2.4 cm (Silurian)

Cryptolithus. Length 2.7 cm (Ordovician)

Eoharpes. Length 1.8 cm (Ordovician)

Cerarus. Length 2.2 cm (Ordovician)

Lichas. Length 5.1 cm (Ord.-Sil.)

Figure 11.23 Geologic range of trilobites.

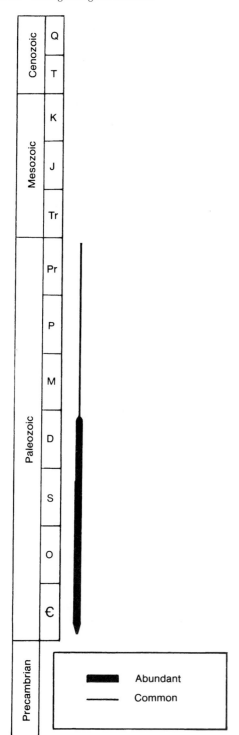

Figure 11.24 Cephalon of *Calliops callicephala* (Ordovician).

Figure 11.25 A living species of ostracode.

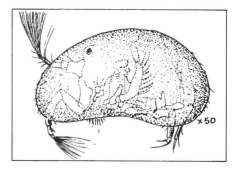

6. What are the three longitudinal primary divisions of the trilobite carapace?

_____

_____

What are the three primary divisions occurring transverse to the longitudinal divisions?

_____

_____

Ostracoda

The **ostracodes** are small, lentil-shaped **crustaceans** with a bivalved shell (**carapace**) enclosing an indistinctly segmented body, from which extend seven pairs of appendages (Fig. 11.25). Adult carapaces are mostly between 0.5 and 4.0 mm in length, are composed of chitin and calcium carbonate, and are articulated along the dorsal margin. Ornamentation and sculpture of the carapace are often intricate and always useful in identification of species.

Ostracodes appeared in the Early Ordovician and are still common today in both marine and freshwater bodies. Like trilobites, they grow by molting. They commonly are found crawling, swimming, and generally swarming about near the bottom of the sea or lake.

Figure 11.26 Paleozoic Ostracoda.

Partly from Collinson, C. W., 1959, Illinois State Geological Survey, Educational Series 4.

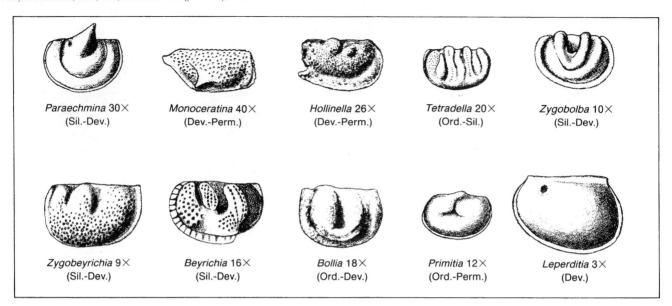

Paraechmina 30× (Sil.-Dev.) Monoceratina 40× (Dev.-Perm.) Hollinella 26× (Dev.-Perm.) Tetradella 20× (Ord.-Sil.) Zygobolba 10× (Sil.-Dev.)

Zygobeyrichia 9× (Sil.-Dev.) Beyrichia 16× (Sil.-Dev.) Bollia 18× (Ord.-Dev.) Primitia 12× (Ord.-Perm.) Leperditia 3× (Dev.)

Fossil ostracodes (Fig. 11.26) are valuable aids in stratigraphic correlation. Because of their small size, they can be obtained from samples of well cuttings. In Paleozoic strata where small fossils are not so plentiful, the ostracodes are used in correlating strata from well to well.

Study Questions

1. Examine the prepared slide of ostracodes and sketch one of the specimens.

2. What other fossil invertebrates have you studied that are bivalved? How might you distinguish the bivalve carapace of ostracodes from the valves of other invertebrates that have two valves?

_____

_____

_____

_____

THE ECHINODERMATA

The Phylum Echinodermata (Fig. 11.27) contains highly developed animals with typically pentamerous (five-fold) symmetry and a skeleton often knobby or studded with spines. Another unique characteristic of the phylum is the presence of fluid-filled **tube feet** (Fig. 11.28), which aid in locomotion and obtaining food. The tube feet are only a part of the water vascular system that characterizes members of the phylum.

The Phylum Echinodermata can be subdivided into seven to twelve classes, some of which were short-lived echinoderm classes in the Lower Paleozoic. Although they were clearly echinoderms, their characteristics were so different from other known echinoderms that separate classes were assigned, and controversy in assignment of these classes has caused the variety of number. In this examination of the fossil record, only the more familiar echinoderms are included.

Class Asteroidea

Members of the Class Asteroidea typically have five arms that radiate outward from an ill-defined central disc. There are two to four rows of tube feet in each arm. The body (Fig. 11.28) is encased in a leathery "skin" studded with small calcareous plates. Upon death and decay, the leathery covering is often lost and the plates swept away. As a result, fossil remains of starfish are not common. Asteroids are predators of worms, crustaceans, and mollusks. Although some species inhabit the deeper parts of the oceans, the majority live within the littoral zone. The asteroids range from Ordovician to the present (Fig. 11.29).

Figure 11.27 Diversity of living echinoderms.

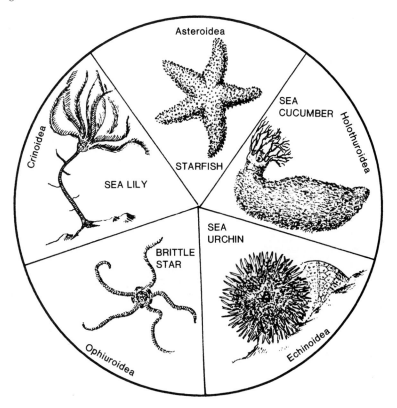

Figure 11.28 Diagram of starfish organs.

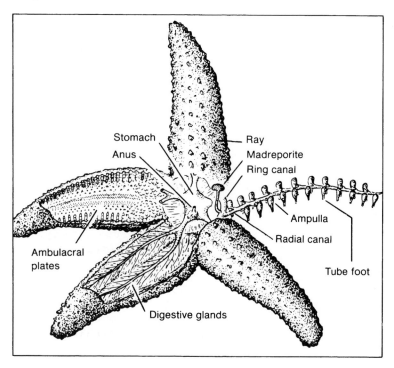

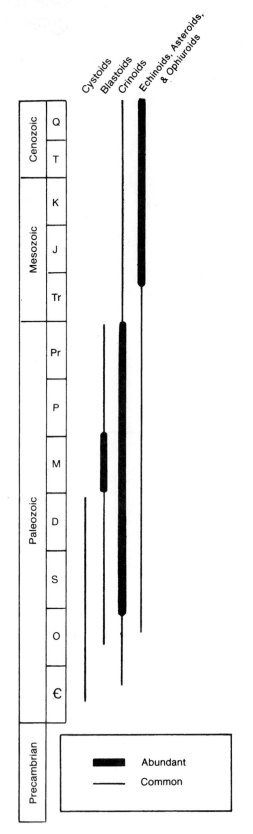

Figure 11.29 Geologic ranges of some important groups within the phylum of Echinodermata.

Class Ophiuroidea

The ophiuroids (brittle stars) differ from the common starfish in their small, sharply delimited central disc and long serpentlike slender arms. Another difference is that the arms lack grooves (**ambulacral** [grooves]) for transporting food particles toward the mouth, and food is taken in only at the mouth on the ventral side of the oral or central disc. The tube feet have sensory and locomotor functions and occur on the ventral sides of the arms. Ophiuroids are known from rocks as old as Ordovician, but they are not common fossils.

Class Echinoidea

The **echinoids** (Fig. 11.30) vary in shape from spheroidal (Fig. 11.31) to flattened discoidal. An echinoid can be compared to a starfish with its arms pulled up over its aboral surface. Five grooves with tube feet, called ambulacral areas, radiate from the mouth up to the opposite (aboral) surface. In flattened forms, the mouth and anus appear on the anterior and posterior periphery. In a typical spheroidal form, the mouth is on the lower surface and may be armed with five sharp teeth as part of a masticatory apparatus called the **"Aristotle's lantern."** Typically, the shell consists of twenty rows of plates. Five of these correspond to the arms of the starfish and are perforated by holes for the tube feet. Regularly arranged tubercles occur on the plates and mark the position at which spines were articulated. Locomotion is accomplished by movement of tube feet and spines.

Echinoids occur at all depths in the sea but prefer warm, shallow waters. They feed largely on seaweed but are known to eat dead animal matter that they come upon. Although echinoids have been found in rocks as old as Ordovician, they did not become abundant as fossils until the Jurassic. Cretaceous and Cenozoic echinoids are common and are widely used as index fossils.

Class Holothuroidea

Perhaps the most bizarre of the echinoderms are the Holothuroidea, exemplified by the "sea cucumber" (Fig. 11.32). The body is sausage-shaped, there are no arms or spines, and the skeleton is composed of numerous microscopic, oddly shaped calcareous ossicles called holothurian sclerites (Fig. 11.32B). Its echinoderm characteristics can be observed most readily in the five longitudinal rows of tube feet. Fossil holothurian sclerites are known from rocks as old as Mississippian. They are sometimes found in rocks being prepared for studies of microfossils such as foraminifers and ostracodes.

Class Crinoidea

The name "sea lily" is descriptively appropriate, though biologically incorrect, for this group of animals. In a way, the **crinoid** (Fig. 11.33) can be regarded as an inverted starfish with a stalk or stem attached to the underside. The crinoid is composed of three main parts: the calyx, the arms, and the stem (Fig. 11.34). Most of the vital organs are encased in the cuplike **calyx** or head of calcareous plates. The mouth is on the upper surface, and the anus is located to one side of

Figure 11.30 Echinoderms.

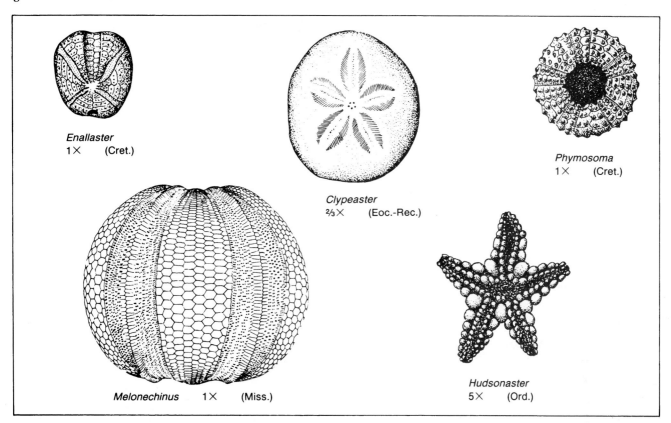

Enallaster
1× (Cret.)

Clypeaster
⅔× (Eoc.-Rec.)

Phymosoma
1× (Cret.)

Melonechinus 1× (Miss.)

Hudsonaster
5× (Ord.)

Figure 11.31 (A) Aboral view of a sea urchin with spines removed. (B) Aristotle's lantern. Redrawn from R. T. Jackson, *Phylogeny of the Echini*, with revision of Paleozoic species.

From *Boston Society of Natural History*, Memoir 7.

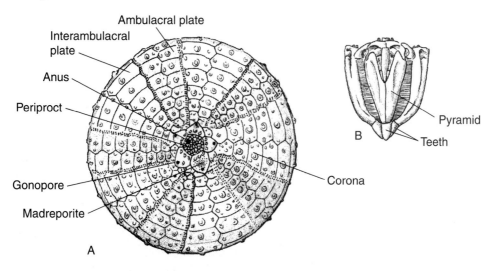

Ambulacral plate
Interambulacral plate
Anus
Periproct
Gonopore
Madreporite
A
Corona

B
Pyramid
Teeth

Figure 11.32 (A) External view of a sea cucumber or holothuroid (x0.4).(B) Holothurian sclerites (x30).

From Levin, H. L., *Ancient Invertebrates and Their Living Relatives,* Upper Saddle River, NJ: Prentice Hall, 1999.

Figure 11.33 Limestone slab containing well-preserved Mississippian crinoids.

Photograph courtesy of Wards Natural Science Establishment, Rochester, New York.

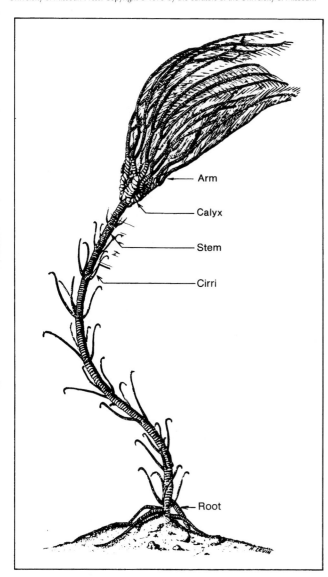

Figure 11.34 Restoration of a crinoid.

Reprinted from *The Common Fossils of Missouri* by A. G. Unklesbay, by permission of the University of Missouri Press. Copyright © 1973 by the curators of the University of Missouri.

Arm

Calyx

Stem

Cirri

Root

Crinoids first appeared in Ordovician rocks, but their golden age was the Mississippian (Fig. 11.36). During that period, crinoid skeletal plates littered the sea floor, forming extensive layers of crinoidal limestone. The Mississippian was truly an "Age of Crinoids." These echinoderms have never since been as numerous nor as varied.

Class Blastoidea

The stalked echinoderms known as **blastoids** (Fig. 11.37) exhibit more readily apparent pentaradiate symmetry than the crinoids. Typically, the budlike calyx (Fig. 11.38) is pentagonal, with five ambulacra radiating symmetrically from the mouth. Thirteen plates (three basals, five radials, and five interradials) enclose the space not occupied by the ambulacral areas. Each ambulacrum possesses a centrally located food groove that branches laterally into side food grooves. Threadlike brachioles extend from the lateral margins of the ambulacra and function in food gathering in a fashion similar to that of the arms of crinoids.

The geologic range of blastoids is Ordovician through Permian (see Fig. 11.29). As with the crinoids, they were most numerous during the Mississippian. Because blastoids have been totally extinct since the Permian, their paleoecology is based largely on organisms found associated with them in rock strata. Such evidence suggests that blastoids lived in clear, shallow marine areas. Almost all are found in limestones and limy shales.

Class Cystoidea*

The pentamerous symmetry that characterizes many classes of echinoderms is poorly shown or even absent in the cystoids (see Fig. 11.38). The calyx has a dorsally located mouth, and the anus is situated on the oral side in an eccentric position. The plates of the calyx may be symmetrically or asymmetrically arranged and are pore-bearing. **Cystoids** (see Fig. 11.29) are found chiefly in Ordovician and Silurian rocks, although their complete geologic range is from the Cambrian to the Upper Devonian.

Study Questions

1. Examine a preserved specimen or a dried specimen of the common starfish *Asterias*. Locate the ambulacral grooves, madreporite (see Fig. 11.28), and position of the mouth. Note the tube feet on the preserved specimen. Prepare a sketch and label features visible on oral and aboral *sides*.

the mouth. Five branching arms extend upward from the calyx. They bear food grooves, barblike pinnules, and cilia, which aid in food gathering. In some living and most ancient forms, a long stem composed of calcareous disclike plates extends downward from the bottom of the calyx for attachment to the sea floor. Disaggregated plates of crinoid stems, arms, and calyx are common fossils. Unfortunately, the complete calyx, which is important in species identification, is rarely found.

Although not so numerous and varied as their Paleozoic ancestors (Fig. 11.35), the crinoids of present seas are notably present in both shallow and deep waters between about 50°S and 80°N latitude. The majority of modern forms are free living, are gregarious, and prefer relatively clear, well-aerated water with an abundance of minute planktonic organisms for food.

---

* The class Cystoidea has been recently dismantled and its members placed in either the class Rhombifera or Diploporita. The designation Cystoidea is retained here for simplicity.

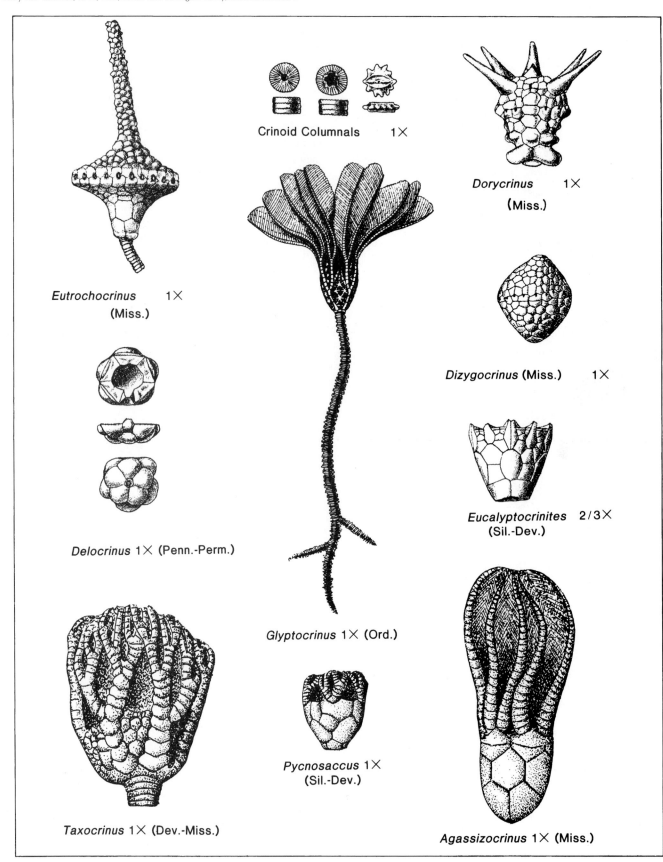

Crinoid Columnals   1✕

Dorycrinus   1✕
(Miss.)

Eutrochocrinus   1✕
(Miss.)

Dizygocrinus (Miss.)   1✕

Delocrinus 1✕ (Penn.-Perm.)

Eucalyptocrinites   2/3✕
(Sil.-Dev.)

Glyptocrinus 1✕ (Ord.)

Pycnosaccus 1✕
(Sil.-Dev.)

Taxocrinus 1✕ (Dev.-Miss.)

Agassizocrinus 1✕ (Miss.)

Figure 11.36 Reconstruction of a Mississippian sea floor with abundant crinoids.

2. Examine a dried specimen of an echinoid, or a well-preserved fossil echinoid from your study set. Sketch the oral and aboral surfaces and label ambulacral areas, mouth, anus, and pores for tube feet.

Figure 11.37 Restoration of a blastoid.

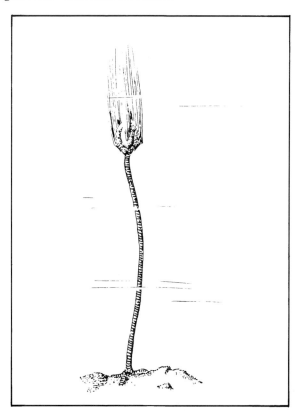

3. Lightly shade in the ambulacral areas on the illustrations of *Melonechinus* and *Phymosoma* in Figure 11.29 and of *Pentremites* in Figure 11.38.

4. What skeletal features of crinoids are also present in blastoids?

_____

_____

_____

_____

_____

_____

5. Label the brachioles, calyx, and stem on the illustration of the blastoid in Figure 11.37.

6. Very intricate relationships exist between living animals, but only rarely do fossils indicate to us the complexity of some of these relationships as they existed long ago. Two basic types of relationships are **antagonism,** in which one species suffers through the action of another, and **symbiosis,** in which one or both species benefit, but neither is really harmed. Figure 11.39 shows two views of a relationship in which a snail shell is in position over the anal opening of a blastoid.

Figure 11.38 Cystoids (top row) and blastoids.

Partly from Collinson, C. W., 1959, Illinois State Geological Survey, Educational Series 4.

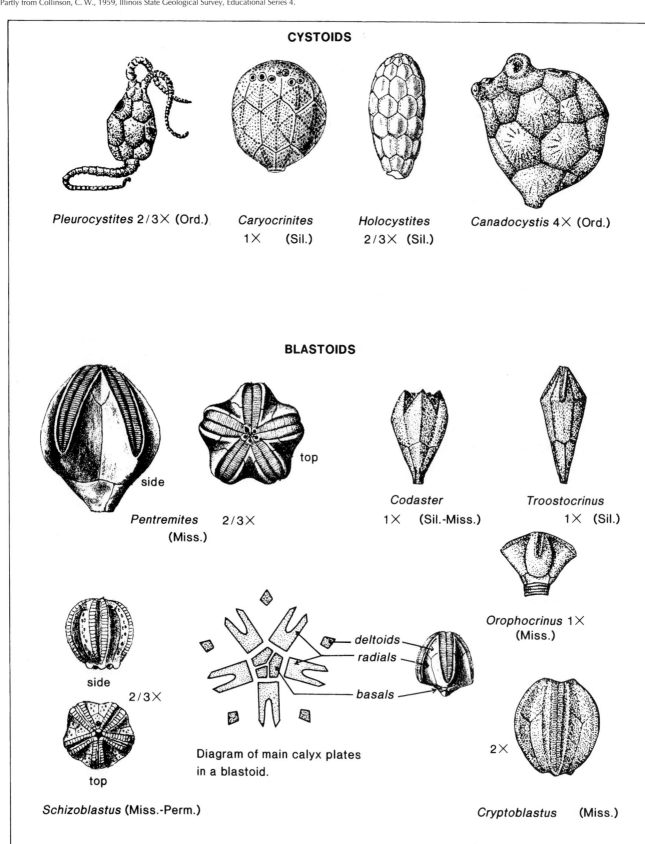

CYSTOIDS

Pleurocystites 2/3✕ (Ord.)

Caryocrinites
1✕ (Sil.)

Holocystites
2/3✕ (Sil.)

Canadocystis 4✕ (Ord.)

BLASTOIDS

side

top

Pentremites 2/3✕
(Miss.)

Codaster
1✕ (Sil.-Miss.)

Troostocrinus
1✕ (Sil.)

Orophocrinus 1✕
(Miss.)

side

2/3✕

top

deltoids
radials
basals

Diagram of main calyx plates
in a blastoid.

Schizoblastus (Miss.-Perm.)

2✕

Cryptoblastus (Miss.)

Figure 11.39 Coprophagous relationship between blastoid and snail.

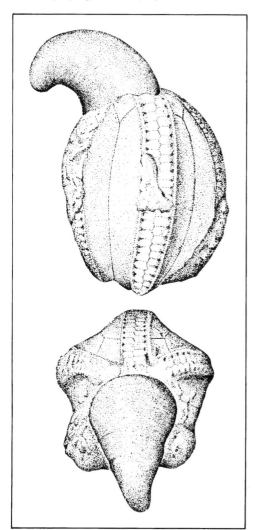

The fossils were collected from Mississippian strata near St. Louis, Missouri. Is this an example of symbiosis or antagonism? If this is symbiosis, is it possible that both organisms benefited, and if so, how?

_____

_____

_____

_____

_____

7. On Figure 11.39, label a central food groove, a side food groove, a radial plate, and a deltoid plate.

GRAPTOLITES

Graptolites (Fig. 11.40) are a group of extinct organisms once thought to be plants and later considered coelenterates, then bryozoans; most recently they have been classified by some paleontologists as primitive members of the Phylum Chordata. They, like the bryozoa, were microscopic, suspension-feeding individuals grouped together into colonies. Their skeletons were of a chitinous rather than calcareous material.

Graptolites are typically preserved as carbon films in black shales. They resemble pencil marks that have a saw-tooth appearance; hence the name _graptos_ (written) _lithos_ (rock). The preserved colonies are termed rhabdosomes and consist of one or more narrow branches called **stipes.** The stipe results from the uniserial or biserial arrangement of cup-like **theca** composed of a type of chitin.

Most graptolites can be identified as belonging to one of two major categories. The first comprises the _dendroid graptolites,_ which built complex, branching, fan-shaped rhabdosomes. Some dendroid graptolites may have attached their bases to the sea floor. Others are believed to have suspended themselves from some floating object by a threadlike tube known as the nema.

The second category of graptolites are termed _grap-toloid._ They were planktonic forms with distinctive patterns of thecae. The graptoloid rhabdosome generally was composed of far fewer stipes than the rhabdosome of dendroids. In their evolution, they underwent a progressive reduction in the number of stipes. In general, Lower Ordovician beds contain graptolites with more stipes than Middle and Upper Ordovician strata. Silurian graptolites characteristically are constructed of only one stipe.

Graptolites are known from rocks of Cambrian through Mississippian age, but they are of greatest importance during the Ordovician and Silurian. Their worldwide distribution and rapid evolution have made them ideal index fossils of special importance in studies of Early Paleozoic strata.

Study Questions

1. Where on the graptolite colony did the individual animals live?

_____

_____

_____

_____

_____

_____

2. How do graptolites differ from corals in living habit and skeleton?

_____

_____

_____

_____

_____

_____

Figure 11.40 Graptolites.

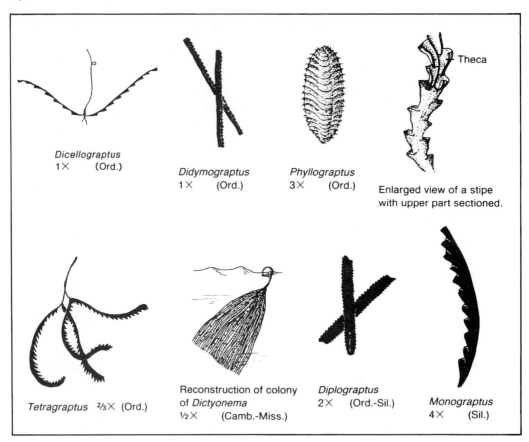

Dicellograptus
1× (Ord.)

Didymograptus
1× (Ord.)

Phyllograptus
3× (Ord.)

Theca

Enlarged view of a stipe
with upper part sectioned.

Tetragraptus ⅔× (Ord.)

Reconstruction of colony
of Dictyonema
½× (Camb.-Miss.)

Diplograptus
2× (Ord.-Sil.)

Monograptus
4× (Sil.)

3. How did the living habits of graptolites contribute to
their being good index fossils?

_____

_____

_____

_____

_____

_____

FOSSIL PLANTS

Plants are multicellular organisms that typically live on land
and carry out photosynthesis. They possess photosynthetic
pigments, such as chlorophyll, located in cell organelles
called chloroplasts. All plants can be placed in one of two
groups: the *nonvascular plants* and the **vascular plants.**
Nonvascular plants include liverworts, mosses, and horn-
worts. Such plants have a scant fossil record. The fossil
record for vascular plants, however, is noteworthy and warrants examination.

Vascular plants have special tissues and canals to
transport moisture and nutrients from beneath the ground to
the chlorophyll-bearing leaves where photosynthesis is ac-

complished. In addition to a vascular transport system, most
vascular plants also have features to prevent desiccation, to
support the weight of the plant, and to facilitate reproduction
on land. There are three major groups of vascular plants:
plants that do not bear seeds and utilize spores in reproduction (such as true ferns), plants that bear naked seeds (such as
pine trees), and plants with protected seeds and flowers.

Our knowledge of the evolution of plants is vastly
improved by the study of **spores** and **pollen** grains (called
palynology). Produced in vast numbers, these reproductive
cells can be transported far and wide by wind and water.
Eventually, many are deposited on the floors of lakes,
seas, and oceans where they are preserved and become
valuable as paleoclimatic indicators and for biostratigraphic correlation.

Although a detailed study of plant evolution and palynology is beyond the scope of this introductory manual,
some recognition of the more common plant fossils is desirable. The following descriptions will be useful in identifying
the more frequently encountered plant fossils.

Seedless Vascular Plants

The first plants to invade the land were spore-bearing, seedless, vascular plants. Precisely when this plant invasion occurred is somewhat uncertain (Fig. 11.41). The first scraps of
vascular tissue containing strands of wood are Early Silurian

Figure 11.41 Geologic ranges, relative abundances, and evolutionary relationships of vascular land plants.

From Levin, H. L., *The Earth Through Time,* Philadelphia: Saunders College Publishing, 1992.

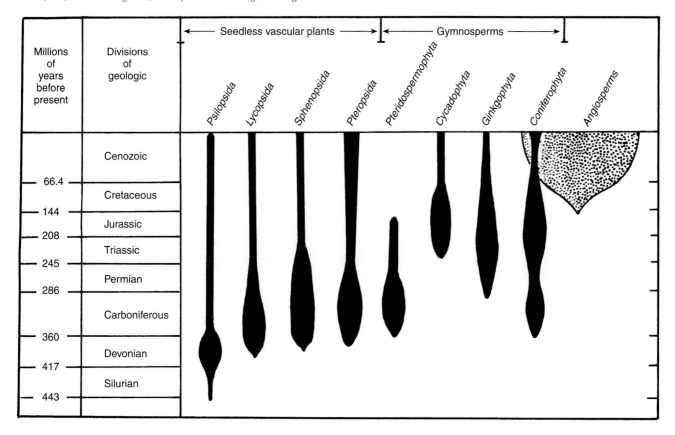

in age. However, spores having three radiating scars (so-called trilete spores), resembling those of seedless vascular plants, become relatively abundant in rocks of Late Ordovician age, strongly suggesting the presence of these land plants at that time.

Among spore-bearing plants, the adult plant *(sporophyte)* produces the reproductive cells called spores. The spores do not require fertilization. When spores are dropped to the moist ground, they develop into small, leafless plants called *gametophytes.* A gametophyte, in turn, produces egg and sperm. When the egg is fertilized by contact with sperm, it develops into a new sporophyte. Such reproduction requires a moist environment to prevent dehydration of the egg or sperm.

The **psilophytes** are a group of Early Paleozoic spore-bearers that represent the vanguard of the land invasion. Psilophytes were small plants (rarely more than 30 cm in height) characterized by horizontal stalks (**rhizomes**) that grew just under the surface of the ground in moist soil. Short, slender stems bearing spore sacks (**sporangia**) grew upward above the soil (Fig. 11.42). The plant had neither true leaves nor roots. From these relatively unimpressive plants, however, huge trees developed during the Late Paleozoic. In swampy regions, the accumulation of dead vegetation from these trees provided the material that was eventually converted to coal. Among the principal plants of these "coal

forests" were *lycopsids, sphenopsids, true ferns,* and *progymnosperms.*

Today, lycopsids are represented by relatively small plants like the "ground pine" and club mosses. In the late Paleozoic, however, lycopsids with robust trunks and extensive root systems towered 40 m above the ground. Most had long, slender leaves attached directly to the limbs in a spiral arrangement. When these leaves were released, they left diamond-shaped leaf scars (Fig. 11.43). For this reason, lycopsids are dubbed "scale trees."

Growing side by side with the scale trees in late Paleozoic forests were the joint-stemmed *sphenopsids.* Scouring rushes and horsetails are living sphenopsids. Extinct forms such as *Calamites* and *Annularia* had unbranched, longitudinally ribbed stems (Fig. 11.43). Circlets of slender leaves grew at each transverse joint of the stems.

True ferns *(pteridophytes),* like *Pecopteris,* were also present among the lycopsids and sphenopsids. Unlike modern ferns, which are generally small, the ferns of the late Paleozoic grew to lofty heights. Spores often were carried in sporangia located on the undersides of leaves.

The final group of seedless plants with a significant fossil record are the *progymnosperms.* As one might surmise from their name, this group of seedless vascular plants are the probable ancestors of the early seed-bearing plants. Indeed, both groups have similar wood tissue.

Figure 11.42 A psilophyte from the Lower Devonian. These primitive small plants rarely exceeded 30 cm in height.

From Levin, H. L., *The Earth Through Time,* Philadelphia: Saunders College Publishing, 1992.

Soil

Vascular Plants with Naked Seeds

Vascular plants with naked seeds are informally called **gymnosperms** (from the Greek meaning "naked seed"). The seeds of this group are considered "naked" because they are not completely enclosed by the tissues of the parent at the time of pollination. The divisions of these plants with naked seeds that have a good fossil record are the *pteridospermophytes, glossopterids, coniferophytes, cycadophytes,* and *ginkgoes.*

Pteridospermophytes are the seed ferns. They had fernlike foliage but bore naked seeds on their leaves. The earliest remains of pteridospermophytes are found in Devonian rocks. They were abundant during the late Paleozoic, declined somewhat during the Triassic and Jurassic, and became extinct during the Cretaceous.

During the Carboniferous, gymnospermal plants called *glossopterids* grew luxuriantly across the southern supercontinent of Gondwana. The group takes its name from *Glossopteris* (Fig. 11.44), fossils of which typically consist of thick, tongue-shaped leaves. *Glossopteris* and associated plants of the glossopteris flora were adapted to more temperate climates, as compared to the more tropical conditions that prevailed in the coal forests of North America and Europe.

Conifers, which include pines, spruces, firs, redwoods, and cedars, are the most familiar of the gymnosperms. In conifers, distinctive male and female cones develop. The smaller male cones produce pollen grains containing sperm. Eggs develop in the protective ovules located within the scales of the larger female cones. After pollen has been transported by wind to the female cone, fertilization is completed by movement of sperm (in relative safety) through a moist tube that grows from the pollen grain to the embryonic seed. Under suitable conditions, the fertilized seed may then grow into an adult conifer. Conifers made their appearance during the late Carboniferous (Mississippian and Pennsylvanian). Perhaps the most famous fossil conifers are those preserved as silicified logs in Petrified Forest National Park, Arizona.

The conifers are members of a larger group of conifer-like plants designated as the *coniferophytes.* Coniferophytes include a group of plants called *cordaites* that flourished during the Pennsylvanian and Permian. Cordaites were cone-bearing trees with long straplike leaves. They were particularly abundant in moist, swampy terrains.

During the Mesozoic, a group of plants known as *cycadophytes* achieved global distribution. Among the cycadophytes, the group known as cycadeoids are extinct, but those called **cycads** are still living and commonly known as sago palms. Many of the Mesozoic cycads were lofty trees surmounted by palmlike foliage. Others grew close to the ground. Whether large or small, they had rough trunks and limbs resulting from bark covered by spirally arranged leaf scars.

The *ginkgoes* are a group of gymnosperms readily recognized by their distinctive bilobed fan-shaped leaves. Ginkgoes made their appearance during the Permian. They spread widely during the Triassic and Jurassic but began to decline in the Cretaceous and Tertiary. Today, only a single species, *Ginkgo biloba,* survives.

Vascular Plants with Protected Seeds and Flowers

The explosive proliferation of vascular plants with protected seeds and flowers was one of the most dramatic events in plant evolution. These plants are called **angiosperms** (Fig. 11.45), and most of the plants we see about us today are angiosperms. In angiosperms, seeds develop in a chamber of the flower called the ovule. Organs on the flower called stamens produce pollen. Pollen is transferred from the stamens to another organ on the flower known as the pistil. As in gymnosperms, a pollen tube is produced by the pollen grain after it has come to rest on the pistil. The pollen tube penetrates to the ovule, and sperm moves through the tube to the seed where fertilization takes place. Colored leaves on the flower are useful in attracting insect pollinators, and the proliferation of plants with flowers was accompanied by a striking parallel evolution of insects. In addition, the enclosed seed permitted the growth of edible coverings, enhancing dispersal by animals that feed on seeds and fruit.

Figure 11.43 Plants of the coal swamps.

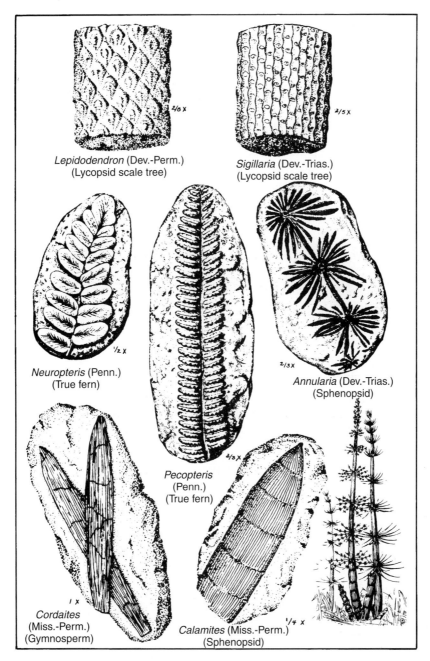

Lepidodendron (Dev.-Perm.)
(Lycopsid scale tree)

Sigillaria (Dev.-Trias.)
(Lycopsid scale tree)

Neuropteris (Penn.)
(True fern)

Annularia (Dev.-Trias.)
(Sphenopsid)

Pecopteris
(Penn.)
(True fern)

Cordaites
(Miss.-Perm.)
(Gymnosperm)

Calamites (Miss.-Perm.)
(Sphenopsid)

The earliest evidence of angiosperms is found in Cretaceous rocks. The group diversified rapidly, and by the end of the Cretaceous, all of the major groups had appeared. Many of these plants were very similar to those observed today. By the mid-Cenozoic, the appearance of prairie grasses supplied the stimulus for the dramatic increase in the numbers of plains-dwelling herbivorous mammals and their predators.

Study Questions

1. Refer to the figures in this section on plants and identify the fossil plants in your study set. Beside each,

   indicate if the plant is a seedless plant, a gymnosperm, or an angiosperm.

2. What types of fossilization are seen in the fossil plants provided in your study set?

   _____

   _____

   _____

   _____

   _____

   _____

Figure 11.44 *Glossopteris* leaf, associated with coal deposits formed from the glossopterid forests of the Permian. This fossil was found on Polestar Peak, Ellsworth Land, Antarctica.

Courtesy of J. M. Schopf, C. J. Craddock, and the U.S. Geological Survey.

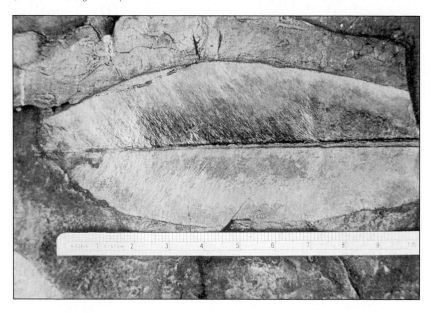

Figure 11.45 Leaves of angiosperms found fossilized in Cretaceous rocks.

From Levin, H. L., *The Earth Through Time,* Philadelphia: Saunders College Publishing, 1992.

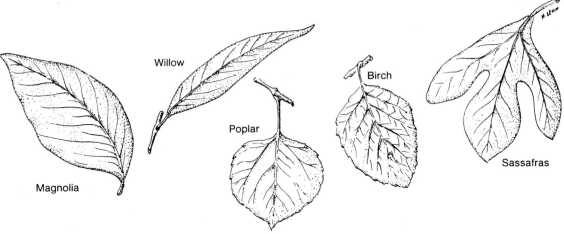

3. What advantages, in terms of assurance of success in reproduction, do gymnosperms have over seedless, spore-bearing plants?

_____

_____

_____

4. A traditional interpretation of the relationship between insects and flowering plants is that the flowering plants evolved first, stimulating the expansion of flower-dependent insects. Some scientists suggest the reverse may have been the case. How might this question be resolved?

_____

_____

TERMS

ambulacral   Pertaining to the rows of openings through which the tube feet of echinoderms are extended, or along which food grooves are located.

amphineuran   A group of mollusks having shells that are composed of eight plates on the dorsal side of the body. The chitons are amphineurans.

angiosperm   A plant with true flowers, in which the seeds are enclosed in the fruit (the fertilized and developed ovary). Examples include grasses, orchids, elms, and roses.

antagonism   The relationship that exists between two different organisms that live in close association, in which one organism suffers through the actions of the other.

Aristotle's lantern   A system of calcareous elements that surround the mouth and function as jaws in echinoids.

beak   The projection at the initial point of shell growth in pelecypods.

belemnites   Members of the molluscan Class Cephalopoda having straight internal shells.

blastoid   Sessile (attached) Paleozoic echinoderm, members of which have a stem and an attached cup or calyx composed of relatively few plates.

calyx   That portion of the skeleton that surrounds the viscera in crinoids, blastoids, and cystoids.

camerae (cephalopods)   The chambers in the conch of a cephalopod.

carapace   A bony or chitinous case or shield covering the whole or part of the back of certain animals.

cephalon   The major anterior body segment (the "head") of a trilobite.

cephalopods   Mollusks characterized by a chambered conch.

chitin   A resistant organic compound that is the common constituent of various invertebrate skeletons such as insect exoskeletons and the inner tests of foraminifers.

columnella   The medial pillar surrounding the axis of a spiral gastropod shell, formed by the coalescence of the inner walls of the whorls.

conch   Any of various marine shells of invertebrates, including bivalve mollusks and brachiopods.

crinoid   A stalked echinoderm with a calyx composed of regularly arranged plates from which arms radiate for gathering food.

crustacean   A member of the Class Crustacea of the Phylum Arthropoda that includes such animals as lobsters and crayfish.

cycads   A gymnosperm having compound leaves and naked seeds borne separately on leaves or in simple cones. Cycads were abundant during the Triassic and Jurassic and declined during the Cretaceous.

cystoid   An attached echinoderm with an irregular arrangement and number of plates in the calyx and with calyx perforated by pores or slits.

echinoids   Globular or discoidal members of the Phylum Echinodermata, having a firm shell (test) composed of symmetrically arranged plates, including such animals as sand dollars, sea urchins, and heart urchins.

gastropoda   A class of mollusks having a well-developed head region, ventral foot for creeping, and unchambered conch. Includes snails and limpets.

graptolites   Extinct colonial, marine invertebrates frequently found as fossils in dark shales of Paleozoic age.

gymnosperm   A plant whose seeds are commonly in cones and never enclosed in a fruit. Examples include cycad, ginkgo, and the conifers (pine, fir, and spruce).

mantle   The pair of fleshy folds that secrete the shell in mollusks, brachiopods, and certain other invertebrates.

monoplacophoran   Primitive marine mollusks with simple, cap-shaped shells.

operculum   A calcareous or chitinous plate that develops on the posterior dorsal surface of the foot of a gastropod and that serves to close the aperture.

ophiuroidea   A class of the Phylum Echinodermata characterized by the presence of five slender, jointed, flexible arms, as in the so-called brittle stars.

ostracode   A small, bivalved crustacean.

pallial line   Linear depression on inside of pelecypod. Shell marking inner margin of thickened mantle edges.

pallial sinus   An indentation in the pallial line in pelecypods. The pallial sinus marks the position of the siphon.

palynology   The study of pollen of seed plants and spores, both living or fossil, including their dispersal and applications in stratigraphy and paleoecology.

pollen   Tiny reproductive bodies produced in the anthers of flowering plants.

psilophyte A primitive vascular plant, generally without roots or leaves, having spore-bearing organs at the stem tips. Geologic range: Late Silurian to Early Devonian. Also called psilopsids.

pteridospermophytes Extinct seed-bearing plants with fernlike leaves.

pygidium The major posterior portion of a trilobite.

rhizome A horizontal, subterranean stem that permits a plant to grow laterally.

scaphopod Mollusks characterized by a simple, tusklike, unchambered shell that is open at both ends.

septa Vertical partitions in the theca of a coral.

siphuncle A long membranous tube extending all through the camerae and septa from the protoconch to the base of the body chamber of a cephalopod shell.

sporangia Structures that produce spores.

spore Any single-celled body, produced as a means of propagating a new individual, often adapted to survive unfavorable environmental conditions.

stipe A branch of a graptolite colony, along which are arranged the thecae.

symbiosis The relationship that exists between two different organisms that live in close association, with at least one being helped without either being harmed.

theca The external skeleton of an echinoderm. Formed of calcium carbonate (corals) or chitin (graptolites).

tube feet The locomotor organs in starfish and echinoids.

umbo Elevated and convex area of the valves of pelecypods and brachiopods that is adjacent to the beak.

valves One of the two usually convex plates that form the shell in brachiopods and pelecypods.

vascular plant A plant with a well-developed conductive system and structural differentiation. The majority of visible terrestrial plants are vascular.

whorl One of the turns of a spiral or coiled shell.

12

Fossil Indicators of Age, Environment, and Correlation

FOSSILS AND AGE DETERMINATIONS

Perhaps the most important value of fossils to geology is the means they provide for differentiating and recognizing units of rock deposited during particular increments of geologic time. Because fossils change or evolve through time, they serve as distinctive and nonrepetitive markers in vertical successions of sedimentary rocks. However, in order to make use of this attribute of fossils, geologists must determine the total time span or **geologic range** during which particular organisms lived. The geologic range of an organism is not known *a priori,* but is determined by experience. Only after collections of fossils are made from successively higher (and thus younger) beds in many geologic sections will the total geologic range of organisms be disclosed. In practice, the determination of geologic ranges can become quite involved. However, the general principle is illustrated by the situation depicted in Figure 12.1.

Study Questions

1. Using the "Law of Superposition," arrange the letters representing the geologic systems in proper vertical order in the geologic column on the left side of Figure 12.1.

2. By means of vertical arrows or heavy lines, plot the vertical ranges of fossils F-1 and F-2 in the space to the right of the standard geologic column.

3. During which of the geologic periods were the rocks in unexplored region IV deposited?

_____

_____

4. During which geologic periods were the rocks in unexplored region V deposited?

_____

_____

Figure 12.1 Problem illustrating the determination and use of geologic ranges of fossils. Capital letters on the sections of different regions represent rocks deposited during hypothetical geologic periods analogous to, for example, the Silurian Period.

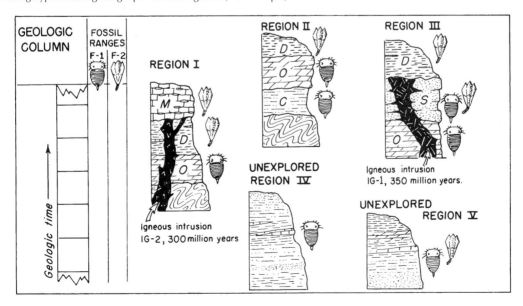

5. What statement can be made about the age of rocks in unexplored region IV?

_____

_____

6. What statement can be made about the age of rocks in unexplored region V?

_____

_____

7. The following questions refer to Figure 12.2. The figures in Chapters 10 and 11 will be useful in answering these questions.

    a.   On limestone slab A, note the branching twig bryozoa, trilobite pygidia, and strophomenid brachiopods. What is the age of this limestone?

_____

_____

_____

Figure 12.2 Geologic age and depositional environment. Identify as many of the fossils as you can on each rock slab shown. Beneath each slab, write the geologic age (era or period) during which the fossils lived and the probable environment (marine, nonmarine, water depth, climate).

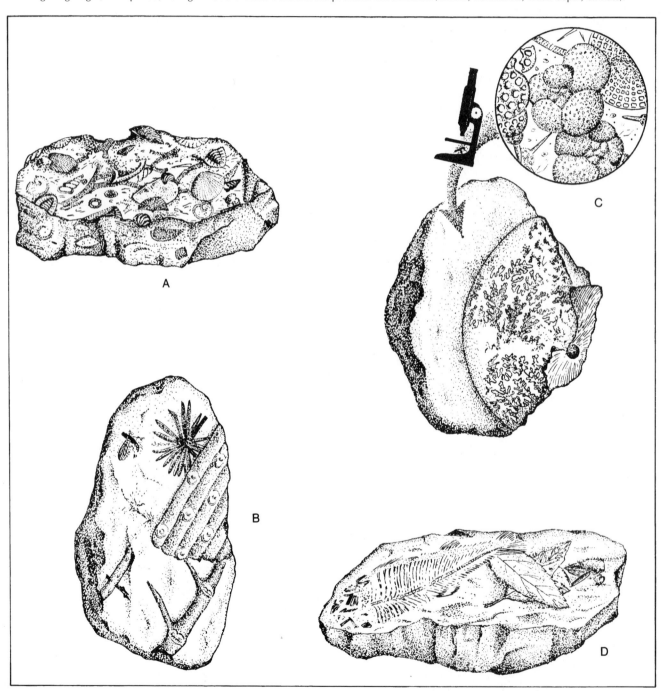

b. Two fossil plants (and an obscure insect) are visible on the surface of slab B.

　　1. What are the names of the two plants (see Fig. 11.43)?

_____

_____

_____

_____

　　2. What is the age of the rock?

_____

_____

_____

_____

c. Rock C is a limestone containing a large cephalopod and various microfossils (visible on microscopic examination).

　　1. What groups of microfossils are present?

_____

_____

_____

_____

_____

　　2. What type of sutures are visible on the fossil cephalopod?

_____

_____

_____

　　3. What is the age of this rock?

_____

_____

_____

d. Rock D is a shale.

　　1. Define the age or age range of this rock as closely as you can. State what criteria you used to make this determination.

_____

_____

　　2. What is the probable environment of deposition?

_____

_____

_____

_____

The geologic ranges of major groups of invertebrate fossils that have left good fossil records are illustrated in Figure 12.3. Knowledge of the time interval during which even such large taxonomic groups lived can serve to determine the geologic system in which particular strata belong. Smaller taxonomic divisions, such as orders, families, genera, or species, can provide the necessary information for identifying correspondingly smaller time-rock units, such as series or stages.

Study Questions

Answer the following questions by referring to Figure 12.3.

1. If the belemnites and trilobites (not identified as to genera or species) are found together in a rock unit, within what limits may the age of the rock be stated? (That is, what is the maximum and minimum age of the rock?)

_____

_____

_____

_____

2. If belemnites, trilobites, and graptolites (not identified as to genera or species) are found together in a rock unit, within what limits may the age of the rock be stated?

_____

_____

_____

_____

3. If trilobites (not identified as to genera and species) are found in a rock unit, within what limits may the age of the rock be stated?

_____

_____

_____

4. For age determination of a rock unit, what are the advantages of an assemblage of **index fossils** over a single kind of index fossil?

_____

_____

_____

_____

BIOZONES AND BIOSTRATIGRAPHY

Recognizing and correlating rock units on the basis of the fossils they contain is called **biostratigraphy,** and the fundamental unit used in biostratigraphy is the **biozone.** A **biozone**

Figure 12.3 Geologic ranges of some major groups of marine invertebrates and the relative importance of each group as currently used for correlation and age determination.

Updated from Teichert (1958), *Geol. Soc. Amer. Bull.,* **69**:99–120.

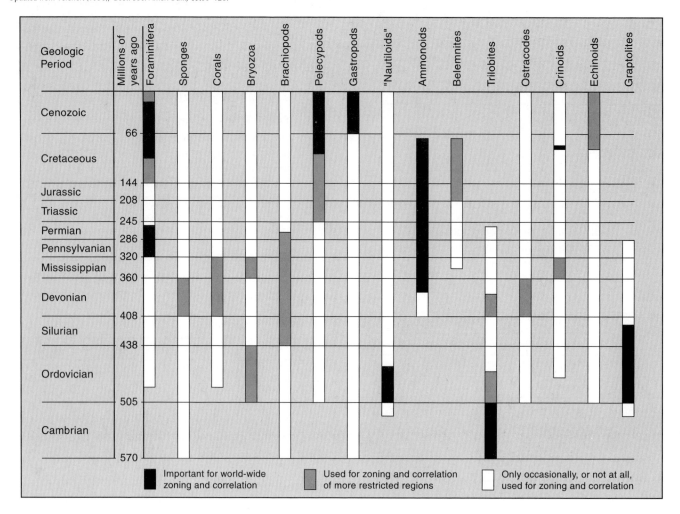

is a body of rock that is identified strictly on the basis of its contained fossils. Without naming them, we have already described two kinds of biozones: the range zone and the assemblage zone. The **taxon range biozone** is a body of rock representing the total geologic life span of an organism. The range zone consists of the interval between the first (lowest) occurrence of the organism and its point of extinction and is named after the fossil itself. Thus, the total range of the organism *Robulus crassus* would be called the *Robulus crassus* range zone.

One may also designate an **assemblage biozone,** selected on the basis of several coexisting species and named after one common and abundant member. A variation of the assemblage zone is the **concurrent range biozone,** recognized by the overlapping ranges of two or more taxa (sing. *taxon*). (The term *taxon* refers to a group of organisms that constitute a particular taxonomic category, such as species, genus, or family.)

Some other frequently used kinds of biozones are the **acme biozone,** which consists of a body of rock repre-

sented by the maximum abundance of some species, genus, or other taxon (and named after that abundant taxon), and the **intrazone,** which represents an interval that lacks diagnostic fossils between two identifiable biozones. The interval zone would have a hyphenated name, including the organism that became extinct below the barren interval and the organism that made its first appearance above the barren interval (e.g., the *Globigerina sicanus-Orbulina suturalis* interval zone).

Once biozones have been identified in sections of rock at many different locations, one can correlate the biozones from place to place. Such correlations are the ultimate reason for designating biozones.

Study Questions

1. On Figure 12.4, identify, name, and draw in the boundaries of a *concurrent range biozone,* an *acme biozone,* an *intrazone,* and an *assemblage biozone.*

Figure 12.4 Designating biozones.

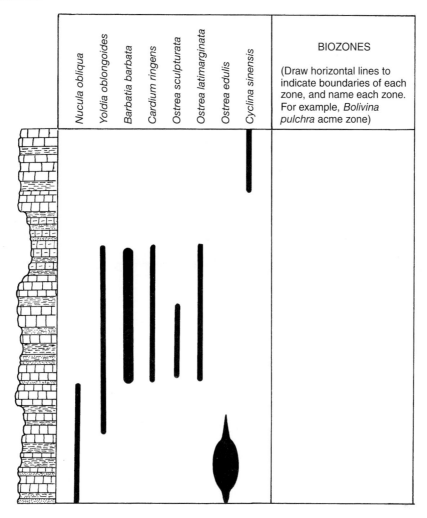

2. On attempting to correlate the biozones you have recognized, you find that you cannot identify the same biozones in rock sections of the same age at distant locations. What may be the reason?

_____

_____

_____

_____

_____

A GRAPHIC METHOD
OF CORRELATION

Even where fossils are present, it is sometimes difficult to precisely correlate a point in one section to a point in another section at a distant location. One interesting way to solve this

problem was devised by geologist Alan B. Shaw. The Shaw method involves the following steps:

a. A geologic column is prepared for the two sections that require correlation. One of these, designated section X, is the standard section to which others are to be compared. The second section is designated section Y (Fig. 12.5A).

b. Geologic ranges of the fossil species are drawn next to each section, with lines that extend from the first appearance of the species to the last, as measured from the base of each section.

c. A graph is then prepared with the standard (X) section on the horizontal axis, and the section being compared on the vertical axis (Y). Points representing the first appearance of a species (use small open circles) and the last appearance of a species (use small filled circles) are plotted on the graph.

Figure 12.5A Graphic method of biostratigraphic correlation.

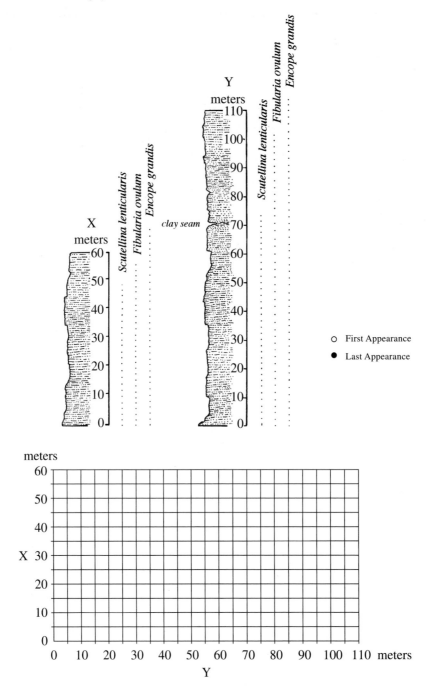

d. A line is drawn through the points plotted on the chart. It is called the *line of correlation.* By using the line of correlation, any point in one section can be correlated to a point in the other. (The example used here is greatly simplified. In practice, there would be many species and a cluster of points, so that one would have to statistically determine the best fit for the line of correlation.)

Correlation Exercise

1. Using the data provided on the accompanying chart (Fig. 12.5B), draw vertical lines next to the geologic columns for sections X and Y, indicating the geologic range of each of the three species used in this example.

2. Plot the correlation line indicated by first and last occurrences by using X and Y data for each point.

Figure 12.5B First and last appearances of species in sections X and Y, Figure 12.5A.

	Section X		Section Y	
	First Appearance	Last Appearance	First Appearance	Last Appearance
Scutellina lenticularis	5 m	30 m	10 m	60 m
Fibularia ovulum	10 m	42 m	20 m	85 m
Encope grandis	25 m	50 m	50 m	100 m

3. Section Y has a 10-cm seam of clay 70 m above the base of the section. To what point above the base of section X would this seam correlate?

_____

_____

_____

4. In which section, X or Y, was the rate of deposition greatest?

_____

INTERPRETATION
OF AN OUTCROP
IN SOUTHERN ILLINOIS

The following questions provide an opportunity to interpret an actual outcrop in southern Illinois. Answer these questions by referring to Figure 12.6.

1. Draw the upper and lower contacts of the Tar Springs Formation on the photograph (Fig. 12.6A).

2. The rocks in this outcrop were deposited during the same geologic period. What was the geologic period?

_____

_____

3. Name two fossils present in the rock slabs that are the basis for your answer to the previous question.

_____

_____

4. Describe the probable depositional environments of the Vienna Formation.

_____

_____

_____

5. What can be inferred about the movement of the shoreline during the time that the formations in this bluff were being deposited?

_____

_____

_____

_____

_____

6. In reference to Question 5, for the movement of the shoreline, would reasons other than sea-level fluctuation explain the assemblages and sediment types? Explain.

_____

_____

_____

_____

INTERPRETATION
OF AN OUTCROP
IN CENTRAL ILLINOIS

For the following questions, you are asked to interpret an outcrop in central Illinois. Answer the questions by referring to Figure 12.7.

1. By comparison with drawings in "Fossils and Their Living Relatives" (Chapters 10 and 11), identify each fossil as to major group (phylum or class) and subgroup (genus and/or species). Write the name of the fossil beneath it on Figure 12.7, and plot its range by a vertical arrow on the range chart.

2. What is the geologic age of the outcrop, assuming that all units belong to the same geologic system?

_____

A

Vienna Fm.

25'

Clastic
limestone

Tar Springs Fm.

4'

Shale,
med. dark gray

2'

Sandstone, silty, mica
ripple marks

2'

Shale, med. gray

1'

Sandstone, calcareous

4'

Claystone, med. gray

5–10'

Sandstone, fine grain,
mica; local cross bedding

3'

Shale, silty, laminated

Glen Dean Fm.

8'

Limestone, clastic
Shale, med. dark gray

B

Figure 12.7 Columnar section of an outcrop in central Illinois.

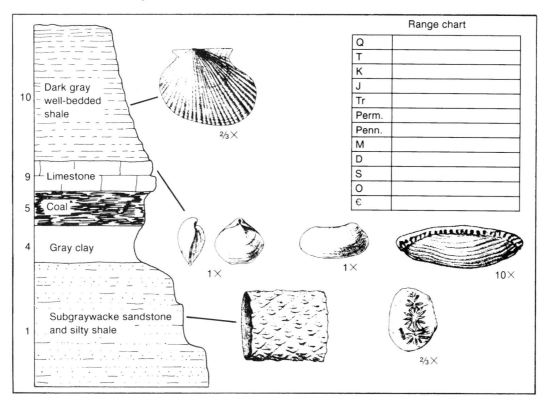

Range chart

Q	
T	
K	
J	
Tr	
Perm.	
Penn.	
M	
D	
S	
O	
Є	

3. Which of the numbered units are transgressive marine?

_____

4. Which of the numered units are terrestrial?

_____

FOSSILS AND PALEOENVIRONMENTS

In addition to correlating and determining the age of strata, fossils are also useful in reconstructing the environment in which ancient organisms lived. In younger strata, the fossil animals and plants can be compared directly with their living relatives (Fig. 12.8), and the assumption is made that both require(d) the same environmental conditions for life. If there are no similar living counterparts for the comparison, then deductions about the environment must be based on the nature of the entire fossil assemblage, evidence from general comparative anatomy, and clues obtained from the enclosing sedimentary rock. In general, fossils of plants, insects, some fish, terrestrial vertebrate bones, and pollen grains and spores indicate a nonmarine environment. Brachiopods, crinoids, bryozoans, corals, and trilobites are certainly marine, and

most species of these probably lived in relatively shallow water. Marine species can give further insight into the environments in which they lived based on their tolerance of salinity, their feeding techniques (suspension versus detritus feeders), and, as in the case of corals, light, water depth, and temperature requirements.

Study Questions

1. What group of living and fossil protists are being compared in Figure 12.8?

_____

_____

2. What kinds of organisms living on the present sea floor are not likely to be preserved as part of the fossil record?

_____

_____

_____

_____

Figure 12.8 Illustration of how the distribution and environment of an existing organism may be related to those of organisms that lived in the geologic past.

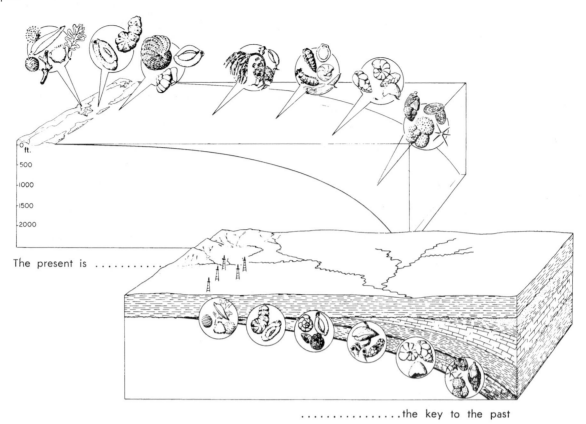

The present is . . . . the key to the past

3. Considerable geologic work has been devoted to determining the depth of water in which particular foraminifers live. Why are petroleum geologists interested in mapping the occurrences of fossil benthic foraminifers to identify trends or directions toward ancient shallow-water areas and shorelines?

_____

_____

_____

_____

_____

_____

 ## THE HABITAT OF MARINE LIFE

To recognize the significance of marine fossils in the history of the earth, it is necessary to study present-day organisms in relation to their natural abode or habitat. The study of those relations is called ecology. Paleoecology deals with the relations between ancient organisms and their environment. Many environmental conditions existing today had their parallels in the geologic past. Thus, if the fossil assemblage in a given rock layer is the counterpart of a community of organisms living today in tropical lagoons, then it can be inferred that the rock layer was deposited in an ancient lagoon. However, caution is required in making such inferences. In fossiliferous beds, the paleoecologist is really studying the environment in which the fossils were buried, which may or may not represent the environment in which they lived.

A simplified classification of marine environments is shown in Figure 12.9. The **littoral zone** (between high and low tide) contains animals and plants that have adapted to withstand periodic exposure, wave action, and sediment movement. The forms of life found in the **neritic-benthic** zone (low tide to about 200 m) require at least some light. That biologic requirement is readily met over most of this shallow zone where sunlight penetrates all the way to the sea floor. Here marine plant life proliferates as do the multitude of organisms that are directly or indirectly dependent on plants. Animals living in the **bathyal-benthic** zone (600 to 6000 ft, or 183 to 1830 m) are adapted to a habitat characterized by high pressure and extreme cold.

Within the environmental zones, organisms can be classified, according to their mode of life, as **planktonic** or **planktic, nektic,** or **benthic.** The **planktonic** (floaters) are predominantly animals and plants too small to be seen without magnification. The unicellular green plants, which make up part of the plankton, are the ultimate food source in the sea. Like their relatives the land plants, they use the energy of sunlight to produce carbohydrates from carbon dioxide by

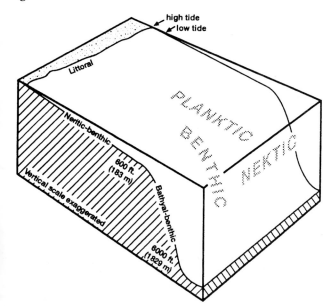

Figure 12.9 Classification of marine environments.

the process known as photosynthesis. **Nektic,** or swimming animals, include fishes, turtles, whales, porpoises, seals, cephalopods, and some arthropods. Most nektic forms are active predators. The bottom dwellers are termed **benthic** and include a great diversity of invertebrate animals as well as microscopic algae, the larger algae (seaweeds), some fungi, and bacteria.

Neritic-benthic invertebrates are among the most common fossils in sedimentary rocks. Based on their feeding habits, they can be divided into plant-eating **herbivores; carnivore**[-scavenger] organisms, which eat other animals both living and dead; the **filter feeders,** which strain tiny organisms and organic detritus from the bottom water; and the *deposit feeders,* which extract nutrition from the organic content of bottom sediment.

The range of water depths preferred by several major groups of fossil and living marine invertebrates is illustrated in Figure 12.10. Similar charts using smaller taxonomic groups such as species or families often provide smaller and more precise depth ranges. Nevertheless, even the larger ranges indicated for major groups can be useful.

Figure 12.10 General depth distribution of some of the groups of marine fossil and living invertebrates discussed in these studies. Similar charts depicting smaller taxa (species, genera, or families) often provide very precise bathymetric information. It should be remembered, however, that depth is not the only factor that influences where organisms live. Temperature, salinity, turbidity, and many other factors are also important.

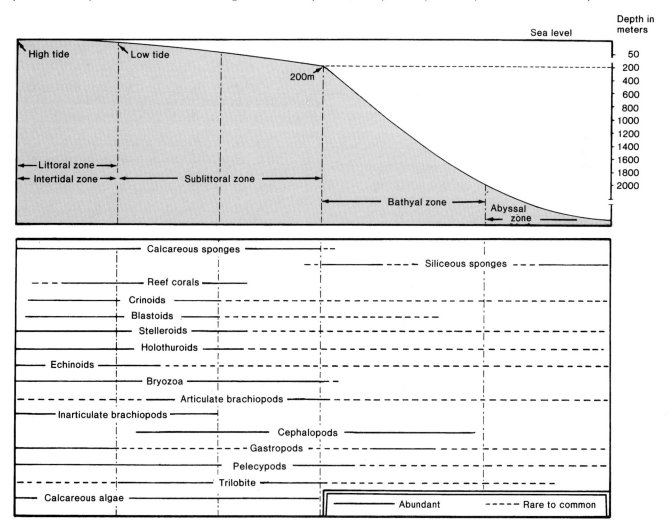

Study Questions

1. At a given locality, a time-rock unit containing abundant fossil calcareous sponges and calcareous algae is conformably overlain by a time-rock unit containing siliceous sponges and cephalopods. What statement can be made about bathymetric change at the locality?

_____

_____

_____

_____

2. In general, why are cephalopods and planktonic foraminifera likely to be less useful than bryozoa in bathymetric analyses?

_____

_____

_____

_____

3. What general statement can be made about the paleobathymetry of a stratum containing abundant crinoids (see Fig. 12.10)?

_____

_____

_____

_____

_____

4. Note that the neritic-benthic environment contains greater numbers and diversity of organisms. What factors may account for this?

_____

_____

_____

_____

_____

5. Many benthic animals burrow into the sea floor and live in the tubes they have excavated. Why are the burrows produced by littoral-benthic animals generally deeper than those produced by animals living in the neritic-benthic zone?

_____

_____

_____

_____

Figure 12.11 The use of fossils as indicators of ancient climates.

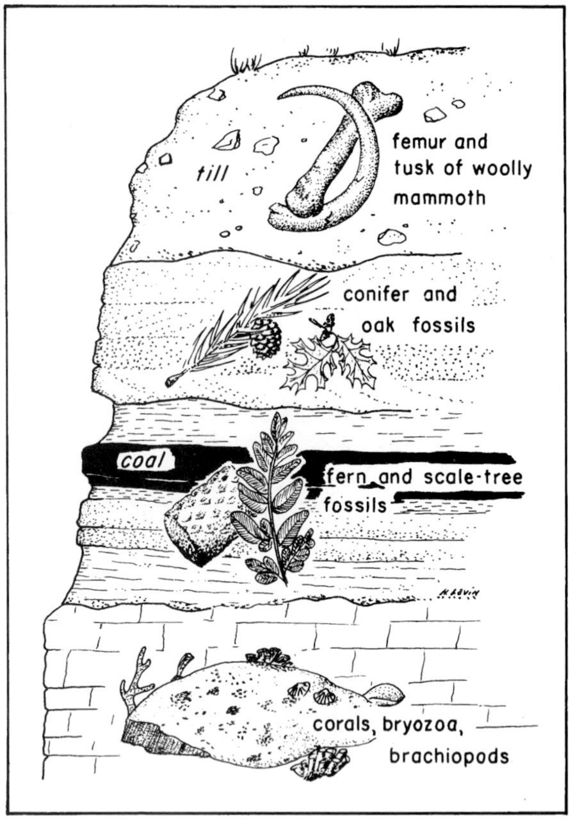

In addition to the information fossils may provide about paleobathymetry, fossils can also be useful indicators of ancient climates. Refer to Figure 12.11 in answering the following questions:

1. Label each of the four time-rock units depicted in Figure 12.11 as either _warm-marine, warm-moist nonmarine, cool-moist nonmarine,_ or _cold-moist nonmarine._

2. Draw an arrow to and label the surfaces of disconformity present at the outcrop. What is the evidence for these disconformities?

_____

_____

_____

_____

3. During which geologic era were the two lower units deposited?

_____

_____

During which geologic era was the uppermost unit deposited?

_____

Figure 12.12 Traces that reflect animal behavior.

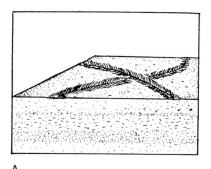

A

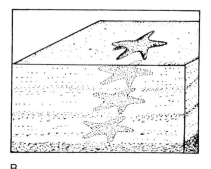

B

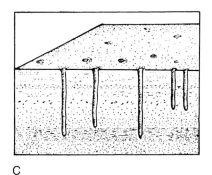

C

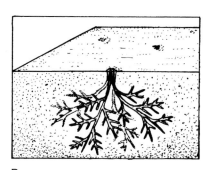

D

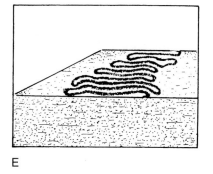

E

What statement can be made about the age of the unit containing fossil remains of hardwood trees?

_____

_____

_____

INVERTEBRATE TRACE FOSSILS

As described in Chapter 10, fossils include not only the petrified remains of ancient organisms but also their **traces.** **Trace fossils** consist of **tracks, trails, burrows** (dwellings in soft sediment), **borings** (dwellings in hard substrates such as **hardgrounds**), fossilized excrement (**coprolites** or **fecal pellets**), and other markings made by former animals and plants. The study of such traces is called **ichnology.** Although trace fossils are often not as attractive as many petrifactions of animals, they are nevertheless very useful to geologists seeking clues about water depth, currents, food supply, and the rates of deposition in ancient seas and lakes. From some invertebrate traces one can sometimes infer body form where no remains of the external skeleton (shell or carapace) exist. One may be able to tell if the animal was resting, crawling, grazing, feeding, or simply living within a relatively permanent dwelling. For example, shallow depressions that more or less reflect the outline of the animal may be **resting traces.** **Crawling traces** are usually linear and show directed movement. **Grazing traces** occur along definite bedding

planes and are characterized by systematic meandering or concentric and parallel patterns that represent the animal's effort to cover the area containing food in an efficient manner. The three-dimensional counterpart of grazing traces are called **feeding traces.** Feeding traces consist of systems of branched or unbranched burrows. Simple or U-shaped structures inclined or perpendicular to bedding are often the **dwelling traces** of various benthic invertebrates.

Study Questions

1. What type of extinct marine arthropod may have produced the crawling trace in Figure 12.12A?

    _____

    _____

2. What kind of marine invertebrate may have produced the trace fossil called *Asteriacites* (Fig. 12.12B)?

    _____

    _____

    _____

    Is it likely the trace represents crawling, resting, or feeding?

    _____

    _____

    _____

Figure 12.13 Trace fossil clues to sedimentation and structure.

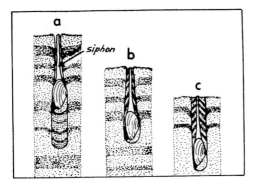

A

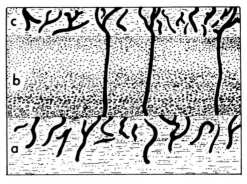

B

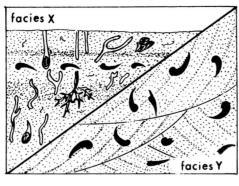

C

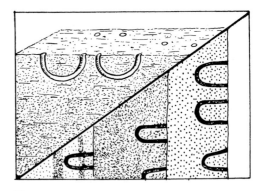

D

Was sediment deposition at this site meager or relatively rapid? Why?

_____

_____

_____

_____

3. What type of behavior is indicated by the trace fossil depicted in Figure 12.12C?

_____

_____

What type of brachiopod might produce a vertical tube such as this?

_____

_____

4. What was the behavior of the animal that formed the trace fossil *(Chondrites)* shown in Figure 12.12D?

_____

_____

5. What type of behavior is indicated by the trace fossil in Figure 12.12E?

_____

_____

In addition to serving as clues to the behavior of animals, trace fossils also provide information about events that have occurred in the environment of deposition. They can be used, for example, to determine if deposition or erosion had taken place at the surface of sedimentation and may even suggest the approximate rate of deposition. In areas where strata have been deformed and overturned, trace fossils may help the geologist distinguish the tops from the bottoms of beds.

1. Figure 12.13A depicts dwelling traces of the pelecypod *Mya.*

    a. Which of the three traces shown was probably made at a time in which the surface of sedimentation was stationary and the animal was growing?

_____

_____

b. Which trace was probably made while the surface of sedimentation was rising as a result of active deposition?

_____

_____

c. Which trace suggests that currents may have been sweeping away sand at the surface of sedimentation?

_____

_____

2. What sequences of events occurred on the ocean floor at the locality depicted in Figure 12.13B? (The burrow traces are made by the marine worm *Nereis*.)

_____

_____

_____

3. Figure 12.13C shows two different facies of a time-rock unit. What biological evidence suggests that the sediment of facies X does not represent a reworked sediment?

_____

_____

_____

_____

4. What conditions other than the reworking of sediment might account for a scarcity of trace fossils within a stratum?

_____

_____

_____

_____

5. The upper part of Figure 12.13D shows the U-shaped dwelling tubes of a species of marine worm. The lower part of the figure depicts dwelling traces of extinct worms presumed to have behaved similarly. The fossil traces occur in vertically dipping strata. Is the original top surface of the strata toward the right or left?

_____

_____

_____

_____

TERMS

acme biozone   Body of strata that contains the maximum abundance of individuals of a species or other taxa.

assemblage biozone   A biozone having three or more taxa of a group that is distinguishable from assemblages in overlying and underlying strata.

bathyal-benthic zone   The ocean-floor environment extending from a depth of 600 to 6000 ft (183 to 1830 m).

benthic (or **benthonic**)   Pertaining to aquatic bottom-dwelling organisms.

biostratigraphy   The study of strata based on the fossils they contain.

biozone   A body of rock defined by the presence, absence, or relative abundance of certain species or other taxa.

boring   Biogenic sedimentary structure excavated into a hard substrate (e.g., rock, shell, or wood), usually to serve as a dwelling.

burrow   Biogenic sedimentary structure emplaced in soft sediment below the sediment surface, usually for the purpose of feeding or dwelling, or as a result of locomotion.

carnivore-scavenger   Pertaining to organisms that can feed on either living or dead animals.

carnivorous   Pertaining to organisms that feed on animals.

concurrent range biozone   A biozone defined by the interval of overlap between the first appearance of one species or other taxon, and the last appearance of a different species or taxon.

coprolite   Large lump of fossilized excrement, generally much greater than 1 cm in length or diameter.

crawling trace   Tracks in sediment or sedimentary rock that are produced by locomotion with the aid of appendages and/or muscular movement of the body.

deposit feeder   An organism that feeds on organic debris on and within sediment covering the floor of a lake or sea.

dwelling trace   Certain burrows and borings in sediment or sedimentary rock that were once more or less permanently inhabited by animals. Most such structures are cylindrical.

fecal pellet   Small spherical, ovoid, or cylindrical lump of excrement; maximum dimensions generally less than 1 cm.

feeding trace   Tracks or burrows in sediment or sedimentary rock that represent the traces of deposit feeders in search of food. Radial patterns in which previously mined sediment is avoided characterize such traces.

filter feeders (invertebrates)   Animals that live by filtering small pieces of organic matter and smaller organisms out of the water. Filter feeders include bryozoa, brachiopods, sponges, and clams.

geologic range (of an organism)   The geologic time span between the first and last appearance of an organism.

grazing trace   Tracks or burrows in sediment in which a feeding organism moves in a symmetrical pattern to efficiently exploit nutritious sediment.

hardground   Substrate composed of indurated rocks at a minor depositional hiatus or erosion surface, in which borings may be produced.

herbivorous   Pertaining to organisms that feed on plants.

ichnology   The study of fossil and recent traces such as tracks, burrows, and borings.

index fossil   A fossil that identifies and dates the strata or succession of strata in which it is found. Also called a **guide fossil.**

intrazone   Unfossiliferous intervals having defined biozones both above and below.

littoral zone   In a simple classification of marine environments, the zone between high and low tide.

nektic (or **nektonic**)   Term pertaining to aquatic organisms that swim, as opposed to those such as plankton, which float.

neritic-benthic zone   The ocean-floor environment that extends from low tide to a depth of 600 ft (about 183 m).

planktonic (or **planktic**)   Organisms that float or drift passively with the movements of a water mass.

resting trace   Impressions formed in sediment or sedimentary rock by an animal temporarily at rest or in refuge; often reflect the morphology of the animal.

taxon range biozone   A body of rock representing the maximum stratigraphic range of a species or other taxon, from its first appearance to its last.

trace   Structure produced in sediment by the activity of an animal or plant; also called a **biogenic sedimentary structure.**

trace fossil   Ancient trace preserved in lithified sediment; also called an **ichnofossil.**

track   Individual footprint.

trackway   Set of multiple, separate footprints that indicate movement.

trail   Continuous locomotion trace without separate footprints.

13

A Brief Survey of the Vertebrates

INTRODUCTION

Because we ourselves are vertebrates, animals with backbones seem to hold particular interest. Vertebrates are a subphylum of the Phylum Chordata. All members of the Phylum Chordata have some sort of longitudinal supportive structure. In you and I it is our vertebral column. In chordates like **tunicate** larvae and **lancelets,** that structure is a flexible rod or *notochord.* In addition to the notochord, a distinguishing feature of chordates, at least at some stage in their development, is a hollow, nerve cord that extends above (dorsal) and parallel to the notochord (Fig. 13.1). Other characteristics include gill slits in the **pharynx,** a post-anal tail, and blood that moves forward in the primary ventral blood vessel and backward in the dorsal.

In chordates belonging to the Subphylum Vertebrata, the longitudinal supportive structure consists of a series of bony segments (**vertebrae)** that form the **vertebral column** or "backbone." The vertebral column in fishes functions in much the same way as a notochord by resisting shortening of the body as muscles acting against it alternately contract in succession so as to provide the sinous motion needed for swimming. In contrast to the notochord with its simple structure, vertebrae offer greater strength and a variety of attachment points for muscles and other skeletal components. Vertebrae alternate with flexible disks made of cartilage. These **intervertebral disks** are partly remnants of the notochord and are held in place by muscles and ligaments extending from vertebra to vertebra. The vertebral column is **homologous** with the notochord. Structures of different organisms that appear to have a common evolutionary origin are termed homologous. In general, homologous structures in different organisms show a correspondence in location and form.

Study Questions

1. On figure 13.1, connect by lines the nerve cord in the lancelet, fish, and vertebra. Similarly, connect by lines the notochord or its homologous structure in the fish and the human vertebra. With what feature of the fish skeleton is the human "spinose process" homologous?

2. If you were to drop a small goldfish and an earthworm into a bucket of water, which would swim more efficiently? Why?

FISHES

The first fishes to appear on earth were jawless fishes belonging to the Class **Agnatha.** Their earliest remains are Cambrian in age. The Cambrian fossils are identified predominantly on the basis of the microstructure seen in fragments of hard tissue. By Ordovician, however, fossils of jawless fishes are more abundant and often well preserved. The jawless fishes were followed during the Silurian by the first fish with jaws. The appearance of bone-supported true jaws was a milestone in the evolution of vertebrates. The ability to grasp, cut, and hold food led to new and more active ways of life and resulted in many new lineages of fishes. There are not fossils that directly show how the jaws may have evolved from structures in jawless fishes, but a traditional view is that jaws formed from modified, anterior gill supports. The evolutionary history of vertebrates reveals many other examples where a structure having one function is changed to perform a quite different function.

Study Questions

1. In Figure 13.2, what skeletal structures behind the mandibular arch in the dogfish appear to be homologous with the **mandibular arch?**

2. What feature of the dogfish appear to be homologous with the **spiracle?**

3. What supportive function does the **hyomandibular** have in the dogfish?

Figure 13.1 Comparison of lancelet transverse section (A), with that of an actinopterygian bony fish (B), and a human thoracic vertebra (C).

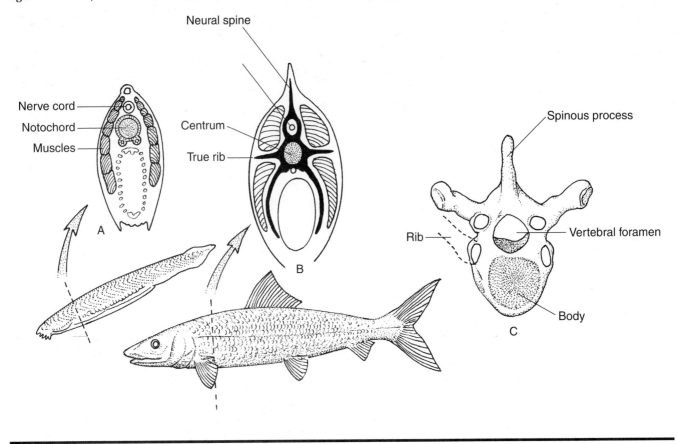

Figure 13.2 Basic skeletal structures of the head of a small shark known as a dogfish.

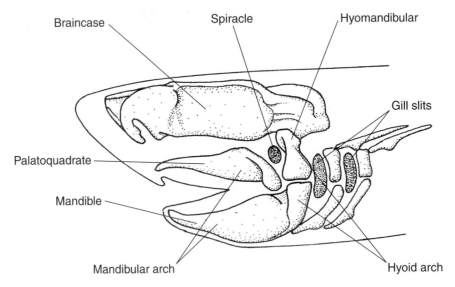

By Late Silurian time, fishes with jaws had branched into two large classes: the **Chondricthyes** (sharks and other fishes with cartilagenous skeletons) and the **Osteichthyes** (fishes with bony skeletons). The Osteichthyes have been the most successful in terms of abundance and diversity. The most numerous bony fishes today are those in the sub-class **Actinopterygii.** In these fishes the fins are supported by parallel cartilagenous rods called radials. Another sub-

class, the **Sarcopterygii,** includes a group of fishes called **crossopterygians.** Crossopterygians had muscular robust "lobe fins" supported by a single basal bone.

Study Question

On Figure 13.3, label the humerus, ulna, and radius on the amphibian and human.

Figure 13.3 Anterior appendage of a crossopterygian fish (A), an early amphibian (B), and a human (C).

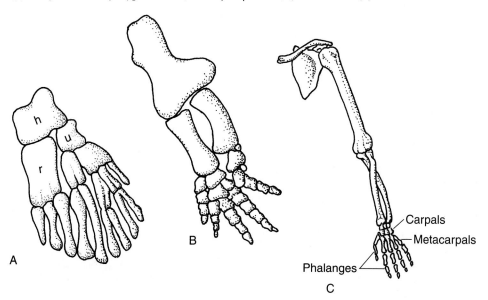

Carpals
Metacarpals
Phalanges

C

AMPHIBIANS

Vertebrates are a diverse group that include both water-dwelling fishes and land animals that walk on all four legs and are therefore called **tetrapods.** It was during the Devonian Period that tetrapods made their appearance. These early tetrapods were amphibians. Prominent among the Devonian amphibians were a group known as ichthyostegans because of the many fishlike characteristics they retained. In fact, many skeletal features of earliest amphibians and crossopterygian fishes are strikingly similar (Fig. 13.4).

Study Questions

1. Many problems had to be solved in making the transition from life in the water to life on land. What changes do you think were made in the way vertebrates:

   a. obtained oxygen?

   _____

   _____

   b. supported their bodies?

   _____

   _____

2. How do amphibians betray their ancestry in their mode of reproduction?

   _____

   _____

3. On Figure 13.4, label the bones on the crossopterygian that are equivalent to those on the icthyostegan.

Figure 13.4 Lateral views of the skull of a crossopterygian fish (A) and an early amphibian (B).

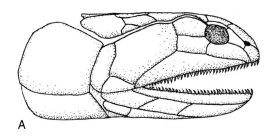

A

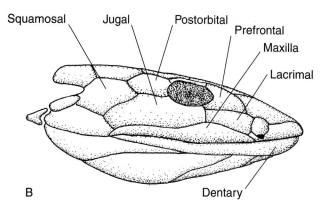

Squamosal   Jugal   Postorbital   Prefrontal   Maxilla   Lacrimal

B   Dentary

4. Note the prominent notch on either side of the back of the ichthyostegan skull. The tympanic membrane ("eardrum") was stretched across this notch. An ear bone, the **stapes,** extended from the tympanic membrane into a hole on the side of the braincase. Sound waves striking the tympanic membrane were transmitted by the stapes to the inner ear. What bone, formerly serving as a prop to support the braincase (see Fig. 13.2) was transformed to form the stapes?

AMNIOTES

Many millions of generations were required before tetrapods evolved from crossopterygian fishes. Even so, the conversion was not complete, for the amphibians that were the first tetrapods needed to return to water bodies in order to lay fishlike, naked eggs that developed into fishlike larvae (tadpoles). In contrast to the fishlike eggs of the amphibians, more advanced tetrapods had evolved an egg called an **amniotic egg.** Within the protective outer covering of this egg, a membrane, the **amnion,** retains water and protects the developing embryo from drying out. Vertebrates having such an egg are termed **amniotes.** Because they do not require standing bodies of water for reproduction, amniotes have been able to roam freely across the lands.

During the Carboniferous Period, amniotes diverged into two branches. The first can be termed the **Reptilia.** This branch includes the **Anapsida** (e.g., turtles), **Diapsida** (e.g., lizards and snakes), and **Archosauria** (e.g., crocodiles, dinosaurs, and birds). The second branch consists of the Synapsida. The anapsids, synapsids, and diapsids can be recognized by the positions and number of openings (temporal fenestra) on the sides of their skulls (Fig. 13.5). Anapsids lack such openings, whereas diapsids have two on each side of the skull. A single opening bordered above by the squamosal and postorbital bones identify the synapsids.

The synapsids were particularly abundant and widespread during the Permian Period. The most important synapsid groups were the **therapsids** and the **pelycosaurs.** The therapsids are particularly interesting because they possessed several mammalian skeletal traits. Pelycosaurs are recognized by the distinctive erect sail along their backs (Fig. 13.6). The sail was supported by greatly elongated bones extending vertically from the vertebrae, called the **neural spine.** It is likely that the sail acted as a heat receptor and at other times as a radiator.

Figure 13.5 The three major skull types in tetrapods. **1.** anapsid; **2.** synapsid; and **3.** diapsid. (p, *parietal;* sq, *squamosal;* po, *postorbital;* j, *jugal;* qj, *quadratojugal*)

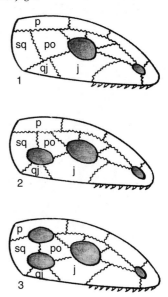

Study Questions

1. What element of the fish skeleton in Figure 13.1 corresponds to the greatly lengthened bone that supports the sail on a pelycosaur's vertebra?

_____

_____

2. The amniotic egg provides for limited exchange of gases necessary for the survival of the embryo. Based on knowledge of your own respiration, which of these gases are taken into the egg, and which are eliminated?

_____

_____

Figure 13.6 *Dimetrodon,* a Permian reptile having a sail-like structure along its back. *Dimetrodon* was one of a group of reptiles termed *pelycosaurs.*

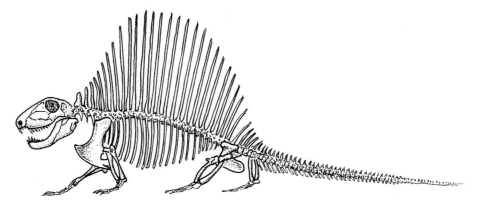

Exercise: Estimating the Weight of a Dinosaur

WHAT YOU WILL NEED

1. Container holding over 1000 ml of water.
2. 1000-ml beaker.
3. Graduated 100-ml beaker.
4. Plastic dinosaur model.

Figure 13.7 Setup for measuring the volume of water displaced by a dinosaur model. It is helpful if the spout on the large beaker is enlarged with plastic modeling clay or putty so that water pours directly into measuring cylinder.

Dinosaur model. The model used here, *Utahraptor,* is one of many in "The Carnegie Collection" produced by Safari Ltd., P.O. Box 63085, Miami, Florida 33163. These models were sculpted under the guidance of paleontologists at the Carnegie Museum of Natural History. They are often available in the shops of science museums.

Study Model		Utahraptor	
1. Begin with length of actual dinosaur in centimeters (provided by instructor)	_____	Length of actual *Utahraptor*	*650 cm*
2. Find length of model in centimeters Cut masking tape into 3-mm wide strips and extend tape along length of the model. Cut or mark tape at total length, remove, place on flat surface, and measure.	_____	Length of model of *Utahraptor*	*16.2 cm*
3. Divide length of actual dinosaur by length of model.	_____	650 cm divided by 16.2 cm Thus the model is about 1/40th the size of the actual *Utahraptor.*	*40.12 cm*
4. Determine volume of the model (Fig. 13.6). Submerge model in a graduated beaker filled to a set level with water and measure the amount of water displaced. Alternatively, fill the beaker brimful. Place a graduated 100-ml cylinder beneath pour-spout, and measure "spill-over" when model is submerged.	_____	Volume of *Utahraptor* model	*20.0 ml*
5. Cube the scale of the model	_____	Cube of the scale of the model *Utahraptor* model is 1/40th the size of actual *Utahraptor.* Thus, 1/40 × 40 × 40 (or 40³)	*1/64,000*
6. Find volume of the actual dinosaur Otained by multiplying the inverse of the above value times the volume of the model.	_____	64,000 × 20 ml	*1,280,000 ml*
7. Convert milliliters to liters	_____	1,280,000 divided by 1,000	*1,280 liters*
8. Determine weight of actual dinosaur A liter of water weighs 1 kg. An average-size crocodile weighs 0.9 kg/L. Thus, 0.9 kg/L times the volume of actual dinosaur gives the weight of the dinosaur.	_____	0.9 kg/L × 1,280 L	*1,115 kilograms*
9. Convert kilograms to pounds To convert kilograms to pounds multiply kilograms by 2.205 lbs/kg.	_____	1,115 × 2.205 lbs/kg	*2,458 pounds or 2.46 metric tons.*

DINOSAURS

Dinosaurs are extinct archosaurians with legs that extend directly under their bodies so as to provide an upright posture. That upright stance differed from the sprawling posture of more primitive tetrapods like the amphibians. Dinosaurs dominated the landscapes of the Mesozoic Era for about 180 million years. From relatively small *bipedal* (running on two legs) forms, dinosaurs branched into two great groups; the **Ornithischia,** recognized by their birdlike pelvic structure, and the **Saurischia,** with pelvic structures more closely resembling those of lizards. In either group, the pelvic structure is composed of three *pairs* of bones fused together (Fig. 13.8). Viewed from the side of a mounted skeleton one would see one bone of each pair or three in all. The uppermost of these bones is the *ilium.* It is firmly clamped to the vertebral column. The bone extending downward and toward the rear is the *ischium.* The remaining *pubis* joins the other two and projects downward and forward to help support the visceral mass. Where the three bones come together, there is a hollow area which affords a socket for the rounded end of the upper leg bone or femur.

Although dinosaurs are widely marveled for the immense size many of them attained, others were small creatures that were as small or smaller than a chicken. Paleontologists can sometimes infer the approximate weight of dinosaurs from the thickness of their limb bones and characteristics of their skeletons. Another method that has been used to provide reasonable estimates of dinosaur weight is based on the volume of accurate scale models. Edwin Colbert (1962), and more recently, Spencer Lucas (1944)* described how such weight estimates of dinosaurs might be ascertained (Fig. 13.7).

MAMMALS

The Mesozoic was not the exclusive domain of the dinosaurs, for by Triassic time mammals had made their appearance. Early mammals were small shrewlike creatures that might have been inconspicuous in a landscape dominated by reptiles. They were the descendants of mammal-like reptiles that lived during the Permian and Triassic periods. Mammals are recognized by the possession of fur or hair (except in whales), a four-chambered heart, and, as implied by their name, mammary glands used in suckling their young. The first of these characteristics is associated with the fact that mammals are warm-blooded **(endothermic),** and the second with mammalian postnatal care of their young. Nei-

---

* Lucas, S.G. (1994) *Dinosaurs: The Textbook.* Dubuque, IA, Wm C. Brown Publishers.

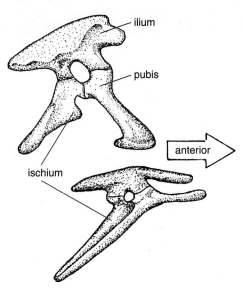

Figure 13.8 Pelvic structures in dinosaurs. Saurischian (above) and ornithischian (below).

ther characteristic, however, is very useful to the paleontologist, who must differentiate mammals from reptiles on the basis of skeletal characteristics. Fortunately, there are several characteristics of the mammalian skeleton that are diagnostic. One of these is the presence of a single bone, the **dentary,** in the lower jaw. In mammals, the old reptilian **quadrate** and **articular** jaw hingement has been modified to the mammalian **squamosal**-to-dentary type of hinge. In the process, the articular became the **malleus** and the [quadrate] became the mammalian incus. Thus, once again, bones serving one function (hingement) are transformed and pressed into service for a different function (transmitting sound vibrations). Other skeletal characteristics of mammals include a braincase considerably larger than that in reptiles and a double knob of bone (double **occipital condyle**) that forms the articulation between the skull and first cervical (neck) vertebra. There is ordinarily a functional differentiation of teeth in mammals. In human dentition, this differentiation consists of incisors, canines, molars, and premolars. Other differences in the mammalian skeleton are found below or posterior to the skull. Ribs are usually absent on neck vertebrae. The posterior lumbar region is also ribless and forms a contrast to the great rib basket of the thorax, and in all but four genera of living mammals, the number of cervical vertebrae is seven. The breastbone **(sternum),** which was relatively unimportant in reptiles, becomes an important base of attachment for the longer ribs. In mammals, one also finds a strong ridge (the *scapular spine*) extending down the middle of the shoulder blades. This feature provided a place for attachment of muscles from the upper forelimb and is related to the mammalian trait of positioning the legs more directly under the body. Bones of the pelvis **(ilium, ischium,** and **pubis)** are fused into a single sturdy structure.

Figure 13.9 Skeletons of the Oligocene dog *Hesperocyon* (A) and the Carboniferous reptile *Hylonomus* (B). The drawings are not at the same scale.

Hesperocyon after Matthew, W. D., 1909, The Carnivora and Insectivora of the Bridger Basin, Middle Eocene, Mem. Amer. Mus. Natu'l. Hist. **9:**291–576. *Hylonomus* after Carroll, R. L. and Baird, D., 1972. Carboniferous Stem Reptiles of the Family Romeriidae. Bull. Mus. of Comparative Zoology **143:**321–363. President and Fellows of Harvard College. Reprinted by permission.

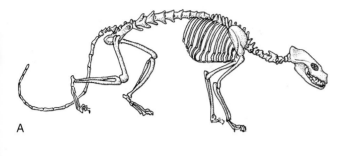

A

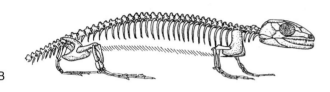

B

Study Questions

1. Examine the skeletons depicted in Figure 13.9. Which of the figures represents the skeleton of a mammal?

_____

_____

2. List below four features of the skeleton you have selected, that are more mammal-like than reptile-like.

    a. _____

    _____

    b. _____

    _____

    c. _____

    _____

    d. _____

    _____

THE TEETH OF MAMMALS

Upon finding the jaw or skull of an extinct mammal, the paleontologist immediately examines the teeth, for these give direct evidence of the kind of food the animal ate as well as certain other information about the animal's behavior. Unlike the teeth of fish, amphibians, and reptiles, the teeth of all but a few mammalian groups are differentiated, or *heterodont*. This means that the teeth in the forward part of the jaws differ in shape and function from those farther to the rear. Typically, these differentiated teeth consist of *incisors* at the front, followed by *canines, premolars,* and *molars*. An expression termed the *dental formula* describes the number of teeth of each type. For example, the dental formula for a coyote is:

$$\frac{3142}{3143}$$

The formula indicates the coyote has a total of 42 teeth. There are ten teeth in *each* of the two upper jaw sides and 11 teeth in *each* of the two lower jaw sides. In the coyote, as in other heterodont mammals, there is never more than one canine in each of the four sides.

Study Questions

1. Figure 13.10 illustrates the teeth in an adult human. What is the dental formula for *Homo sapiens?*

_____

2. How many permanent teeth are there in the entire mouth of *Homo sapiens?*

_____

Figure 13.10 Permanent human dentition.

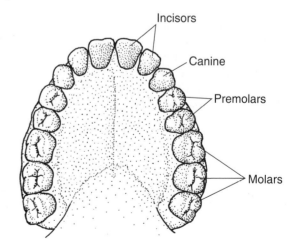

The incisor teeth of mammals function in cutting, grasping, and gnawing. In large grazing animals like horses, they are flat and chisel-like so as to effectively bite and crop grasses and associated plants. Incisors occur in both the upper jaw and the lower jaw in horses, with upper incisors anchored in the premaxilla bones. Those in the lower jaw have their roots in the large dentary bone. Upper incisors are not present in deer and cattle. Instead, these animals cut plant food by pinching it off between the lower incisors and a horny plate on the upper jaw.

The incisor teeth in rodents are wonderfully suited for gnawing. They have the form of curved, keen-edged chisels. Hard enamel covers the front or convex surface of the tooth, whereas softer dentine lies behind the enamel. Because of the difference in hardness between the enamel and the dentine, the enamel layer is able to maintain a razor-sharp edge. In rodents, lateral incisors and canines are missing, leaving a gap or *diastema* between the remaining incisors and the cheek teeth.

The most spectacular of all incisor teeth occur in elephants and their tusked ancestors the mastodonts and mammoths. The immense tusks of these *proboscideans* are actually enormously enlarged and modified second incisors.

Behind the incisors lie the canines. Although many plant-eaters have canines that are used in fighting and defense, these teeth are particulary prominent among the flesh-eaters. In the famous sabre-tooth cat *Smilodon,* the upper canines have the shape of huge daggers and were apparently very effective in cutting out chunks of flesh from the large herbivores on which they preyed. The great tusks of walruses are also canines used for pulling the animal up onto the ice and also for dislodging shellfish from hard surfaces on the ocean floor. In the insect-eating bats, canines help to snare insects, whereas in vampire bats they are used to pierce skin and release a small flow of blood.

The cheek teeth (premolars and molars) have the important function of breaking food apart into small particles, thereby increasing the total surface area that can be coated with digestive enzymes. The flesh-eaters (carnivores) possess laterally compressed cheek teeth with sharp, bladelike cutting edges. They are called *carnassials.* Carnassials in the lower jaw shear like scissor blades across those in the upper jaw to efficiently cut flesh and sinew.

Mammals that feed on the coarse and gritty vegetation of grasslands require cheek teeth that can both withstand wear and serve as grinding mills. To accomplish the grinding, the occlusal surfaces of the premolars and molars are broad and flat. Grinding, however, could not be accomplished if the surfaces were smooth. To provide the necessary roughness, the hard enamel is folded up and down in the softer cement and dentine. Because of enamel's greater resistance to wear, the layers of enamel form ridges, whereas the softer dentine and enamel wear down as troughs between the ridges. In some herbivores, the ridges have the form of concentric, "half moons," forming a pattern known as *selenodont.* Deer, for example, have selenodont molars. Teeth in which the enamel runs transverse, as in elephants, are termed *lophodont.*

To provide for the long wear required for a diet of harsh vegetation, the roots of the molars in large grazing animals like cattle, camels, deer, and horses extend deep into the jaws. In addition, the crown (that part of the tooth above the root) is exceptionally long. As the wear surface of the tooth is gradually reduced, the roots rise to compensate for the wear, and thereby expose more of the crown. Bone fills the space vacated by the rising root. By the time most of the crown has been worn away so that the roots can be seen at the gum line, the animal has reached old age. Malnutrition often follows. Deep-rooted, high-crowned teeth such as those described here are termed *hypsodont.*

Many mammals, including pigs and humans, are capable of eating both plant and animal food. They are omnivorous. In most omnivores, the incisors are bladelike. Canines tend to be less prominent except where they are used in defense or have a secondary sexual function. The cheek teeth have low cusps, short crowns, and well-developed roots. They are spoken of as *bunodont.*

As noted earlier, most mammals are *heterodont* in that their teeth can be divided into functional types, such as the incisors, canines, premolars, and molars. A few groups, including porpoises and armadillos, have reverted to the *homodont* condition, in which all the teeth have a more or less similar appearance. Also, teeth may not be present at all in some mammals, as seen in baleen whales and South American anteaters.

Study Questions

1. Which of the drawings in Figure 13.11 depict the following:

    _____a. Molars characteristic of a flesh-eating mammal.

    _____b. Incisors specialized for efficiency in gnawing.

    _____c. Hypsodont molars with selenodont patterns.

    _____d. Homodont dentition.

2. Among the choices *deer, beaver, cat,* and *porpoise,* what is the most likely animal depicted in figure:

    A. _____   C. _____

    B. _____   D. _____

3. Refer to Figure 13.10. Are the molars depicted in this figure carnassial, hypsodont, or bunodont?

_____

_____

 BIRDS

The final group of vertebrates to be visited in this exercise are the birds. Like the mammals, birds are endothermic or "warm blooded." Feathers, a distinctive feature of birds, not only provide for flight but also are an ideal insulation to help maintain high avian body temperature. Birds have been fashioned by nature for high power and low weight. Feathers,

Figure 13.11 Examples of mammal dentition.

Redrawn from Lawlor, T. E. *Handbook to the Orders and Families of Living Mammals*. Eureka, California: Mad River Press, 1979.

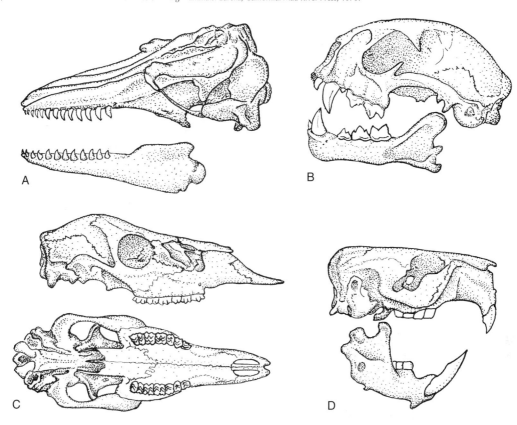

hollow bones, efficient wings powered by sturdy breast muscles, a large robust heart, and an efficient respiratory system all combine to produce a remarkable flying machine.

From the time of Charles Darwin, naturalists have been aware of the skeletal similarities between birds and reptiles. These similarities prompted a remark from Thomas Huxley that "birds are only glorified reptiles." Indeed, birds appear to be related to a subgroup of theropod dinosaurs known as coelurosaurs (the familiar flesh-eater *Velociraptor* is a coelurosaur). These reptiles were already birdlike in their bipedal stance and in the structure of the forelimbs, hindlimbs, shoulder girdle, and skull. In fact, a new classification of vertebrates has been proposed in which the close relationship between dinosaurs and birds is indicated by the inclusion of both groups within the Archosauria.

Study Questions

1. Figure 13.12 is a depiction of the skeleton of the Jurassic bird *Archaeopteryx* and a modern pigeon.

   Name three skeletal features of *Archaeopteryx* that are more characteristic of reptiles than of birds.

   a. _____

   b. _____

   c. _____

2. Describe the differences in the following features between the two skeletons.

   skull _____

   forelimbs _____

   tail _____

   pelvic structure _____

3. What is the function of the sternum in modern birds?

   _____

   _____

   _____

4. What features of the pigeon skeleton indicate greater capability as a flyer?

   _____

   _____

   _____

   _____

VERTEBRATE ICHNOLOGY

The tracks and other traces of vertebrate animals are no less interesting and useful than those of the invertebrates discussed

Figure 13.12 Skeletons of *Archaeopteryx* (A) and a pigeon (B). The drawings are not at the same scale.

After Heilmann, G., *The Origin of Birds,* New York. Appleton-Century-Crofts Inc., 1927.

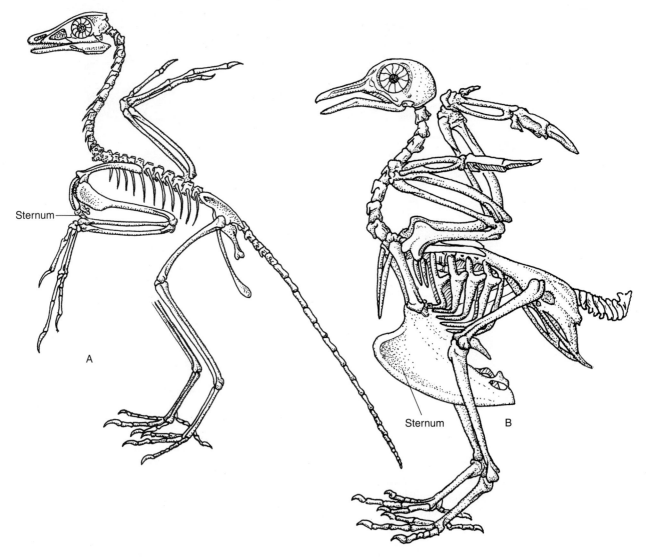

Sternum

A

Sternum

B

Figure 13.13 Vertebrate tracks in siltstone, similar to those found in Triassic sedimentary rocks of Connecticut. These tracks record the passage of reptiles across soft, wet sediment before it became rock.

0   10   20   30 cm

earlier in this chapter. They are often particularly fascinating because they record the dynamic and sometimes dramatic record of the action of ancient animals (Fig. 13.13). **Tracks** and **trackways** (continuous series of tracks made by a single animal) inform us whether the track maker was **bipedal** (walked on two legs) or **quadrupedal** (walked on four legs), **digitigrade** (walked on toes) or **plantigrade** (walked on the flat of the foot); whether it had an elongate or short body; if it was lightly built or ponderous; and sometimes whether it was aquatic (perhaps with webs of skin between toes), **carnivorous** (with claws), or **herbivorous** (tracks of hooves would indicate Cenozoic grazing mammals, for example).

Study Questions

1. How many different kinds of vertebrates left tracks in the large slab of siltstone (Fig. 13.13)?

   _____

2. In what chronological order did the animals cross the area?

   _____

   _____

3. Draw circles around the tracks of the plantigrade quadruped.

4. Which are the tracks of a digitigrade biped?

   _____

5. What event is recorded by the tracks in the center of the slab?

   _____

   _____

6. Figure 13.14 is a sketch of vertebrate tracks discovered in Germany in 1834. The tracks occur in a Triassic red sandstone. The smaller tracks are about 10 cm long, and the larger about twice that length. They occur along a single line, with what *appear* to be the thumb impressions directed alternately to the left and right of each pair. The following hypotheses were advanced as interpretations of these tracks:

   a. The larger tracks were made by a cave bear, and the smaller by a monkey, which followed the bear about.

   b. In 1851, Sir Richard Owen proposed that the tracks were made by a large amphibian or reptile having larger rear than front feet, and that the unusual location of the thumb resulted from the animal's gait, which consisted of crossing one leg over the other. Thus, at each step the left foot was placed to the right of the right foot, followed by the right foot being advanced to the left of the left foot.

   c. In 1925, Wolfgang Sorgel proposed that the tracks were made by a bipedal reptile that occasionally walked on all four legs. Professor Soegel stated that the presumed impression of a thumb was not

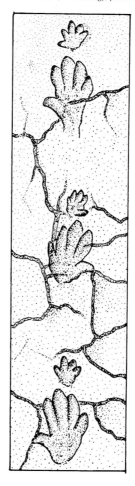

Figure 13.14 Sketch of footmarks discovered in 1834 in a Triassic red sandstone in Germany. The animal presumed to have made the tracks was named *Cheirotherium,* meaning "hand beast."

After a drawing in Sir Richard Owen's book, *Paleontology,* published in 1861.

that of a thumb at all but a finger pointed outward and away from the other fingers.

Discuss the three hypotheses, providing arguments against those you consider untenable. (Also, on your next visit to the zoo, examine the toes of lizards.) Was professor Soergel correct in stating that the position of the presumed thumb impression was an unnecessary cause of confusion?

_____

_____

_____

_____

_____

_____

_____

_____

_____

_____

TERMS

Actinopterygii   Bony fishes with fins supported by cartilageonous rods ("ray-finned" fishes).

Agnatha   Jawless fishes.

amnion   The membrane forming a fluid-filled sac in which the embryo resides.

amniote   A vertebrate employing an amniotic egg in reproduction.

amniotic egg   Type of egg in which the developing embryo is maintained and protected by an arrangement of membranes.

amphibian   Vertebrates that lead a dual life, in that adults are terrestrial and breathe by means of lungs, whereas larvae (tadpoles) are aquatic and breathe by gills.

Anapside   Reptile group characterized by a skull that lacks temporal fenestra.

Archosauria   A major branch of tetrapodal amniotes that include dinosaurs, crocodiles, and birds.

articular bone   Bone in the posterior portion of the lower jaw that forms the surface for the joint.

bipedal   Vertebrates that walk and run on their two hind legs.

carnivorous   Feeding on other animals, flesh-eaters.

Chondrichthyes   The cartilagenous fishes, including sharks, skates, and rays.

crossopterygians   Progressive air-breathing fishes of the subclass Sarcopterygii.

dentary bone   In mammals, the single bone of the lower jaw. In reptiles and amphibians, the lower jaw includes bone in addition to the dentary.

diapsids   Tetrapodal amniotes with two temporal openings in the skull.

digitigrade   Condition of walking on digits (toes) with the sole of the foot raised off the ground.

endothermic   Term describing animals that use metabolic energy to maintain constant body temperature despite variations in environmental temperature, such as birds and mammals.

herbivorous   Feeding upon plants.

homologous   Pertaining to structures or organs that are similar in origin and basic structure.

hyomandibular   The modified upper bone of the hyoid arch, which functions as a connecting element between the jaw and the braincase in certain fishes.

ichthyostegans   Group of early amphibians of the subclass Labyrinthodontia, exemplified by *Ichthyostega.*

ilium   The uppermost of the three bones located on each side of the pelvic girdle.

intervertebral disks   Disk-shaped structures containing cartilage that provide flexibility and shock absorption between vertebrae.

ischium   The most posterior of the three bones located on each side of the pelvic girdle.

lancelet   Small chordate having a body that tapers at both ends, segmented muscles, and a full-length notochord.

malleus   One of the three middle-ear ossicles that transmit sound in mammals. The stapes is the innermost, the incus is the middle element, and the malleus has one surface against the eardrum.

mandibular arch   The bones forming the upper and lower jaw in vertebrates, the lower element of which is the mandible.

neural spine   Dorsal projection of bone on vertebrae that serves for attachment of muscles and ligaments in some vertebrates and for the support of the dorsal "sail" in pelycosaurs.

occipital condyle   Rounded knob or knobs of bone at the rear of the skull by which the skull is attached to the first vertebra.

Ornithischia   Dinosaurs having bird-like pelvic structure.

Osteichthyes   The bony fishes.

pectoral fins   The anterior pair of fins in fishes.

pelycosaurs   Permian reptiles, many of which possessed elongated neural spines supporting a web of skin down the middle of the back.

pharynx   The throat region or canal through which food passes from the mouth into the gut.

plantigrade   The condition of walking with the soles of the feet flat against the ground.

pubis   One of the three bones of the pelvic girdle, which is composed of three bones on each side: the ilium above, and the pubis and ischium below, with the pubis anterior to the ischium.

quadrate   Bone at the rear of the skull in reptiles, amphibians, and crossopterygians for hingement of the jaw.

quadrupedal   The condition of walking on all four legs.

Reptilia   The first tetrapodal vertebrates to employ an amniotic egg in reproduction.

Sarcopterygii   The lobe-finned bony fishes, including the progressive air-breathing crossopterygians.

Saurischia   Dinosaurs having a tri-radiate reptile-like pelvic structure.

spiracle   Remnant of an anterior gill slit seen in modern sharks.

squamosal   One of the temporal bones of the skull, typically located posterior to the postorbital.

stapes   The main bone associated with hearing in living amphibians and reptiles. In mammals the stapes is the innermost of the three ear ossicles.

sternum   A midventral bone common in tetrapods to which the distal ends of the thoracid ribs are attached.

synapsids   Tetrapodal amniotes having a hole in each side of the skull behind the eye.

tetrapod   Four-footed vertebrates.

therapsids   A group of advanced mammal-like reptiles.

tracks   Individual footprints.

trackways   Sets of multiple separate footprints that indicate movement.

tunicate   Chordates of the subphylum Urochordata that are exemplified by the sea squirts.

vertebrae   The short bone or cartilage segments making up the vertebral column in vertebrates.

vertebral column   The series of vertebrae extending from the skull to the end of the tail.

14

Geologic Maps and Geologic Structures

INTRODUCTION

If all the surficial material (e.g., vegetation, soil, water, and rock) were removed from the earth's surface so that the underlying rocks were exposed, and then the exposed surface of each rock formation were painted a different color and photographed from an aircraft, such a photograph would be a **geologic map** (see colorplates following this chapter). In actual practice, a geologic map is constructed by locating in the field the contact lines between different rock formations and plotting these contacts accurately on a base map, preferably a topographic base map. Occasionally, the surficial material (called **cover** or **alluvium**) may be very thick and can be mapped as a separate unit and assigned a color or design to differentiate it on the geologic map.

Outcrop patterns on geologic maps are analogous to outcrop patterns on the upper surfaces of block diagrams, except that geologic maps are not distorted by the perspective that characterizes block diagrams. Most geologic maps are accompanied by structure sections that are analogous to the sides of block diagrams.

The legend or explanation of a geologic map consists of information about the designs, colors, and symbols used on the map. The stratigraphic units, which may be either rock units or time-rock units, are arranged from oldest to youngest in the legend. Igneous and metamorphic units are commonly grouped separately from the sedimentary units. There is no universal usage of a particular color or design for rocks of a certain type or age, although the United States Geological Survey (USGS) uses standard colors and symbols for sedimentary rocks of different periods.

ATTITUDE, STRIKE, AND DIP

Geologists employ the term **attitude** in referring to the way a bed or fault surface is oriented relative to a horizontal plane. For example, a particular stratum may have a horizontal attitude, or it may be tilted. If the bed is tilted, it is important to know how much and in what direction. The two terms *strike* and *dip,* and their map symbols (Fig. 14.1), are essential for conveying this information. **Strike** is the compass direction of a line formed by the intersection of the surface of an inclined feature, such as a bed, with an imaginary horizontal plane (Fig. 14.2). **Dip** refers to the maximum angle, perpendicular to strike, of an inclined stratum measured directly downward from a horizontal plane. On the strike and dip map symbols (Fig. 14.1), strike is indicated by the longer line. The shorter line, at right angles to the strike line, represents the direction toward which the bed is inclined. The amount of that inclination or dip is recorded in degrees next to the symbol.

Study Questions

1. A bed has a strike of N 70° E and is inclined toward the northwest. What is the actual compass direction of dip?

   _____

   _____

   _____

2. A bed has a direction of dip of N 45° E. What is the strike?

   _____

   _____

   _____

Figure 14.1 Symbols used on geologic maps.

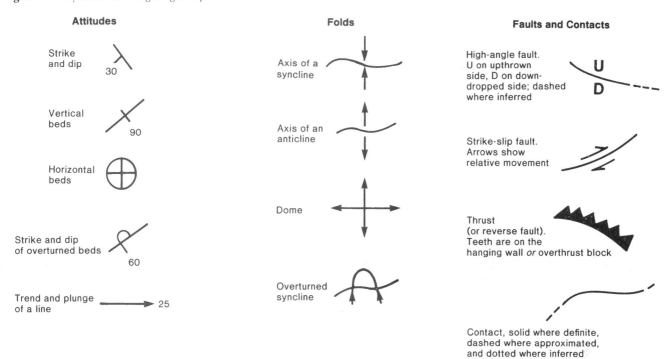

Attitudes

Strike and dip — 30

Vertical beds — 90

Horizontal beds

Strike and dip of overturned beds — 60

Trend and plunge of a line — 25

Folds

Axis of a syncline

Axis of an anticline

Dome

Overturned syncline

Faults and Contacts

High-angle fault. U on upthrown side, D on down-dropped side; dashed where inferred — U D

Strike-slip fault. Arrows show relative movement

Thrust (or reverse fault). Teeth are on the hanging wall *or* overthrust block

Contact, solid where definite, dashed where approximated, and dotted where inferred

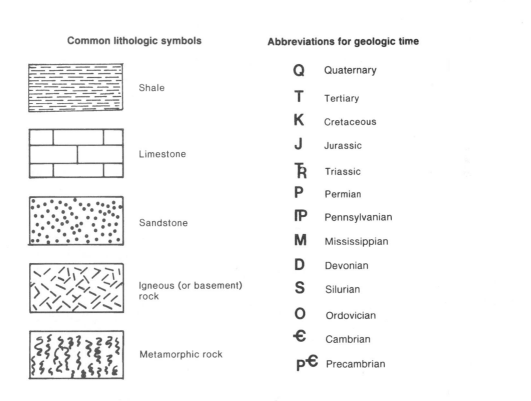

Common lithologic symbols

Shale

Limestone

Sandstone

Igneous (or basement) rock

Metamorphic rock

Abbreviations for geologic time

Q — Quaternary

T — Tertiary

K — Cretaceous

J — Jurassic

TR — Triassic

P — Permian

IP — Pennsylvanian

M — Mississippian

D — Devonian

S — Silurian

O — Ordovician

Є — Cambrian

pЄ — Precambrian

Figure 14.2 Strike and dip of tilted strata. (A) The strike is represented by the line of intersection of a horizontal plane and the inclined (dipping) bed. In this illustration, the horizontal plane is represented by the water surface of a lake. The dip is perpendicular to the strike. The symbol for strike and dip is shown in (B).

From Levin, H. L., *Contemporary Physical Geology,* 2nd ed., Philadelphia: Saunders College Publishing, 1986.

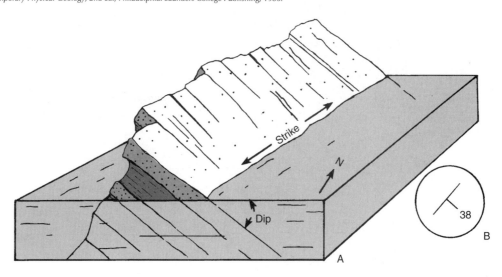

3. Why is it necessary to measure the dip angle perpendicular to the strike of the inclined bed? (Use the surface of your textbook as the inclined plane and the direction of strike as the bound edge. Imagine taking angular measurements not perpendicular to the strike. Does it make sense that the dip perpendicular to strike is the maximum angle?)

_____

_____

_____

_____

 # FOLDS

Folds are common features of deformed rocks. They are usually produced by crustal compression and can range in size from microscopic wrinkles to huge folds involving thicknesses of rock up to several thousands of meters and extending for distances of hundreds of kilometers (Fig. 14.3).

Figure 14.4 shows the features of two simple folds. Note that the **limbs** of an **anticline** dip *away from* the **fold**

axis, and the limbs of a **syncline** dip *toward* the fold axis. Many folds, however, are not as symmetrical as those depicted in Figure 14.4A. Strong deformational forces and variations in the strength of the rock may create complex shapes (Fig. 14.5) and may cause the axis of the fold to be inclined relative to the horizontal. The angle between the fold axis and the horizontal is called the **plunge** of a fold and is used in the same manner as strike and dip to define the attitude of a fold (Fig. 14.4B).

Two other large-scale structural forms can be related to folds. A **dome** is the result of forces acting upward to produce an uplifted portion of the crust where the beds dip *away from* the center in all directions (Fig. 14.6A). A **basin** (Fig. 14.6B) is an area of depression, usually caused by tectonic activity such as downwarping of the crust. Basins can be divided into two general groups. The *structural basin,* formed by tectonic activity, is a depression that may form a catchment area in which sediments accumulate. A *sedimentary basin* may be formed in this manner or by the accumulation of large amounts of sediments accompanied by some form of subsidence of the crust. The beds of sediment accumulated in these basins dip toward the center. Both domes and basins range widely in size, and their overall shape varies from circular to elongate.

Figure 14.3 Tight minor fold of Triassic-Jurassic age, bedded tuffaceous rocks. Ritter Range, central Sierra Nevada Mountains, California.
Courtesy of R. S. Fiske.

Figure 14.4 Features of simple folds. (A) Examples of simple anticline and syncline. Note that the limbs of the anticline dip *away* from the fold axis, and the limbs of the syncline dip *toward* the fold axis. (B) Example of a plunging fold. The angle between the horizontal and the axis of the fold is the plunge. The direction in which the fold axis plunges is called the *trend* of the fold.

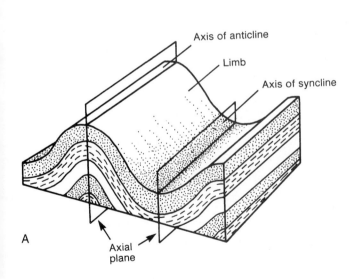

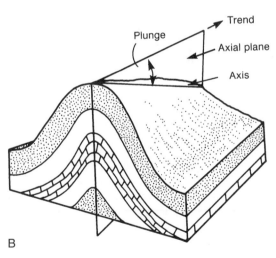

Figure 14.5 Illustrations of symmetrical, asymmetrical, overturned, and recumbent folds.

Fold type and description	
Symmetrical folds Axial plane is vertical	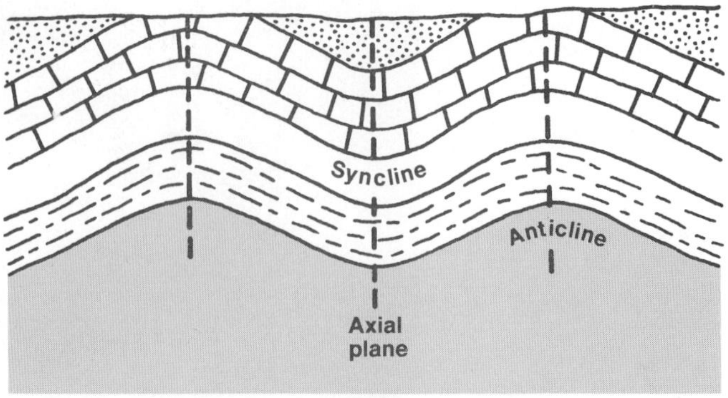
Asymmetrical folds Beds in one limb of the fold dip more steeply than those in the other limb.	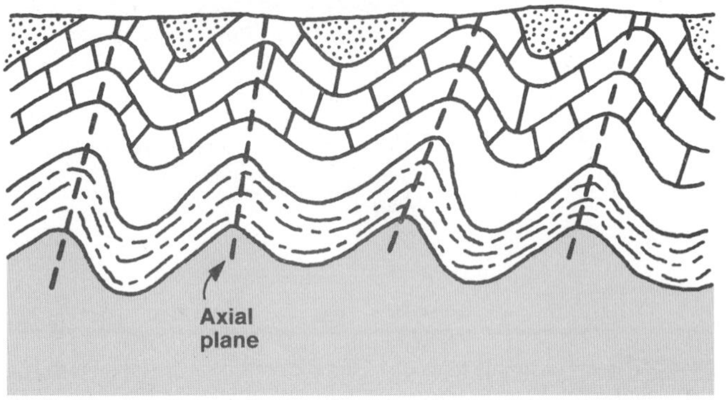
Overturned folds Upper limb of syncline and lower limb of anticline are tilted beyond vertical and dip in the same direction.	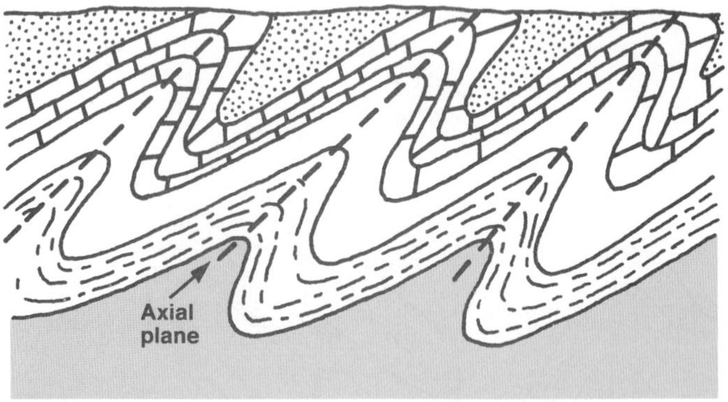
Recumbent folds Beds in lower limb of anticline and upper limb of syncline are *upside down.* Note that the axial plane is nearly horizontal.	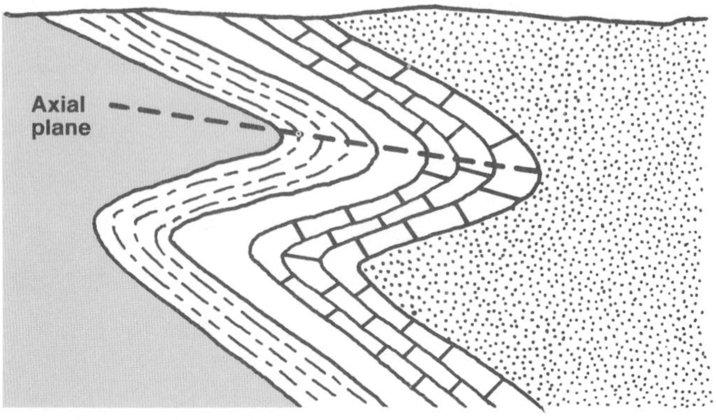

Figure 14.6 Diagrammatic illustration of a dome and basin showing the general dip directions and bedding orientation. Note that the outcrop pattern will be expressed by a vertical view of a plane that cuts through the structure. See Figure 14.30 for an interpretation of the map pattern.

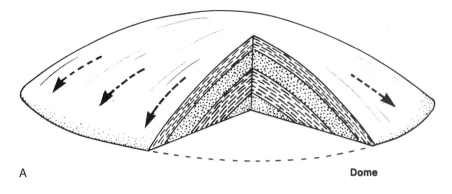

A   **Dome**

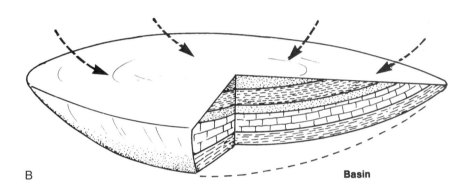

B   **Basin**

Figure 14.7 Map pattern of folds that are not plunging. Syncline, at center, is bordered by two anticlines. (A) Pattern on an idealized flat erosion surface; position of fold before erosion shown in dotted lines. (B) Pattern on surface crossed by stream channel.

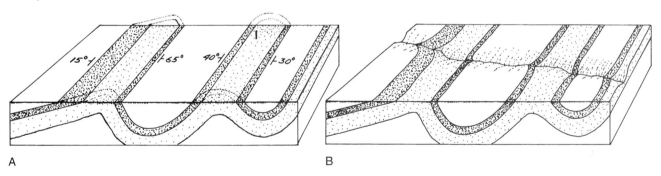

A   B

Figure 14.8 Map pattern of folds, plunging to the left. (A) Pattern on idealized flat erosion surface; position of fold before erosion shown in dotted lines. (B) Pattern on surface crossed by stream channel.

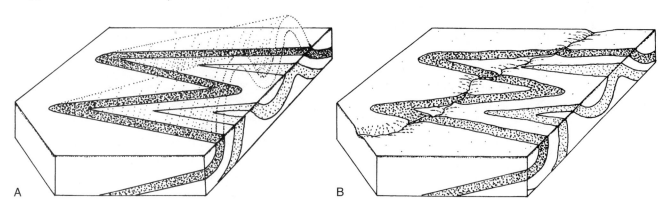

Study Questions

Using Figures 14.7 and 14.8, answer the following questions that relate to map patterns of folds.

1. In Figure 14.7, why is the outcrop band of the dark bed wider in some places than others?

   _____

   _____

   _____

   _____

   _____

   _____

2. In Figure 14.7A, how can the strike-and-dip symbols be used to distinguish the outcrop pattern of an anticline from that of a syncline?

   _____

   _____

   _____

   _____

   _____

3. Note that an outcrop band takes a V-shape where it crosses a stream channel (Figs. 14.7B and 14.8B). What is the relation between the dip direction of the beds and the direction in which the V is pointing? See Figure 14.13 for additional details.

   _____

   _____

   _____

   _____

4. The relative ages of the beds shown in different patterns can be found from their order of superposition on the front of the blocks. Are the beds in the center of a syncline older or younger than the beds on its flanks?

   _____

   _____

   _____

   _____

   _____

   _____

   Are the beds in the center of an anticline older or younger than the beds on its flanks?

   _____

   _____

   _____

   _____

   _____

5. To summarize, list three ways in which the pattern of an anticline on a geologic map can be distinguished from the pattern of a syncline.

   _____

   _____

   _____

   _____

   _____

   _____

   _____

FAULTS

Faults are breaks, or ruptures, in rock, where the rock on one side of the break moves relative to the rock on the other side. Like folds, they come in all sizes, from microfaults in crystals to extensive faults such as the San Andreas Fault in California (Figs. 14.9 and 16.1).

A useful basis for the classification of faults is the nature of the relative movement of the rock masses on opposite sides of the fault plane.

Figure 14.10 shows the various types of faults. A method of remembering which type of fault you are examining is to imagine walking down the inclined slope of the fault. The block under your feet is called the *footwall*, the one over your head is the *hanging* wall. If the hanging wall has moved downward relative to the footwall, it is a **normal fault;** if the opposite, it is a **reverse fault.** When the dip of the fault plane is low (much less than 45°) and the hanging wall has moved forward for many kilometers, the reverse fault is described as a **thrust fault.** A fault in which the

Figure 14.9 Reverse fault in the quartzofeldspathic layer of augen gneiss from Whipple Wash, southeastern California. Lens cap for scale.

Courtesy of A. L. Heatherington.

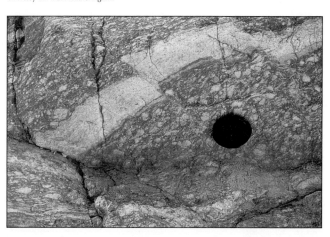

Figure 14.10 Types of faults. On unfaulted block at left, the position of a potential fault is shown by a dashed line; horizontal bed is stippled, and vertical dike is black. (A) Normal fault; lower block shows pattern after surface is leveled by erosion. (B) Reverse, or thrust fault; barbs pointing toward upper or overthrust plate are a map symbol, usually reserved for low-angle thrust faults. (C) Strike-slip fault.

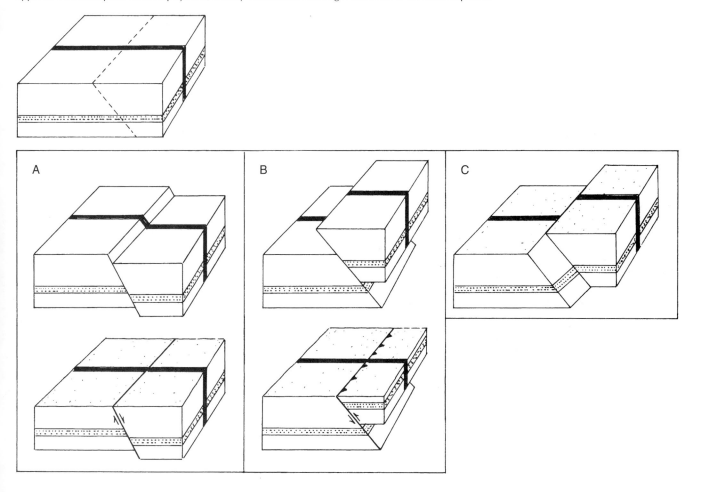

blocks have moved mainly sideways is called a **strike-slip fault.** This type of fault is designated as either a *right-lateral* or a *left-lateral* fault, depending on how the ground opposite you appears to have moved when you stand facing the fault.

Study Questions

1. Can a reverse fault cause older beds to overlie younger beds?

_____

_____

_____

Can a normal fault cause older beds to overlie younger beds?

_____

_____

_____

2. In Figure 14.10C, is the horizontal bed displaced vertically by the strike-slip fault?

_____

_____

_____

_____

3. The surface pattern of the black dike in Figure 14.10A and B shows no surface displacement by either the normal or the reverse faults. If it were dipping, would its surface pattern show displacement (see also Fig. 14.11)?

_____

_____

_____

4. The surface pattern of the vertical black dike is displaced by the strike-slip fault (Fig. 14.10C). If you were standing on the top of the block facing the fault, would the opposite side of the fault be displaced to your right or to your left?

_____

_____

_____

_____

How would you name this fault (Fig. 14.10C)?

_____

_____

_____

Would it matter which side of the fault you were standing on? Explain.

_____

_____

_____

_____

GEOLOGIC MAPS AND GEOLOGIC STRUCTURE SECTIONS

The tops of the blocks in Figures 14.10 and 14.11 represent geologic maps. As you can see in Figure 14.11A, faulting results in the displacement of the exposed dipping bed (black) on either side of the fault. From this outcrop appearance the geologist can interpret the type of fault as well as the motion along the fault. Figure 14.11B shows how faulting can also result in a repeated pattern of the same rock type. Repeated beds are often a result of faulting or asymmetrical to overturned (folded) beds (see Figure 14.5).

Figure 14.11 Effects of faulting on a dipping bed. (A) Fault trending in dip direction of bed. (B) Fault trending in strike direction of bed.

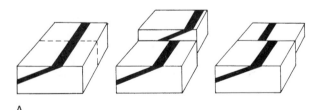

A

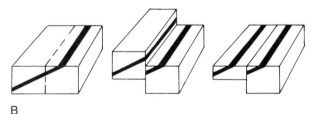

B

The outcrop pattern on a geologic map can also give information about the dip (and dip direction) of the strata. The width of an outcrop on a geologic map is related to the angle of the strata intersection with the slope of the surface. Figure 14.12 shows how the width of the strata changes as the slope changes on which it is exposed. Further information about the direction of dip can be found by examining stream valleys. When a stream erodes across a landscape, the rate of erosion is often controlled by the underlying geology. Geologists can use the outcrop appearance of strata cut by a stream to interpret the dip and direction of dip. This technique is called the "Rule of Vs" (Figure 14.13). Simply put, if the strata is horizontal (no dip), the outcrop pattern in a stream valley will be parallel to the contour lines. However, if the strata is dipping, the pattern will form a "V" shape where the stream cuts through the rock. If the V points downstream, the beds are dipping downstream. If the V points upstream, the beds are dipping upstream. When you look at the V caused by this form of erosion, the degree of narrowness of the V shape and the length it extends (either up- or downstream) indicates the apparent dip of the rock as well as the dip direction. In Figures 14.14 and 14.15, sketch the outcrop pattern and follow the ideas of outcrop width and the "Rule of Vs."

On the block diagrams, the front and sides represent **geologic structure sections.** Structure sections are our interpretation of the subsurface geology and are often distorted by the perspective view of a block diagram. In Figure 14.16, construct a general geologic structure section for each block diagram. A more complete structure section can be made

---

Figure 14.12 Width of outcrop of a bed in relation to its angle of intersection with the slopes of the earth's surface. The width of outcrop equals the width of bed only where the angle of intersection is 90°. To visualize the outcrop widths as they would appear on a geologic map, the drawings must, in effect, be viewed from above. Thus, beds cropping out on a nearly vertical cliff have very narrow outcrop bands and are difficult to represent on a map.

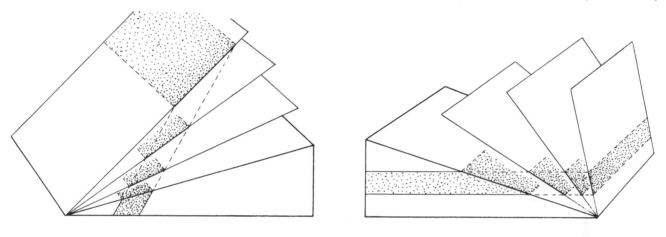

Figure 14.13 "Rule of Vs." (A) Outcrop pattern of horizontal beds intersected by valleys. Outcrop bands are parallel to contour lines because both represent the intersection of horizontal planes with the ground surface. (B) Beds are dipping downstream at a steeper angle than the valley gradient, and the outcrop bands form Vs pointing downstream. (C) Two of the beds are dipping upstream at a steeper angle than the valley gradient, and the outcrop bands form Vs pointing upstream. Because the dike at rear is vertical, its outcrop band is not influenced by the valleys. Observations such as these permit one to infer the dip of a bed by looking at the surface of the ground or a geologic map.

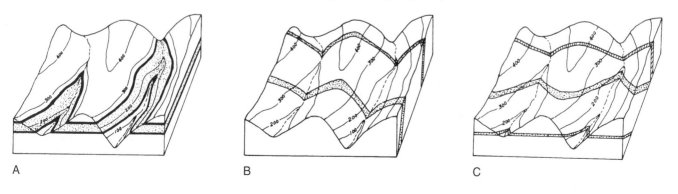

A   B   C

Figure 14.14 On each block, sketch the surface outcrop pattern of the black bed.*

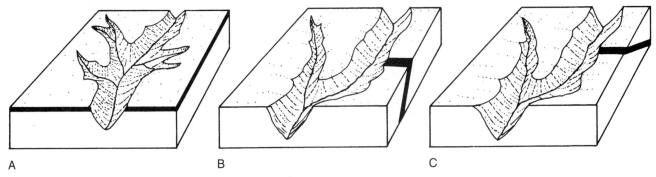

A   B   C

*Use "Rule of Vs" from Figure 14.13 to help place the surface outcrop pattern.

Figure 14.15 On the front and sides of each block, sketch an interpretation of the subsurface structure. The oldest bed is numbered 1.*

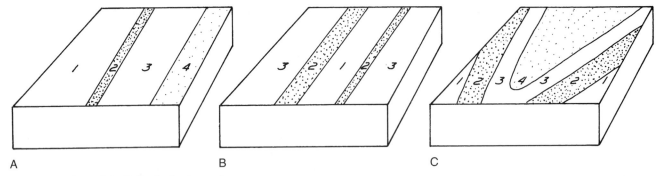

A   B   C

*Use Figures 14.7 and 14.8 to help with this subsurface interpretation.

Figure 14.16 (A) On the front and sides of the block, sketch an interpretation of the subsurface structure. (B) Propose two hypotheses to account for the displacement of the bed shown in dotted pattern. (C) On the front and sides of the block, sketch an interpretation of the subsurface structure.

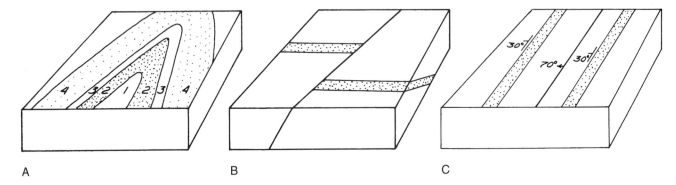

A   B   C

from flat geologic maps by placing the edge of a sheet along the desired line of the structure section, which should be nearly perpendicular to the strike of the beds (Figure 14.17). The subsurface continuation of a bed can then be sketched accurately if strike and dip symbols are on the map or approximately if the direction of dip must be inferred from outcrop pattern.

Figure 14.18 illustrates the procedure used to construct these sections. The edge of the paper is placed along the line of the structure section, and the vertical scale is drawn on the paper. As there is to be no vertical exaggeration, the vertical scale is the same as the horizontal scale on the map, and the dip angles of the beds can be laid off with a protractor. By projecting the dip angles in the beds, the subsurface structure is sketched. Unless data from subsurface drilling is available, the configuration at depth of features such as folds involves some guesswork, and a fault may flatten or steepen with depth. According to the interpretation in Figure 14.18, the base of the sandstone bed lies at an actual depth of about 500 feet at the trough of the fold. Using the information given in Figure 14.17, construct the geologic structure section for the northern Black Hills.

Figure 14.17 Geologic map of the northern part of the Black Hills. In the space provided, sketch a structure section of the Black Hills. At both the east and west ends of the section, the base of the Cretaceous (K) is about 610 m (2000 ft) below the altitude of the outcrop. The beds do not dip uniformly away from the area of surface outcrop shown on the map. On the east, they dip at first more steeply and then flatten. On the west, they are flat near the outcrop, then dip more steeply, then flatten again.

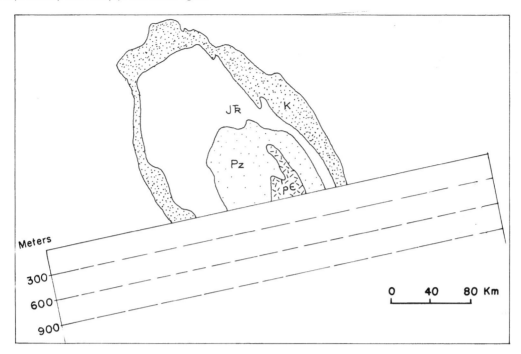

Figure 14.18 Procedure for sketching a structure section for a geologic map, with no exaggeration of the vertical scale (1 ft = 0.3048 m).

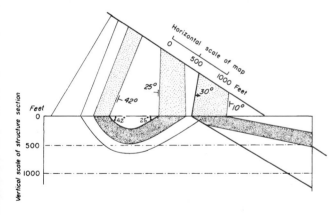

Figure 14.19 Effect of a 2X vertical exaggeration on the structure section of Figure 14.18 (1 ft = 0.3048 m).

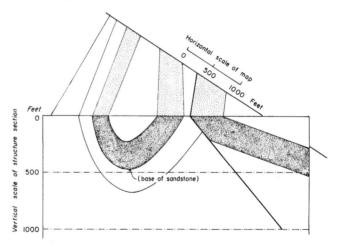

The above procedure is used with geologic maps of large scale, on the order of 1:12,000 (1,000 feet to the inch). But suppose the scale of the map is small, say about 1:1,200,000 or 10,000 feet to the inch. Then the thickness of the structure section to a depth of 1,000 feet is only 0.01 inch, obviously too thin to show the structure. The vertical scale must be stretched or exaggerated. Figure 14.19 illustrates the effect of stretching or exaggerating the vertical scale by a factor of two. As you see, the apparent dip of the beds and of the fault has been greatly increased and the structure has a distorted appearance. When you look at a geologic structure section, always notice whether or not the vertical scale has been exaggerated.

GEOLOGIC MAP OF THE ATHENS QUADRANGLE, TENNESSEE

The geologic map of Colorplate 1 and the accompanying explanation and structure section of Colorplates 2 and 3 are reproduced from a quadrangle map published by the

U.S. Geological Survey. These colorplates are located at the end of Chapter 14. Colorplate 1 is about one-fourth of the original map. Geologic maps of this sort represent the most basic data about the earth, and are therefore well worth learning about. For ease in answering some of the following questions, cut out the structural cross section and align it along the line A-A′ (found on Colorplate 3).

The stratigraphic units represented by boxes in the explanation of Colorplate 2 are **formations.** The first, or geographic, part of the formation name comes from a locality where it is well exposed, and the second part is either a rock type or the word "Formation." Since this map was published, it has become customary always to capitalize the second word. Notice also the term "Knox Group," which includes several formations.

Study Questions

1. Several successive formations consist mainly of dolomite. How did the field geologist find the contact between the Longview Dolomite and the Kingsport Formation?

_____

_____

_____

2. What do the members of the Knox Group have in common? If the group is considered as a unit, how does it differ from the rocks below and above?

_____

_____

_____

_____

3. The symbol for each formation begins with a capital letter. What is the source of this letter?

_____

_____

_____

_____

_____

4. Which of the formations has the most prominent topographic expression as indicated by the contour lines on the map?

_____

_____

_____

_____

_____

5. In reading the description of this formation, can you discern any reason for its prominence?

_____

_____

_____

_____

_____

6. Which of the formations stands at the lowest altitudes on the map? From its description, can you discern any reason for this?

_____

_____

_____

_____

7. How are contact lines distinguished from fault lines on the map?

_____

_____

_____

_____

8. If the structure section was not available, in what two different ways could you identify the large fold in the center of the map as an anticline?

_____

_____

_____

_____

9. In what direction is this anticline plunging? (See Fig. 14.4 for review.)

_____

_____

_____

_____

10. Notice the syncline in the northeast part of the map. In what direction is it plunging?

_____

_____

_____

_____

11. From the structure section, determine the approximate depth that a well at Cunningham Cemetery (on map, near line of section) would encounter the base of the Copper Ridge dolomite?

_____

_____

_____

_____

12. Is the Chestuee a normal or a thrust fault? (See Fig. 14.10 for review.)

_____

_____

_____

_____

13. What is the approximate total thickness of all the formations?

_____

_____

_____

_____

_____

14. To what association do these beds belong (refer to Fig. 5.3)?

_____

_____

_____

_____

_____

15. What are the inferred tectonic setting (refer to Fig. 5.3) and sedimentary environment (refer to Table 4.1)?

_____

_____

_____

_____

_____

MAPPING HORIZONTAL BEDS

Before mapping a previously unstudied region, the geologist will make a reconnaissance of the region, examine all the rock layers exposed, and decide on the rock units (formations) that will be plotted on the geologic map. A sketch or reconnaissance map can then be made of the region, and if the economic or scientific importance of the region warrants it, a more detailed map may be made later. In the exercises that follow, the formations in each region are briefly described and tentatively named at numbered localities, and notes taken at other localities provide sufficient information for the preparation of a geologic map, much in the same manner as maps are prepared in actual practice.

Goose Creek Area Map (Fig. 14.20)

The Goose Creek area is a good exercise to start for the novice geologist. To answer the questions for this mapping area, first imagine that you are walking from station to station in this field area (Fig. 14.20). Each station is a location where a characteristic rock formation is exposed. At each station (described below), you examine the rock and determine its identity, presence or absence of fossils (Fig. 14.21), and any structural information (strike and dip, fold orientation, faults, etc.) that is present. As you walk from station to station you may cross over one rock formation to another. This boundary is called a *contact* and is marked on the map (see Station #2). To answer the study questions that follow the station descriptions, walk through all the stations and determine the contacts between different formations as well as the identity of the formations in the Goose Creek area.

Sta. #1 Gray calcareous shale, fossiliferous, contains a few thin layers of limestone. Crops out on river bank at elevation 560 ft., base not exposed.

Figure 14.20 Geologic map of the Goose Creek Area.

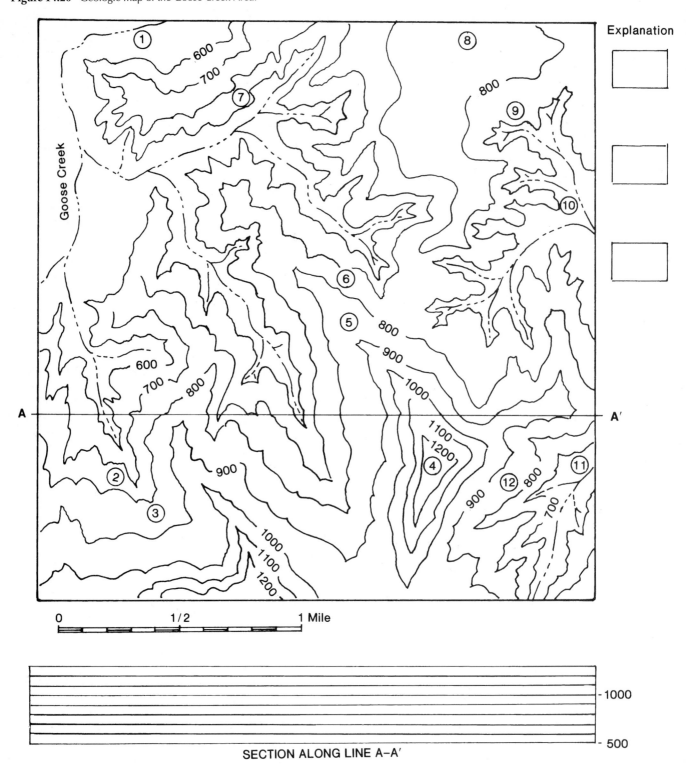

0   1/2   1 Mile

SECTION ALONG LINE A–A'

- 1000

- 500

Traveling from Sta. #1 to Sta. #2 Shale is clayey rather than silty, limestone is very fine grained, both form alternating thin beds that crop out continuously along bank of main stream and of tributaries. Attitude, nearly horizontal.

Sta. #2 Contact between gray shale and white, thick-bedded clastic limestone appears at elevation of about 700 ft. Minimum thickness of shale, 140 ft. Both formations nearly horizontal. Shale and thin-bedded limestone named Dry Creek Fm.

Figure 14.21 Fossils of the Goose Creek Area.

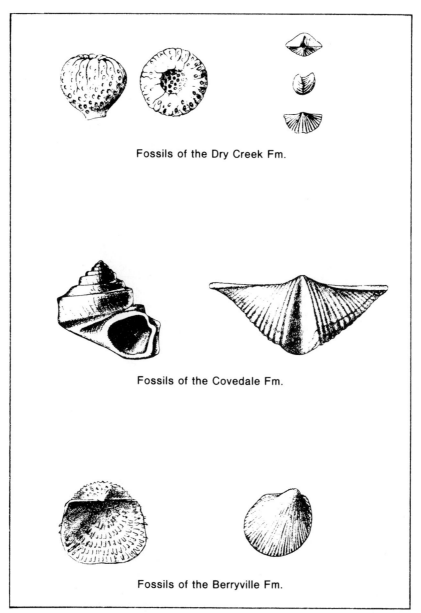

Fossils of the Dry Creek Fm.

Fossils of the Covedale Fm.

Fossils of the Berryville Fm.

Traveling from Sta. #2 to Sta. #3 Massive limestone crops out from place to place upslope, maximum thickness about 150 ft. A few whole fossils collected just downslope from Sta. #3. Unit named Covedale Fm.

Sta. #3 Contact between Covedale Fm. and gray silty sandstone at an elevation of about 850 ft. Both formations are horizontal, but Covedale shows weathering and dissolution at the contact, which appears to be unconformable.

Traveling from Sta. #3 to Sta. #4 Silty sandstone crops out along slope to top of hill. Locally interbedded with thin layers of carbonaceous shale. At about 20 ft. above contact, two layers of bituminous coal are interbedded with the shale and sandstone, the lower about 8 inches in thickness, the upper about 20 inches, separated by 14 inches of shale. Rock unit named Berryville Fm.

Sta. #4 Berryville Fm. crops out at top of ridge.

Sta. #5 Contact at Covedale Fm. and Berryville Fm., lower part of Berryville contains fragments from Covedale. Both formations are nearly horizontal. Elevation of contact 840 ft.

Sta. #6 Contact of Dry Creek Fm. and Covedale Fm., elevation 700 ft.

Sta. #7 Contact of Dry Creek Fm. and Covedale Fm., elevation 705 ft.

Sta. #8   Covedale Fm. crops out at top of hill.

Sta. #9   Contact of Dry Creek Fm. and Covedale Fm., elevation 695 ft.

Sta. #10   Dry Creek Fm. crops out at bottom of creek bed.

Sta. #11   Contact of Dry Creek Fm. and Covedale Fm., elevation 694 ft.

Sta. #12   Contact of Berryville Fm. and Covedale Fm., elevation 845 ft.

Study Questions

1. Use a blue pencil to delineate streams, and choose a color for each formation described above and located on Figure 14.20.

2. At each numbered station, draw in the contact line for a short distance and color the formations for a short distance on either side of the contact line.

3. When all the stations have been completed, sketch in contact lines between stations. In more careful geologic mapping, the contacts are observed and plotted wherever they are either exposed or inferred to be present.

4. Color the formations according to the scheme chosen.

5. Identify the fossils collected from each formation, and use them to determine the age of the formation (see Fig. 14.21). Now complete the explanation legend by coloring each of the boxes, labeling it properly with the formation name and letter abbreviation, and indicating the geologic age of the formations.

6. What is the geologic age of the following formations that are exposed in the Goose Creek Area?

    a.   Dry Creek Fm. _____

    b.   Covedale Fm. _____

    c.   Berryville Fm. _____

7. Draw a topographic profile and construct a geologic structure section (use lined box at bottom of Figure 14.20).

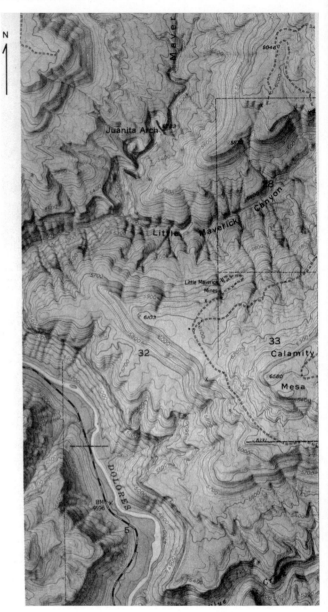

Figure 14.22   Part of the Juanita Arch Quadrangle, U.S. Geological Survey 7.5-minute series, 1950. Same scale as Figures 14.24 and 14.25.
Courtesy of U.S. Geological Survey.

Figure 14.23 *Left,* view upstream along Little Maverick Canyon, Dolores River in foreground. *Right,* Entrada Fm., overlying Kayenta Fm. (below tree) at Colorado National Monument, about 35 miles northeast of Juanita Arch.

Juanita Arch, Colorado

The Juanita Arch area is along an easily accessible canyon (of the Dolores River) in the northeastern part of the Colorado Plateau, south of Grand Junction, Colorado. Because of the arid climate, the formations are well exposed, and many of them are brightly colored. The purpose of this study is to illustrate the pattern of horizontal beds on a geologic map and on a vertical aerial photograph. The formations that crop out in the area of Figure 14.22 are described as follows:

Alluvium (Qal)   Deposits along streams; Quaternary.

Burro Canyon Fm. (Kbc)   White, gray, and red sandstone and conglomerate; green and purplish shale; Cretaceous.

Morrison Fm. (Jm)   Variegated shale and mudstone; red sandstone and conglomerate; Jurassic.

Summerville Fm. (Js)   Thin-bedded red, green, and brown sandy shale; Jurassic.

Entrada Sandstone. (Jec)   Pale brown and white crossbedded mature sandstone; Jurassic.

Navajo Sandstone. (Jn)   Gray, crossbedded, mature sandstone; Jurassic (see Fig. 14.23).

Kayenta Fm. (TrJk)   Unevenly bedded, grayish-orange sandstone, siltstone, and shale; Triassic and Jurassic.

Wingate Sandstone. (Trw)   Reddish brown, crossbedded, mature sandstone; Triassic.

Chinle Fm. (Trc)   Red siltstone, sandstone, shale; Triassic.

Moenkopi Fm. (Trm)   Dark brown, ripple-marked sandstone, shale, siltstone, some conglomerate; Triassic.

Study Questions

1. By referring to Figure 14.22, label the following on the airphoto (Fig. 14.24): Dolores River, Little Maverick Canyon, Calamity Mesa.

2. On the geologic map (Fig. 14.25), Qal is shown in light dotted pattern, Trw in black, and Js in stippled pattern. Identify the areas of outcrop of the other formations by writing their symbols on the outcrop bands.

3. Which formation forms the prominent cliff along the Dolores River?

   _____

   _____

   _____

   _____

4. The Entrada Sandstone appears as a white band on the airphoto (Fig. 14.24). On the photo, draw a line following its contact with the Summerville Fm.

5. Where the upper contact of the Entrada Sandstone crops out on the topographic profile of Figure 14.25, the altitude of the contact is indicated by a tic mark. The upper and lower contacts of the Wingate Sandstone are similarly indicated. *On the profile,* draw

Figure 14.24 Vertical airphoto of part of Juanita Arch Quadrangle, north to the right. Lower boundary corresponds with the right edge of the inset box on the right side of Figure 14.22 and lower boundary of Figure 14.25.

Courtesy of U.S. Geological Survey.

Figure 14.25 Geologic map and topographic profile of part of Juanita Arch Quadrangle. Vertical exaggeration of profile, 2X. Lower boundary of map corresponds with lower boundary of Figure 14.24.

Map traced from Geologic Map of the Juanita Arch Quadrangle, Colorado, by E. M. Shoemaker, 1955, USGS Map GQ 81.

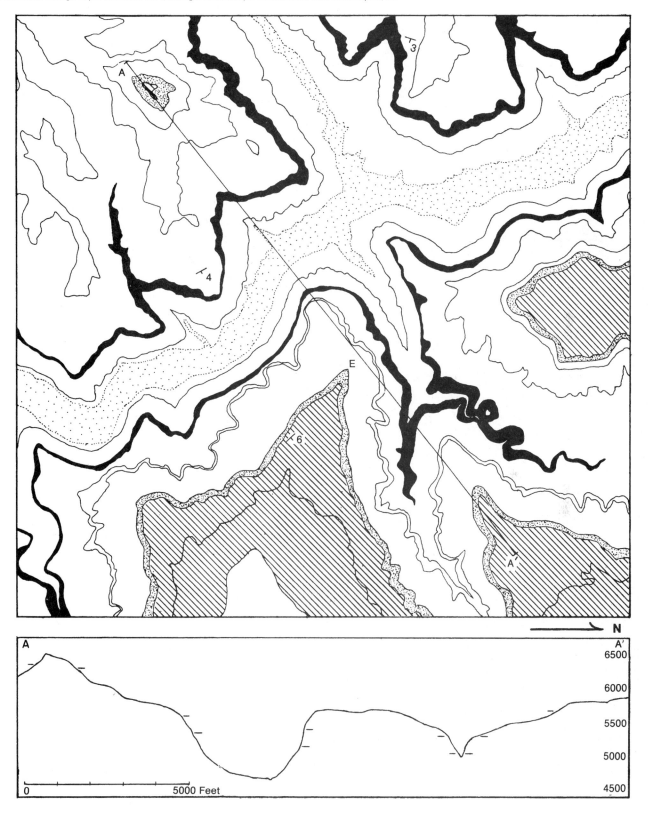

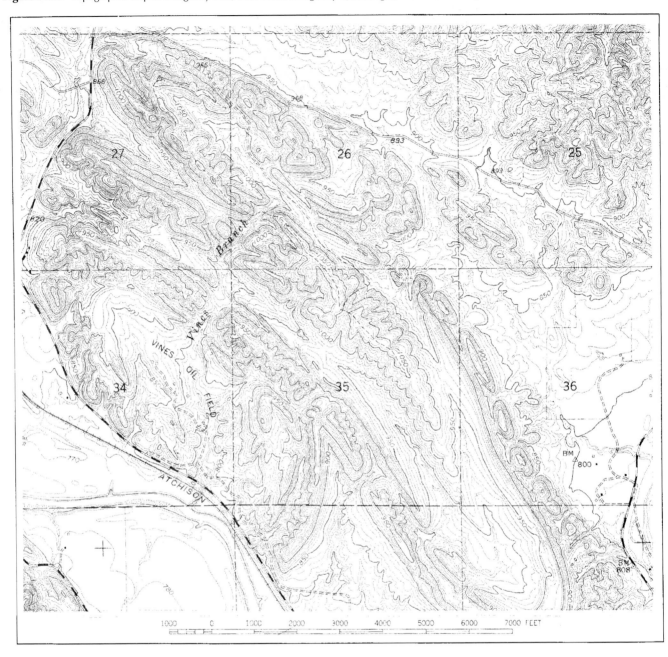

lines between the tic marks to show the probable subsurface location of the Wingate and the top of the Entrada. What is the thickness of rock between the top of the Entrada and the top of the Wingate at the point marked "A" on the geologic map?

_____

_____

_____

_____

_____

At the point marked "A'"?

_____

_____

_____

Judging from the geologic map, which formation has decreased in thickness between these points?

_____

_____

_____

_____

Figure 14.27 Oblique view of Dougherty area, looking southeast.

From F. A. Melton, Oklahoma Geological Survey Guide Book X.

6. In Figure 14.23 (left), label the Chinle Fm. and the Wingate and the Entrada Sandstones.

7. In Figure 14.23 (right), what formation is missing here?

_____

_____

_____

_____

What is the probable sedimentary environment of the Entrada Sandstone?

_____

_____

_____

_____

_____

Of the Kayenta Fm.?

_____

_____

_____

_____

_____

MAPPING FOLDED BEDS

Arbuckle Mountains, Oklahoma

The Arbuckle Mountains of southern Oklahoma are the surface expression of a belt of folded and faulted beds, similar to the Appalachian Ridge and Valley but better exposed because of a somewhat more arid climate. A small part of the Arbuckle Mountains, called the Dougherty area because it is near the small town of that name, is represented on the maps and photographs of this study.

Study Questions

1. By comparing the topographic map (Fig. 14.26) and the vertical airphoto (Fig. 14.28), label on the photo the Wichita River; the Atchison, Topeka, and Santa Fe railroad track; and the Vines Oil Field.

2. What is the altitude of the hill just west of center, Section 26, on the topographic map? Find this hill on the vertical airphoto and indicate its position by writing its altitude across it on the photo.

3. By writing the appropriate symbols (refer to Fig. 14.1), label on the geologic map the formations that are not labeled.

4. Draw a geologic structure section along line A-A′ in Figure 14.29. The topographic profile on which the structure section is to be drawn can be brought into

Figure 14.28 Vertical airphoto of Dougherty area. Same horizontal scale as Figure 14.30. The dark line down the center results from joining two photos that are somewhat differently exposed.

U.S. Dept. of Agriculture, Stabilization and Conservation Service, photos CKG-4DD-205 and 31, Murray College, Oklahoma.

The formations that crop out in the Dougherty area are, from youngest to oldest, as follows:

Alluvium (Qal):	Gravel, sand, silt, and clay; along rivers; Quaternary.	Sylvan Fm. (Osy):	Gray and green clay shale; Ordovician.
Vanoss Fm. (Pvs):	Limestone conglomerate, shale, arkosic sandstone; Late Pennsylvanian.	Viola Fm. (Ov):	Crystalline limestone and fine grained limestone; Ordovician.
Caney Fm. (Mc):	Dark gray fissile shale; Mississippian.	Simpson Fm. (Os):	Clastic limestone and quartz sandstone; Ordovician.
Sycamore Fm. (Ms):	Thin bedded, bluish gray, fine grained limestone; Mississippian.	Arbuckle Fm. (O Єa):	Limestone, dolomite, and quartz sandstone; Ordovician and Cambrian.
Hutton Fm. (D Sh):	Crystalline limestone, marly limestone, shale; Devonian and Silurian.		

Figure 14.29 Geologic map of Dougherty area and topography profile along the line A-A'. Vertical exaggeration 2X, same horizontal scale as Figure 14.30.

Map modified from Ham and McKinley (1954), Geologic map and sections of the Arbuckle Mts., Oklahoma Geological Survey. See also Oklahoma Geological Survey Guidebook III, *Geology of the Arbuckle Mt. Region.*

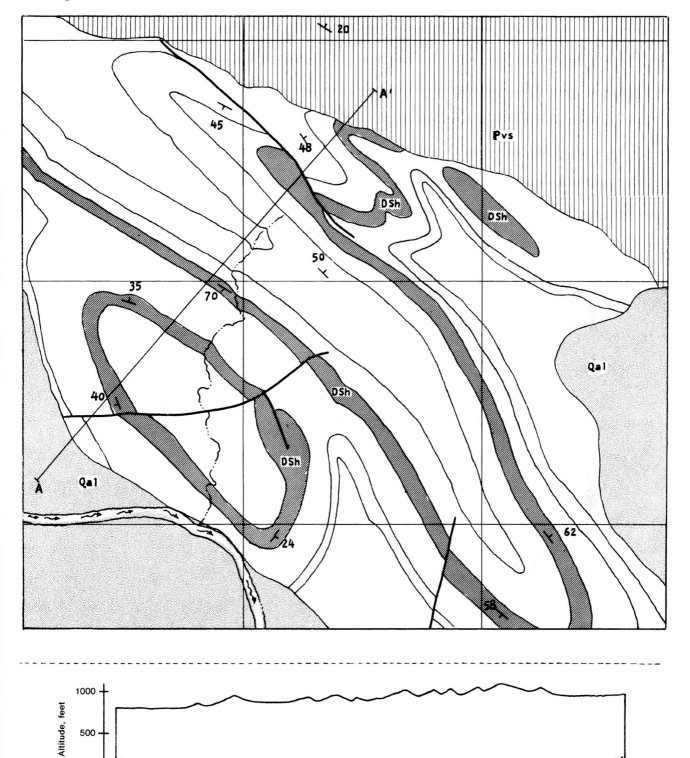

correspondence with line A-A′ on the map by detaching the page at the side, folding along the dashed line, and rotating the profile upward. Where the geologic section crosses a fault, indicate by arrows the upthrown and downthrown sides. Assume that the faults are vertical.

5. Of the formations listed, which have the most prominent (highest) expression in the relief?

_____

_____

_____

6. What is the direction of plunge of the large fold that crosses the area diagonally (see Fig. 14.27)?

_____

_____

_____

7. What is the approximate thickness of the Hunton Fm. where its dip is 70° on the structure section?

_____

_____

    What is the approximate thickness of the Viola Fm., assuming its dip on the north side of the large fold to be 50°? (The thickness can be approximated graphically from the cross section, or more accurately computed by trigonometry from the width of the outcrop; for your information, the sin 70° is 0.94; the sin 50° is 0.76.)

_____

_____

8. What evidence on the geologic map indicates that the Vanoss Fm. lies unconformably upon older formations?

_____

_____

_____

9. Draw on the vertical airphoto the boundaries of the Hunton Fm. and the Viola Fm. Draw in the faults shown on the geologic map.

STRUCTURE CONTOUR MAPS

In addition to geologic maps and structural cross sections, a geologist's understanding of the configuration of rocks in the earth's crust can often be improved by constructing a map that shows the elevation of the surface of a key bed. Such maps are called **structure contour maps.** Structure contour maps depict the structural configurations of rocks at any depth reached by drilling. Unlike structure sections (such as those depicted on the fronts of block diagrams), structure contour maps give a three-dimensional representation of a structure. With the use of structure contour maps, the depth at which an oil well will penetrate a given horizon, or the point at which a mine tunnel will intersect a vein, can be calculated accurately.

In making a structure contour map, a *key horizon* must first be selected for contouring. The key horizon is commonly the upper surface of a persistent and easily recognizable sedimentary bed or formation. The elevation of the key horizon in reference to sea level (the usual datum) is found for each well by algebraically subtracting the depth of the key horizon from the ground-surface elevation at the top of the well. Thus, if the key horizon is at a depth of 200 m in a well that has a ground elevation of 400 m, the elevation of the key horizon is 400 m minus 200 m, or 200 m above sea level. If the key horizon is found at a depth of 500 m in a well with a ground elevation of 300 m, the elevation of the key horizon is –200 m, or 200 m below sea level. Note that when the key horizon is below the sea-level datum (Fig. 14.30A), the contours are negative. Conversely, when the key horizon is above sea level (Fig. 14.30B), the contours are positive. A map having positive contours has the larger contour numbers on the higher parts of the structure, whereas a map having negative contours has the larger numbers on the lower parts of the structure.

Study Questions

Figure 14.31 represents an early stage in the making of a structure contour map. Beside each well location (the small circles), a number is plotted that represents the elevation above sea level at which the key horizon was encountered during drilling.

1. Complete the map by neatly drawing in the structure contours. Like topographic contours, structure contours are lines connecting points of equal elevation. Use a contour interval of 10 m (number lines 160, 150, 140, etc.). You will have to estimate the position of contours between the elevations plotted on the map.

2. What is the major structural feature of the area?

_____

_____

3. If surface elevation is 1000 m, at what depth would a proposed well at locality A reach the key horizon?

_____

_____

    A proposed well at locality B?

_____

_____

4. Which wells on the map would be most favorably located for oil production?

_____

_____

5. What is the strike and direction of dip of wells in the northwestern quadrant of the map?

_____

_____

Figure 14.30 (A) Structure contour map drawn on a key horizon of a basin-shaped structure. Structure is below sea-level datum; therefore, contours are negative, and larger contours are in lower parts of the structure. (B) Structure contour map drawn on a key horizon of a dome-shaped structure. Structure is above sea-level datum; therefore, contours are positive, and larger contours are in higher parts of the structure.

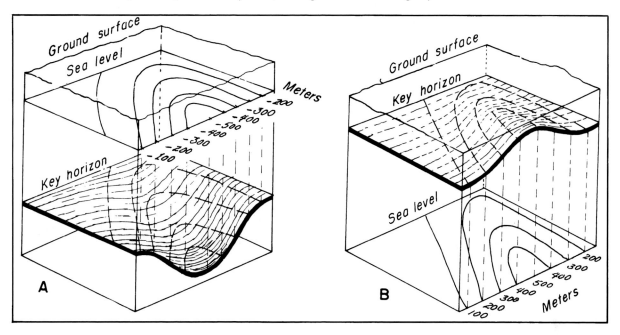

Figure 14.31 Structural contour map. Note that this is not a topographic surface but a key surface. (See Fig. 14.30B or Fig. 17.1 for reference.)

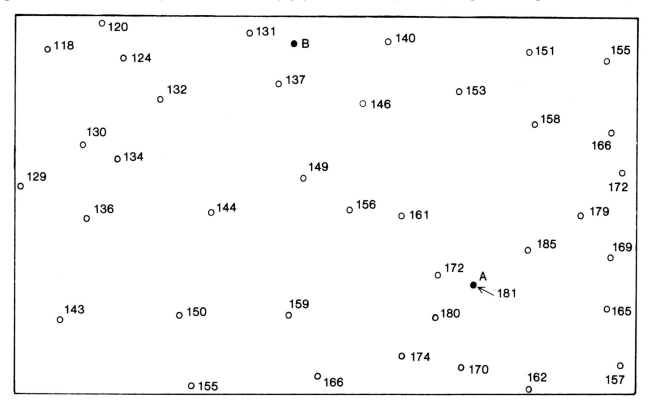

TERMS

alluvium   Recent sediments deposited on existing rocks. These unlithified materials may be considered as a mappable formation if abundant or pervasive. Also called **cover.**

anticline   A geologic structure in which strata are bent into an upfold where the limbs dip toward the fold's axis and where older rocks occur near the center of curvature (fold axis).

attitude   The position of a structural surface (fold, fault, etc.) relative to the horizontal, expressed quantitatively by the strike, direction of dip and angle of dip.

basin   A low area in the earth's crust, of tectonic origin, in which sediments have accumulated. The beds of a basin dip in toward the center.

cover   Another term used to describe a package of recent sediments deposited on existing rocks. See **alluvium.**

dip   The maximum angle formed by the intersection of an inclined bed with a horizontal plane, measured perpendicular to strike.

dome   An uplift or anticlinal structure, either circular or elliptical in outline, in which the rocks dip away from the center in all directions.

fault   A fracture or a zone of fractures along which there has been displacement of the sides relative to one another (parallel to the fracture).

fold axis   A line connecting points along the center of a fold, from which both limbs bend.

formation   A mappable, lithologically distinct body of rock having recognizable contacts with adjacent rock units. Formations may be combined into *groups* or subdivided into *members.* See Chapter 8 for further details.

geologic map   A map on which the underlying geology (such as the rock type, strike and dip, foliation, and age relationships) is depicted.

geologic structure section   A vertical slice along a transect through a geologic map. This cross section allows the geologist to examine the distribution of rock at depth as well as any structural features (such as folds, faults, etc.).

limb (of a fold)   The areas of the fold on either side of the fold axis. The dip of the strata of the limb relative to the fold axis is used to define an **anticline** or a **syncline.**

normal fault   A fault in which the rocks lying above the fault plane (called the *hanging* wall) have been moved down relative to the rocks beneath the fault plane (called the *footwall*).

plunge (of a fold)   The angle of inclination of the axis of a fold measured from a horizontal surface.

reverse fault   A fault formed by compression in which the hanging wall appears to have moved up relative to the footwall.

strike   A direction indicated by the intersection of an inclined bed and a horizontal plane.

strike-slip fault   A fault in which the movement (called the slip) is parallel to the strike of the fault.

structure contour map   A map that depicts the configuration of the surface of a key bed or formation by lines (structure-contour lines) of equal elevation.

syncline   A downfold in which the rock strata dips toward the fold's axis and where younger rocks occur near the center of curvature (fold axis).

thrust fault   A low-angle (much less than 45°) reverse fault where the displacement of the hanging wall is measured in kilometers.

Geologic Maps and Additional Exercises

The following section of color illustration includes geologic maps, geologic structure sections, and the accompanying explanation descriptions reproduced from geologic quadrangle maps published by the U.S. Geologic Survey. These maps and the additional exercises are included so as to follow students in different areas of the country to gain more experience with the topics discussed in Chapter 14 (Geologic Maps and Geologic Structures) and the regional geology discussions of Chapters 15 to 17.

Appalachians

      1–3 Athens Quad
      4–6 Ewing Quad
      7–9 Pine Grove Quad

Cordillera

      10–12 Golconda Quad
      13–15 Waucoba Spring Quad
      16–18 White Mountain Peak Quad

Following Colorplate 18 are a series of exercises keyed to Colorplates 4 to 18. Have the students remove the appropriate exercise and colorplates for use.

    The Geologic Map of Colorplate 1 (and its accompanying explanation and geologic structure section) will be used in the exercise on the Athens Quadrangle (Tennessee) in Chapter 14. Tear carefully the three colorplates from the laboratory manual so as to use them for this exercise.

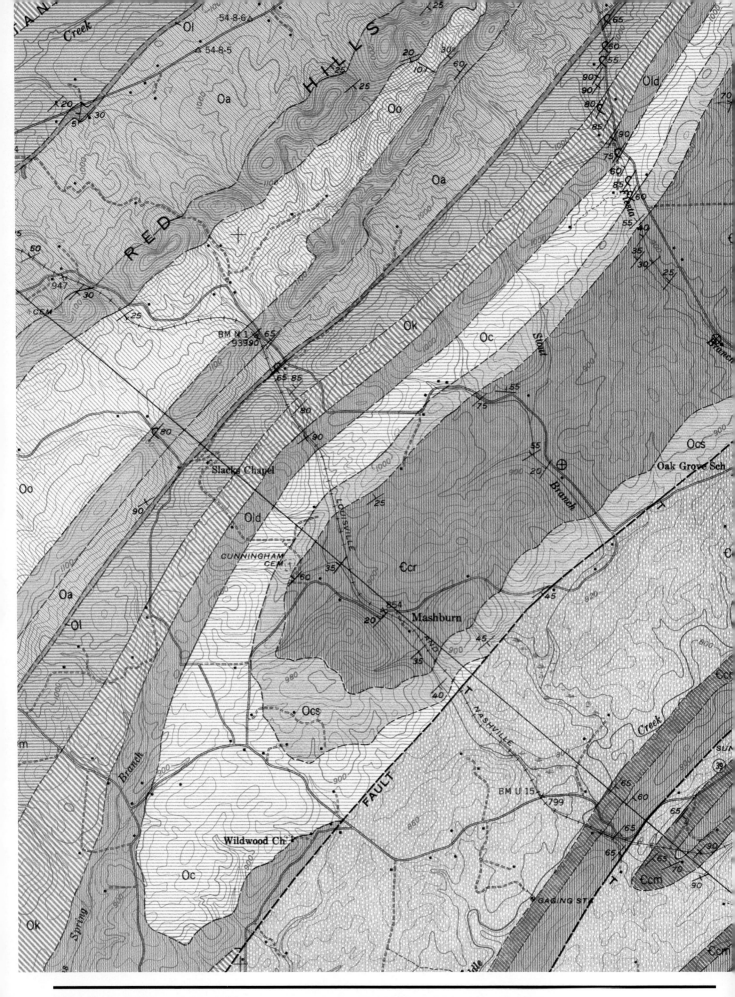

Colorplate 1 Geologic map of part of the Athens Quadrangle, Tennessee. Scale 1:24,000, contour interval 20 ft.

EXPLANATION

Ottosee shale

Yellow-weathering, blue calcareous shale and shaly limestone, with thin layers and lenses of blue and red crystalline limestone; some quartzose in character. Upper member, Oou, red crystalline limestone; some of it quartzose, interbedded with yellow-weathering blue shale. Marble lenses, Oom, white, pink, and red crystalline limestone; little shale.

Holston limestone

Red and blue crystalline limestone, much of it highly quartzose, especially in upper third; formerly called Tellico sandstone. Shale streaks common, forming persistent middle member.

Athens shale

Yellow-weathering, blue, calcareous shale or shaly limestone.

Lenoir limestone

Blue, argillaceous limestone

Mascot dolomite

Gray, well-bedded dolomite, with layers of limestone; thin sandstone layers in lower part.

Kingsport formation

Gray, well-bedded to massive dolomite; some limestone in lower part.

Longview dolomite

Gray, mostly well-bedded dolomite; characterized by much massive chert in weathered residuum.

Chepultepec dolomite

Gray, well-bedded to massive dolomite. Basal sandy member, Ocs, includes several sandstone layers.

Copper Ridge dolomite

Gray to black, mostly massive dolomite. Includes equivalent of Conococheague limestone, and also includes in eastern belts only, much blue limestone in upper and lower thirds.

Conasauga shale

Greenish, noncalcareous shale, commonly with scattered lenses of blue limestone. Maynardville limestone member, Ꞓcm, blue limestone and gray dolomite; the limestone chiefly below.

Middle Ordovician

Lower Ordovician

Knox Group

Upper Cambrian

ORDOVICIAN

CAMBRIAN

Middle (?) Cambrian

Colorplate 2 Explanation section from **Geologic Map of the Athens Quadrangle, Tennessee.** (Map by John Rodgers, 1952, U.S. Geol. Survey.)

EXPLANATION
(Continued)

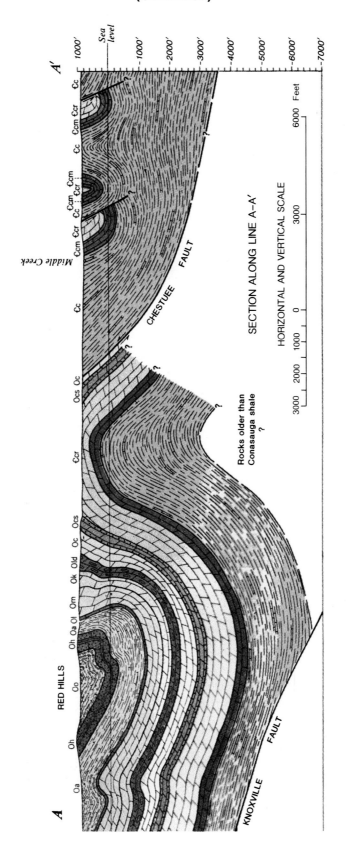

Colorplate 3 Structure section from **Geologic Map of the Athens Quadrangle, Tennessee.** (Map by John Rodgers, 1952, U.S. Geol. Survey.)

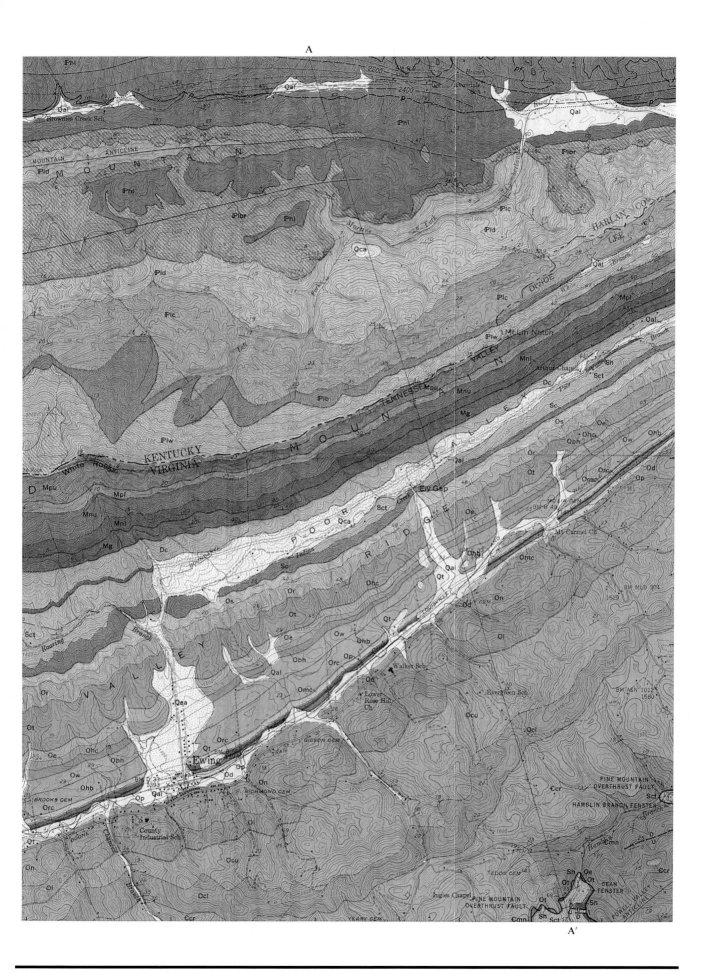

Colorplate 4 Geology of the Ewing Quadrangle, Kentucky and Virginia.

EXPLANATION

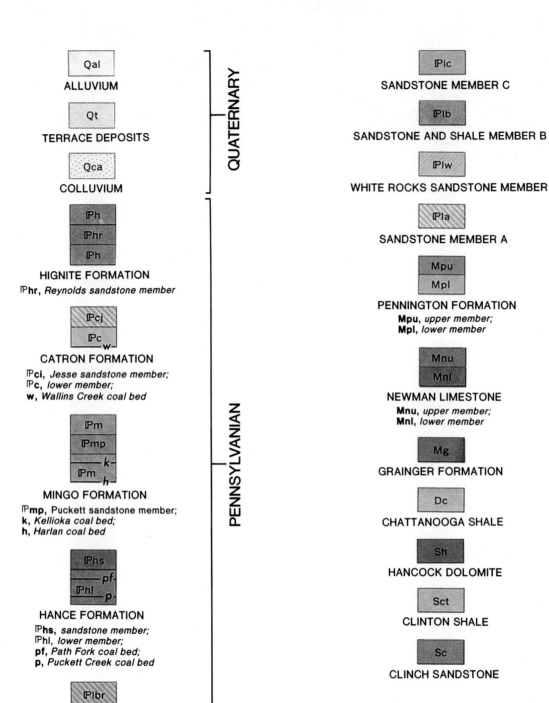

ℙlc
SANDSTONE MEMBER C

ℙlb
SANDSTONE AND SHALE MEMBER B

ℙlw
WHITE ROCKS SANDSTONE MEMBER

ℙla
SANDSTONE MEMBER A

Mpu
Mpl
PENNINGTON FORMATION
Mpu, *upper member;*
Mpl, *lower member*

Mnu
Mnl
NEWMAN LIMESTONE
Mnu, *upper member;*
Mnl, *lower member*

Mg
GRAINGER FORMATION

Dc
CHATTANOOGA SHALE

Sh
HANCOCK DOLOMITE

Sct
CLINTON SHALE

Sc
CLINCH SANDSTONE

QUATERNARY

PENNSYLVANIAN

MISSISSIPPIAN

DEVONIAN

SILURIAN

Qal
ALLUVIUM

Qt
TERRACE DEPOSITS

Qca
COLLUVIUM

ℙh
ℙhr
ℙh
HIGNITE FORMATION
ℙ**hr,** *Reynolds sandstone member*

ℙcj
ℙc
—w—
CATRON FORMATION
ℙ**ci,** *Jesse sandstone member;*
ℙ**c,** *lower member;*
w, *Wallins Creek coal bed*

ℙm
ℙmp
ℙm
—k—
—h—
MINGO FORMATION
ℙ**mp,** *Puckett sandstone member;*
k, *Kellioka coal bed;*
h, *Harlan coal bed*

ℙhs
ℙhl
—pf—
—p—
HANCE FORMATION
ℙ**hs,** *sandstone member;*
ℙ**hl,** *lower member;*
pf, *Path Fork coal bed;*
p, *Puckett Creek coal bed*

ℙlbr
BEE ROCK SANDSTONE MEMBER

ℙld
SANDSTONE AND SHALE MEMBER D

Colorplate 5 Explanation section from **Geology of the Ewing Quadrangle, Kentucky and Virginia.** (Map by K. J. Englund, H. L. Smith, L. D. Harris and J. G. Stephens, 1961, U.S. Geol. Survey.)

EXPLANATION
(Continued)

Os
SEQUATCHIE FORMATION

Or
REEDSVILLE SHALE

Ot
TRENTON LIMESTONE

Oe
EGGLESTON LIMESTONE

Ohc
HARDY CREEK LIMESTONE

Obh
BEN HUR LIMESTONE

Ow
WOODWAY LIMESTONE

Ohb
HURRICANE BRIDGE LIMESTONE

Omc
MARTIN CREEK LIMESTONE

Orc
ROB CAMP LIMESTONE

Op
POTEET LIMESTONE

Od
DOT FORMATION

On
NEWALA DOLOMITE

Ol
LONGVIEW DOLOMITE

Ocu
Ocl
CHEPULTEPEC DOLOMITE
Ocu, *upper member;*
Ocl, *lower member*

ORDOVICIAN

€cr
COPPER RIDGE DOLOMITE

€mn
MAYNARDVILLE LIMESTONE

CAMBRIAN

Contact
Dashed where approximately located, dotted where
concealed

U
D
Fault
Dashed where approximately located. **U,** *upthrown side;* **D,**
downthrown side; **T,** *upper plate of thrust fault*

Anticline
Dashed where approximately located

Overturned anticline

Syncline
Dashed where approximately located

22
Strike and dip of beds

Vertical beds

47
Strike and dip of overturned beds

— 2000 — —
Structure contour drawn on top of Harlan coal bed
Dashed where projected from other beds. Contour interval
40 feet. Datum is mean sea level

— w — —····
Coal bed
Dashed where approximate, dotted where concealed; letters
designate coal bed

Strip mine   **Mine adit**   ✕ **Prospect or outcrop of coal bed**

Abandoned oil well   **Dry hole**

Colorplate 6 Explanation section from **Geology of the Ewing Quadrangle, Kentucky and Virginia.** (Map by K. J. Englund, H. L. Smith, L. D. Harris and J. G. Stephens, 1961, U.S. Geol. Survey.)

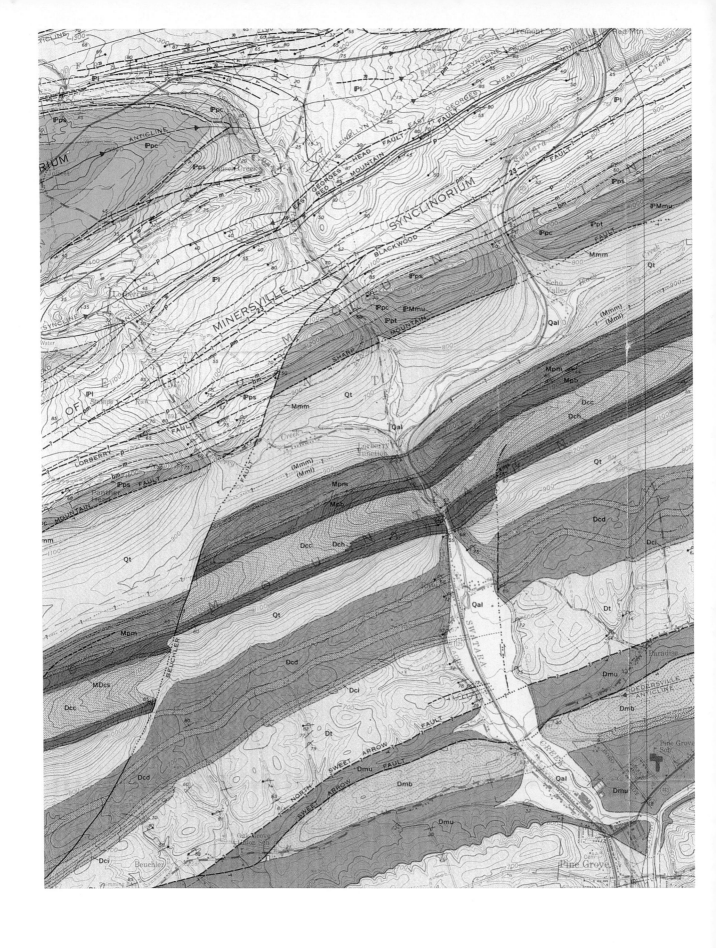

Colorplate 7 Geology Map of the Pine Grove Quadrangle Schuylkill, Lebanon, and Berks Counties, Pennsylvania.

EXPLANATION

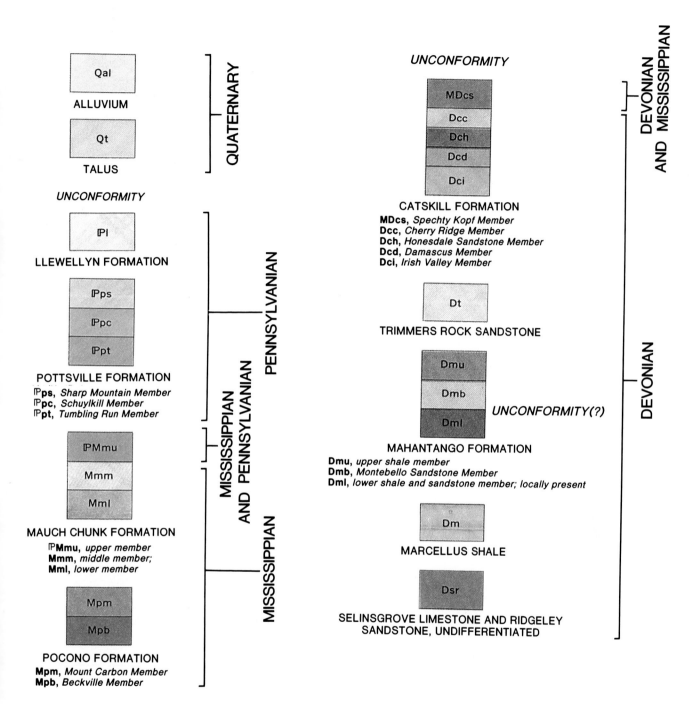

Colorplate 8 Explanation section from **Geology Map of the Pine Grove Quadrangle, Schuylkill, Lebanon, and Berks Counties, Pennsylvania.**
(Map by Gordon H. Woods, Jr. and Thomas M. Kehn, 1968, U.S. Geol. Survey.)

EXPLANATION
(Continued)

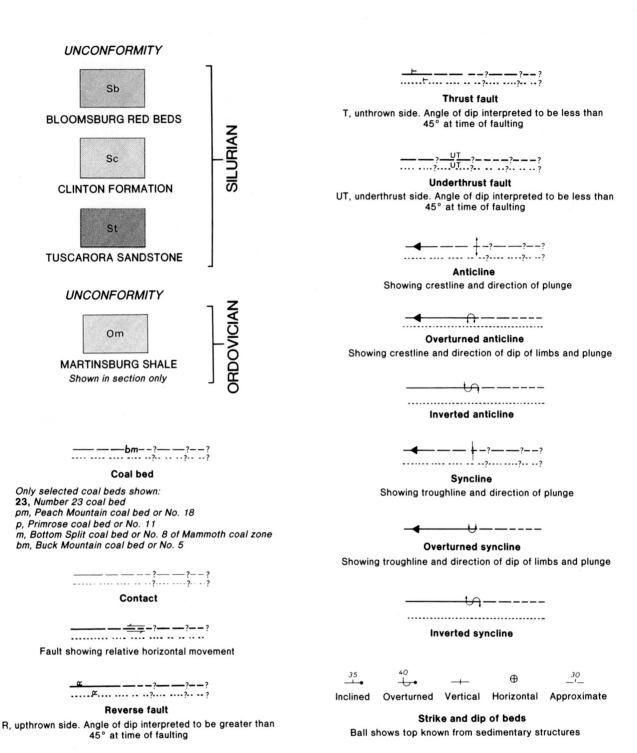

UNCONFORMITY

Sb

BLOOMSBURG RED BEDS

Sc

CLINTON FORMATION

St

TUSCARORA SANDSTONE

SILURIAN

UNCONFORMITY

Om

MARTINSBURG SHALE
Shown in section only

ORDOVICIAN

— — —bm--?— —?— —?

Coal bed

Only selected coal beds shown:
23, *Number 23 coal bed*
pm, Peach Mountain coal bed or No. 18
p, Primrose coal bed or No. 11
m, Bottom Split coal bed or No. 8 of Mammoth coal zone
bm, Buck Mountain coal bed or No. 5

Contact

Fault showing relative horizontal movement

Reverse fault

R, upthrown side. Angle of dip interpreted to be greater than
45° at time of faulting

Thrust fault

T, unthrown side. Angle of dip interpreted to be less than
45° at time of faulting

Underthrust fault

UT, underthrust side. Angle of dip interpreted to be less than
45° at time of faulting

Anticline

Showing crestline and direction of plunge

Overturned anticline

Showing crestline and direction of dip of limbs and plunge

Inverted anticline

Syncline

Showing troughline and direction of plunge

Overturned syncline

Showing troughline and direction of dip of limbs and plunge

Inverted syncline

35	40			30
Inclined	Overturned	Vertical	Horizontal	Approximate

Strike and dip of beds

Ball shows top known from sedimentary structures

Colorplate 9 Explanation section from **Geologic Map of the Pine Grove Quadrangle, Schuylkill, Lebanon, and Berks Counties, Pennsylvania.**
(Map by Gordon H. Woods, Jr. and Thomas M. Kehn, 1968, U.S. Geol. Survey.)

Colorplate 10 Geologic Map of the Golconda Quadrangle, Humboldt County, Nevada.

DESCRIPTION OF MAP UNITS

Qal — ALLUVIUM (QUATERNARY)

Qds — DUNE SAND (QUATERNARY)

Qg — GRAVEL (QUATERNARY)

Qt — HOT-SPRINGS DEPOSITS (QUATERNARY)

Tb — BASALT FLOWS (PLIOCENE)

Ta — ANDESITE FLOWS (MIOCENE)

Tt — ANDESITE TUFF (TERTIARY)

Kgd — GRANODIORITE (UPPER CRETACEOUS)

Kqp — QUARTZ PORPHYRY (UPPER CRETACEOUS)

Kqd — QUARTZ DIORITE (UPPER CRETACEOUS)

Pem — EDNA MOUNTAIN FORMATION (UPPER PERMIAN)—quartzite

PⅠPpu — LOWER PERMIAN AND UPPER, MIDDLE, AND LOWER PENNSYLVANIAN

PⅠPpq — Quartzite and limestone

ⅠPpsc — Interbedded shale and chert

ⅠPpq — Quartzite

ⅠPpg — Greenstone

ⅠPpu — Chert and shale undivided

PⅠPap — ANTLER PEAK LIMESTONE (LOWER PERMIAN AND UPPER PENNSYLVANIAN)

ⅠPhy — HIGHWAY LIMESTONE (MIDDLE AND LOWER PENNSYLVANIAN)

ⅠPhyc — Conglomerate (mélange)

ⅠPb — BATTLE FORMATION (LOWER PENNSYLVANIAN)—quartzite

€p — PREBLE FORMATION (UPPER AND MIDDLE CAMBRIAN)—shale

€pl — Limestone

€o — OSGOOD MOUNTAIN QUARTZITE (LOWER CAMBRIAN?)

Colorplate 11 Description of map units from **Geologic Map of the Golconda Quadrangle, Humboldt County, Nevada.** (Map by R. L. Erickson and S. P. Marsh, 1974, U.S. Geol. Survey.)

Contact

Fault

Dashed where inferred; dotted where concealed. U,
upthrown side; D, downthrown side

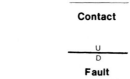

Thrust fault

Sawteeth on upper plate

$\underset{\perp}{\overset{50}{}}$ Inclined

$\overset{}{\underset{81}{\curvearrowright}}$ Overturned

━┼━ Vertical

Strike and dip of beds

⚒ Mine

⊰ Adit

◪ Shaft

⤳ Prospect pit

Mines or prospect pits

Altered area

Rocks are commonly silicified and stained red, orange, or
brown. Some areas contain anomalously high amounts of
metals (Erickson and Marsh, 1971a, b, c, d, 1972)

Colorplate 12 Description of map units from **Geologic Map of the Golconda Quadrangle, Humboldt County, Nevada.** (Map by R. L. Erickson and S. P. Marsh, 1974, U.S. Geol. Survey.)

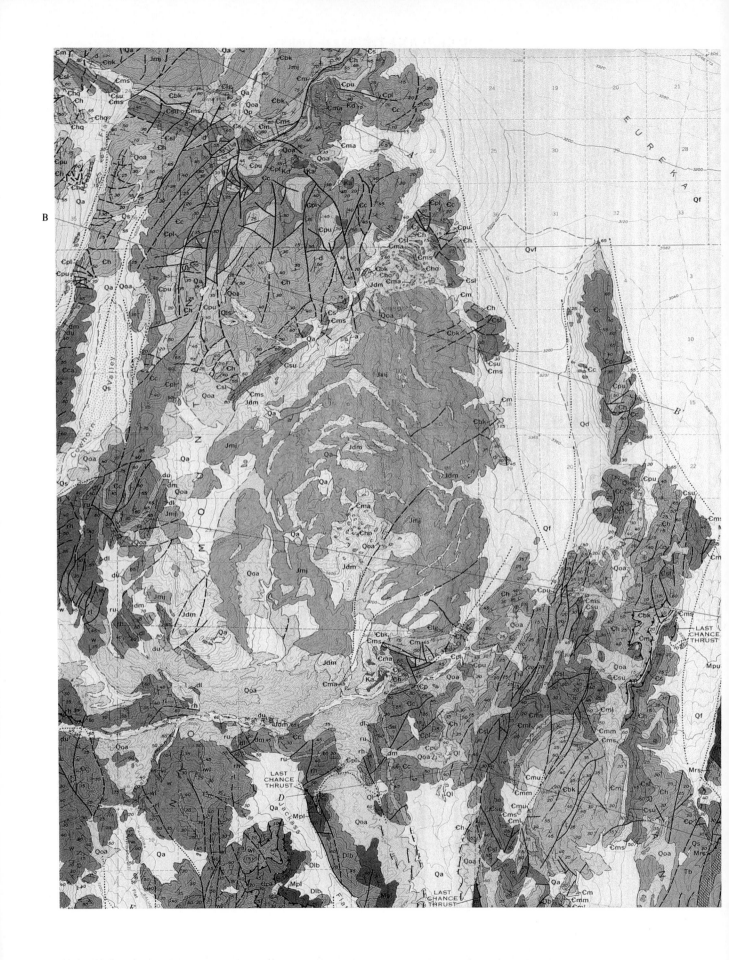

Colorplate 13 Geologic Map of the Waucoba Spring Quadrangle, Inyo County, California.

EXPLANATION

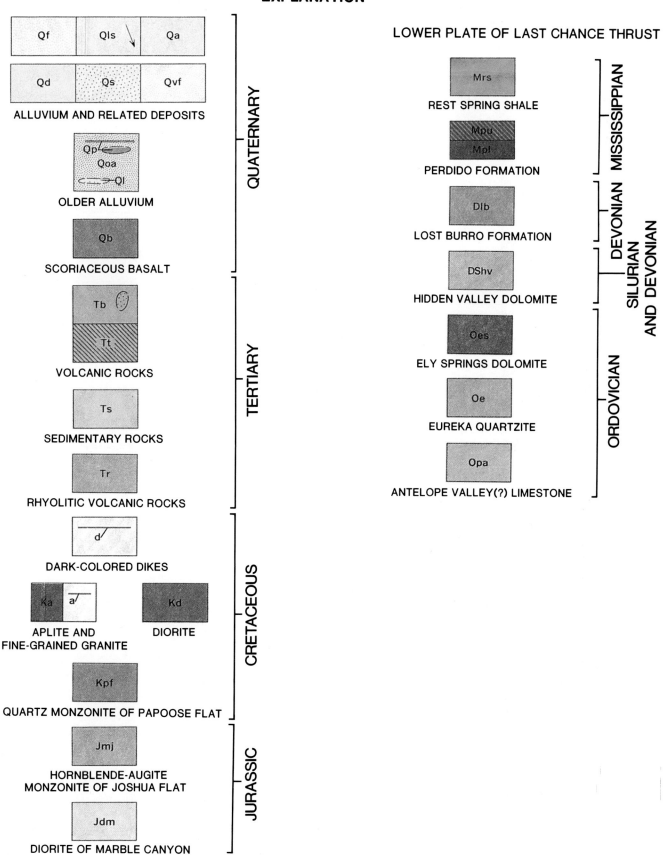

LOWER PLATE OF LAST CHANCE THRUST

ALLUVIUM AND RELATED DEPOSITS — QUATERNARY

Qf | Qls | Qa
Qd | Qs | Qvf

OLDER ALLUVIUM — QUATERNARY

Qp | Qoa | Ql

SCORIACEOUS BASALT — Qb

VOLCANIC ROCKS — TERTIARY — Tb / Tt

SEDIMENTARY ROCKS — Ts

RHYOLITIC VOLCANIC ROCKS — Tr

DARK-COLORED DIKES — d — CRETACEOUS

APLITE AND FINE-GRAINED GRANITE — Ka / a

DIORITE — Kd

QUARTZ MONZONITE OF PAPOOSE FLAT — Kpf

HORNBLENDE-AUGITE MONZONITE OF JOSHUA FLAT — Jmj — JURASSIC

DIORITE OF MARBLE CANYON — Jdm

REST SPRING SHALE — Mrs — MISSISSIPPIAN

PERDIDO FORMATION — Mpu / Mpl

LOST BURRO FORMATION — Dlb — DEVONIAN

HIDDEN VALLEY DOLOMITE — DShv — SILURIAN AND DEVONIAN

ELY SPRINGS DOLOMITE — Oes — ORDOVICIAN

EUREKA QUARTZITE — Oe

ANTELOPE VALLEY(?) LIMESTONE — Opa

Colorplate 14 Explanation section from **Geologic Map of the Waucoba Spring Quadrangle, Inyo County, California.** (Map by C. A. Nelson, 1971, U.S. Geol. Survey.)

EXPLANATION
(Continued)

UPPER PLATE OF LAST CHANCE THRUST

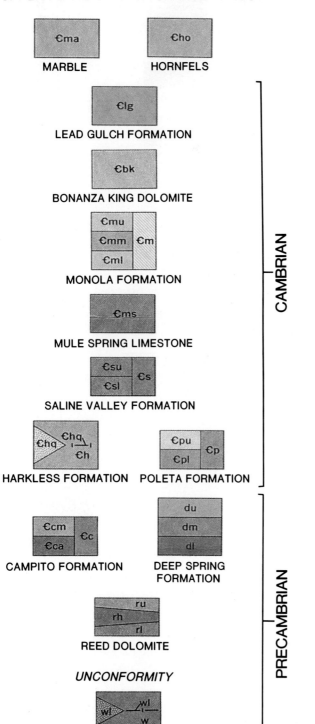

MARBLE (€ma)

HORNFELS (€ho)

LEAD GULCH FORMATION (€lg)

BONANZA KING DOLOMITE (€bk)

MONOLA FORMATION (€mu, €mm, €ml, €m)

MULE SPRING LIMESTONE (€ms)

SALINE VALLEY FORMATION (€su, €sl, €s)

HARKLESS FORMATION (€hq, €h)   **POLETA FORMATION** (€pu, €pl, €p)

CAMBRIAN

CAMPITO FORMATION (€cm, €ca, €c)   **DEEP SPRING FORMATION** (du, dm, dl)

REED DOLOMITE (ru, rh, rl)

UNCONFORMITY

WYMAN FORMATION (wl, w)

PRECAMBRIAN

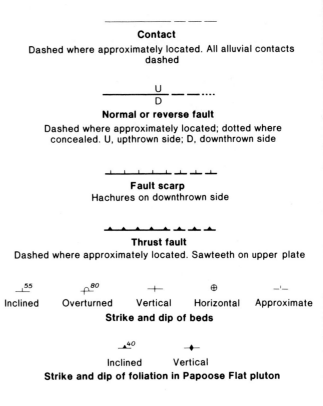

Contact
Dashed where approximately located. All alluvial contacts dashed

U
D
Normal or reverse fault
Dashed where approximately located; dotted where concealed. U, upthrown side; D, downthrown side

Fault scarp
Hachures on downthrown side

Thrust fault
Dashed where approximately located. Sawteeth on upper plate

Inclined (55)   Overturned (80)   Vertical   Horizontal   Approximate
Strike and dip of beds

Inclined (40)   Vertical
Strike and dip of foliation in Papoose Flat pluton

Colorplate 15   Explanation section from **Geologic Map of the Waucoba Spring Quadrangle, Inyo County, California.** (Map by C. A. Nelson, 1971, U.S. Geol. Survey.)

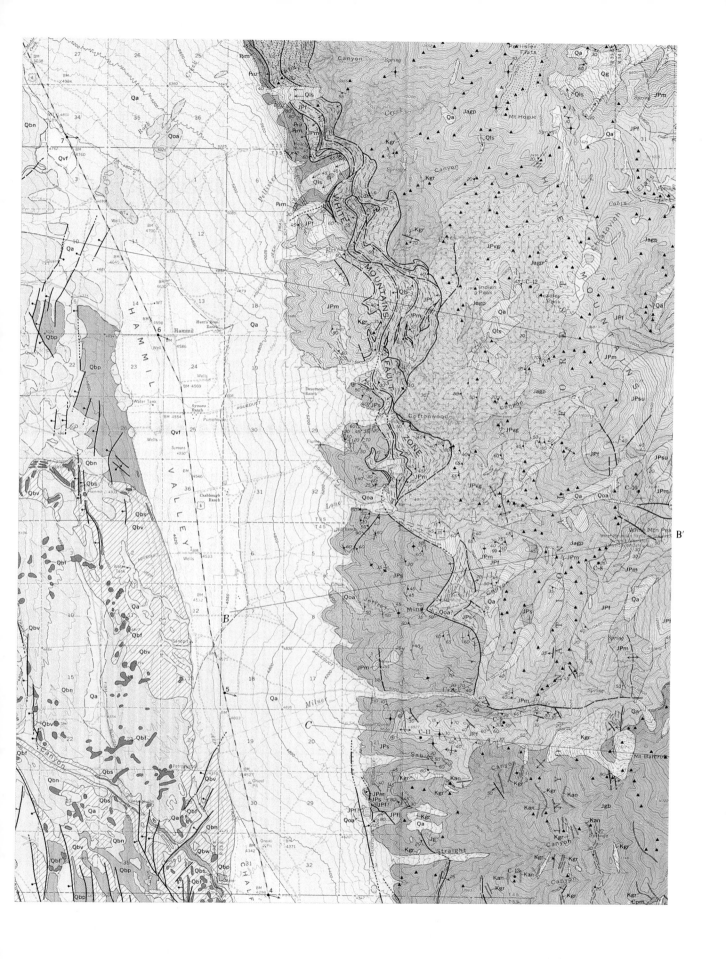

Colorplate 16 Geologic Map of the White Mountain Peak Quadrangle, Mono County, California.

EXPLANATION

SURFICIAL DEPOSITS

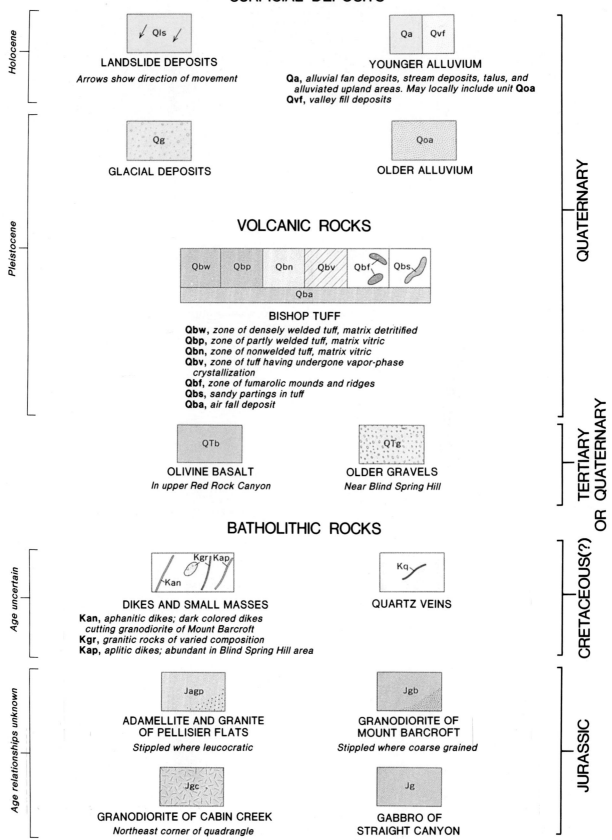

Holocene

LANDSLIDE DEPOSITS

Arrows show direction of movement

YOUNGER ALLUVIUM

Qa, *alluvial fan deposits, stream deposits, talus, and alluviated upland areas. May locally include unit* **Qoa**
Qvf, *valley fill deposits*

Pleistocene

GLACIAL DEPOSITS

OLDER ALLUVIUM

VOLCANIC ROCKS

BISHOP TUFF

Qbw, *zone of densely welded tuff, matrix detritified*
Qbp, *zone of partly welded tuff, matrix vitric*
Qbn, *zone of nonwelded tuff, matrix vitric*
Qbv, *zone of tuff having undergone vapor-phase crystallization*
Qbf, *zone of fumarolic mounds and ridges*
Qbs, *sandy partings in tuff*
Qba, *air fall deposit*

OLIVINE BASALT

In upper Red Rock Canyon

OLDER GRAVELS

Near Blind Spring Hill

BATHOLITHIC ROCKS

Age uncertain

DIKES AND SMALL MASSES

Kan, *aphanitic dikes; dark colored dikes cutting granodiorite of Mount Barcroft*
Kgr, *granitic rocks of varied composition*
Kap, *aplitic dikes; abundant in Blind Spring Hill area*

QUARTZ VEINS

Age relationships unknown

ADAMELLITE AND GRANITE OF PELLISIER FLATS

Stippled where leucocratic

GRANODIORITE OF MOUNT BARCROFT

Stippled where coarse grained

GRANODIORITE OF CABIN CREEK

Northeast corner of quadrangle

GABBRO OF STRAIGHT CANYON

QUATERNARY

TERTIARY OR QUATERNARY

CRETACEOUS(?)

JURASSIC

Colorplate 17 Explanation section from **Geologic Map of the White Mountain Peak Quadrangle, Mono County, California.** (Map by Dwight F. Crowder and Michael F. Sheridan, 1972, U.S. Geol. Survey.)

EXPLANATION
(Continued)

GRANODIORITE OF BENTON RANGE — TRIASSIC

PRE-BATHOLITHIC ROCKS

MIXED METAVOLCANIC AND GRANITIC ROCKS (JPvg)

METAVOLCANIC AND METASEDIMENTARY ROCKS

JPsu, *upper sedimentary unit; fine- to coarse-grained metasedimentary rocks rich in metavolcanic fragments*
JPf, *felsic metavolcanic rocks*
JPm, *mafic metavolcanic rocks*
JPs, *lower metasedimentary unit; fine- to coarse-grained metasedimentary rocks rich in metavolcanic fragments*

Age relationships uncertain

QUARTZ SERICITE SCHIST OF BENTON RANGE (Pzqsh)

CALC-HORNFELS (PZCH) AND MARBLE (PZM) OF BENTON RANGE

METASEDIMENTARY ROCKS, UNDIVIDED
Along White Mountain front Pzm, marble

PALEOZOIC

POLETA FORMATION
€pm, *major marble beds*

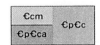

CAMPITO FORMATION
Mostly hornfels. Only locally divisible into:
€cm, *Montenegro Member*
€p€ca, *Andrews Mountain Member*

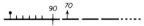

CONTACT
Dashed where approximately located; dotted where concealed

FAULT SHOWING DIP-DIRECTION, AND AMOUNT, WHERE KNOWN
Dashed where approximately located, dotted where concealed or inferred. Ball indicates downthrown side; hachures indicate scarp and direction it faces (only shown for faults along front of White Mountains—many faults in Bishop Tuff also have scarps). In the White Mountains shear zone only the most conspicuous faults are shown

SHEARED ROCKS
In overpatterned areas the dominant lithology is shown, although there is considerable mixing of metavolcanic types and sheets of flaser gneiss formed from the Pellisier Flats rocks in the fault zone. The fault zone contact is generalized and in places lenses of unsheared rocks are unmapped near the contact. Alined shear symbols indicate concentrated shear and possible fault (generally marked by a gully)

Anticline   **Syncline**

Inclined   Vertical   Horizontal
Strike and dip of beds

Inclined   Vertical   Horizontal
Strike and dip of foliation
Generally due to late shear

Inclined   Vertical
Strike and dip of cleavage

Homogeneous granitic rock
No conspicuous foliation or lineation

•C-1
Sample locality
Chemical analysis

7
Point of geologic interest
Arrow shows direction of view—see text

Colorplate 18   Explanation section from **Geologic Map of the White Mountain Peak Quadrangle, Mono County, California.** (Map by Dwight F. Crowder and Michael F. Sheridan, 1972, U.S. Geol. Survey.)

1. This geologic map quadrangle, located at the southeastern border of Kentucky and Virginia, is in what physiographic province (use Plate 2 in the endpapers)?
2. Identify the tectonic significance of this area. What clues are evident in the geologic map? (Use Plate 1 at the end of manual to aid in answering this question.)
3. Contrast this region with that of the Pine Grove (PA) Quadrangle (Colorplate 7). What immediate differences do you observe? What might this imply with regard to the timing and duration of the tectonic activity?
4. Construct a topographic profile and geologic cross section from A to A′. How does this tie in with your responses to Questions 2 and 3? What differences are brought out by examining a cross section than could be immediately noticed in the geologic map? (Use Plate 3 at the end of the manual to compare your cross section.)
5. Identify the major geologic structures depicted in this region. Why is Cumberland Mountain a prominent feature?
6. Using the "rules of V's" (as explained in Chapter 14), in what direction are the formations **Єcr, Ohb,** and **Pm** dipping?
7. With regard to your interpretation of the geologic structures in Question 5, what reasons can you give for the variety of attitudes given for the outcrop of **Єcr?**

SCALE 1:24 000

CONTOUR INTERVAL 20 FEET
DATUM IS MEAN SEA LEVEL

KENTUCKY

QUADRANGLE LOCATION

Colorplate 19 Questions to accompany Colorplate 4, Geology of the Ewing Quadrangle, Kentucky and Virginia.

1. Locate this region on the United States physiographic province map (Plate 2) and the tectonic map (Plate 1).

2. Construct a topographic profile and geologic cross section from the Pine Grove School to the town of Lorberry. What are the major geologic structures in this region?

3. The Beuchler Fault cuts through the center of this quadrangle. What is the type of fault and how would you describe its motion (review fault terminology in Chapter 14 if necessary)?

4. What type of geologic structure is the Minersville Synclinorium? Describe it with respect to its attitude and setting.

5. In the Mahantango Formation (Middle Devonian) there is an unconformity separating the Montebello Sandstone from the underlying shale and sandstone strata. Explain what evidence you would look for in the field to confirm this type of contact.

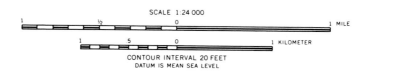

SCALE 1:24 000

1 MILE

1 KILOMETER

CONTOUR INTERVAL 20 FEET
DATUM IS MEAN SEA LEVEL

PENNSYLVANIA

QUADRANGLE LOCATION

Colorplate 20 Questions to accompany Colorplate 7, Geologic Map of the Pine Grove Quadrangle, Schuylkill, Lebanon and Berks Counties, Pennsylvania.

1. Locate this region with respect to the tectonic map (Plate 1) and the physiographic province map (Plate 2).

2. Construct a topographic profile and geologic cross section along D–D'. What geologic structures are depicted in this region?

3. Notice the time gap in strata. What reasons can you give for this jump from the Cambrian to the Pennsylvanian? How about the lack of strata for the Permian to the Cretaceous? What field evidence would you look for in order to substantiate your hypotheses?

4. What are the outcrops of **Kqd** in the mass of **Ɛp?** What is their origin? Are the outcrops of **Ɛpl** caused by the same mechanism? Explainn.

5. What is the direction of thrusting for the Golconda Thrust? What about the unlabelled thrust fault to the east of the Golconda Thrust? Discuss what seems to have occurred here. Can you determine the timing of the thrusting (use cross-cutting relationws and superpostsition)?

6. The numerous faults that cut the Pennsylvanian strata are of what theype? Did the faulting occur during the Late Paleozoic, Mesozoic, or later? What evidence can you find in the geology (use Plate 3 for addtional information)?

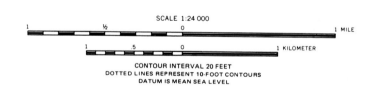

SCALE 1:24 000

1 MILE

1 KILOMETER

CONTOUR INTERVAL 20 FEET
DOTTED LINES REPRESENT 10-FOOT CONTOURS
DATUM IS MEAN SEA LEVEL

NEVADA

QUADRANGLE LOCATION

Colorplate 21 Questions to accompany Colorplate 10, Geologic Map of the Golconda Quadrangle, Humboldt County, Nevada.

1. Locate this region on the United States physiographic province map (Plate 2) and the tectonic map (Plate 1).
2. Construct a topographic profile and geologic cross section from B-B'. What geologic structures are found in this area?
3. How would you describe the Jurassic igneous rocks in the central portion of this geologic map? What is their stratigraphic relation to the surrounding rock types? What type of contact relations are found?
4. How would you explain the presence of Cambrian-age marble (**€ma**) and Harkless Fm. siltstone and shale in the interior of the Marble Canyon diorite? What might this imply about the tectonic and structural events surrounding the emplacement of these Jurassic igneous rocks?
5. Notice that although the Jurassic rocks are seldom faulted, the surrounding rock types have experienced numerous faulting events. What is the relative timing of the faulting? Are the major faulting events related to the emplacement of the Jurassic igneous rocks? Can you support this hypothesis? Explain.

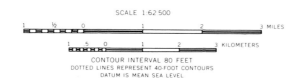

CALIF

QUADRANGLE LOCATION

Colorplate 22   Questions to accompany Colorplate 13, Geologic Map of the Waucoba Spring Quadrangle, Inyo County, California.

1. Locate this region with respect to the tectonic map (Plate 1) and the physiographic province map (Plate 2).
2. Construct a topographic profile and geologic cross section from B-B'. What geologic structures are found in this area? What are the major rock types found in this region? What conclusions can you develop about the tectonic activity in this region with regard to both the geologic structures and the revealed strata? Be cautious: Examine the ages of the strata and the types of strata before answering this question.
3. Bounding the eastern margin of the Hammil Valley is an enormous deposit of Holocene alluvium. How would you describe this deposit? Outline the geologic and environmental conditions that could result in this feature. List the field evidence that you would look for to substantiate your hypothesis. (Examine the topographic relief of the region to help answer this question.)
4. Describe the White Mountain Fault Zone using the terminology of Chapter 14. Notice the presence of Pleistocene alluvium exposed above the Holocene alluvium. What can you hypothesize has caused this stratigraphic relation? Explain the reasoning for your answer. (Examine the cross-cutting relations to help answer this question.)

16°

TRUE NORTH

MAGNETIC NORTH

APPROXIMATE MEAN
DECLINATION, 1972

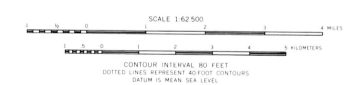

SCALE 1:62 500

1   ½   0   1   2   3   4 MILES

1   5   0   1   2   3   4   5 KILOMETERS

CONTOUR INTERVAL 80 FEET
DOTTED LINES REPRESENT 40-FOOT CONTOURS
DATUM IS MEAN SEA LEVEL

CALIF

QUADRANGLE LOCATION

Colorplate 23 Questions to accompany Colorplate 16, Geologic Map of the White Mountain Peak Quadrangle, Mono County, California.

15

Canadian Shield and Basement Rocks of North America

THE BASEMENT ROCKS

The most tectonically stable region of most continents has not undergone significant orogenic deformation for hundreds of millions of years. However, these regions often have a subtle change in the topography as you move across the continent. This gentle warping of the continental lithospheric plate, which does not involve intense folding, faulting, or seismic activity, is called **epeirogeny.** Although this type of deformation of topography is common to the central stable regions of the continents, it can also occur in orogenic belts and ocean basins.

The most stable portion of the continent is called the **shield.** Shields are those areas in which the oldest crystalline basement rocks are exposed at the surface. The shield is surrounded by generally flat-lying packages of sedimentary rock that were derived by erosion from the shield. These younger rocks that border the margin of the shield are called the **platform.** Thus, the continent is actually composed of two lithologically and chronologically different portions, the shield and the surrounding platform. Together these are called the **craton.** When the craton interacts with another lithospheric plate by collision or subduction, the platform sedimentary rocks are the "bumper" that experiences the deformation and orogenic activity.

The oldest rocks of North America, which constitute the crystalline basement of the continent, are exposed at the earth's surface in an area of about two million square miles in the Canadian Shield and in smaller areas of the Cordilleran and Appalachian mountain belts. Elsewhere, the basement is buried beneath layers of sedimentary rock, as shown on Plate 1 (found after Chapter 20). Our information on the structure and composition of these basement rocks depends mainly on samples taken from wells drilled through the sedimentary cover and images of the underlying rock revealed by the use of seismic reflection data. The dominant rock type of the Canadian Shield is **granite gneiss** (see Chapters 19 and 20 for additional information on this rock type), but most other rock types occur there, including carbonate rocks. According to radiometric age determinations, most basement rocks range in age from a maximum of about 2.5 billion years to about 0.5 billion years. However, there are regions in the Canadian Shield, and in the accreted orogenic belts, that are much older. A small patch of rocks 3.5 billion years old has been found in Minnesota, and recently, rocks as old as 3.96 billion years have been found in the region north of the town of Yellowknife in the Northwest Territories (Wopmay Orogen) of Canada. These rocks (**tonalite gneisses**) suggest that the continental crust has been forming since the Early Archean and that the preservation (and discovery) of such relicts are, in most cases, a matter of serendipity.

When the locations of dated rock samples are plotted on a map such as Figure 15.1, rocks of approximately the same age tend to occupy belts or areas, which constitute Precambrian (Archean to Proterozoic) geologic provinces. Each province is inferred to represent an orogenic belt, in which the rocks were recrystallized during metamorphism and associated igneous (commonly granitic) intrusions. These processes reset the "radiometric clocks" within the minerals (and rocks) to zero (see Chapter 8 for review of absolute age determination methods) and define the time of the *last* metamorphic or plutonic event. When the ages of basement rocks are plotted on a graph according to frequency of occurrence, fairly well-defined maxima occur at ages of greater than 2.5, from 2.5 to 1.8, from 1.8 to 1.7, from 1.7 to 1.5, and from 1.5 to 1.0 billion years ago. This suggests that there were, at least, five major orogenic belts formed at different times during the Archean to Proterozoic history of the North America craton. Using this as a working hypothesis, answer the following questions using Figure 15.1.

Study Questions

1. With pencil lines, divide the unshaded part of North America in Figure 15.1 into four areas or belts, each of which is dominated by the rocks of a particular age group and can be regarded as a Precambrian geologic province. Label the provinces, from oldest to youngest, as Superior, Churchill, Central, and Grenville.

Figure 15.1 Ages of basement rocks at specific localities in North America. Each symbol represents the location of one dated rock, or of several rocks from the same general vicinity. Wavy lines, mainly in the Canadian Shield, represent trends of foliation.

Generalized from Engel (1963), *Science,* **146;** Goldrich (1966), *J. Geophys. Res.,* **22:** 5386; and Goodwin (1991), *Precambrian Geology,* Academic Press.

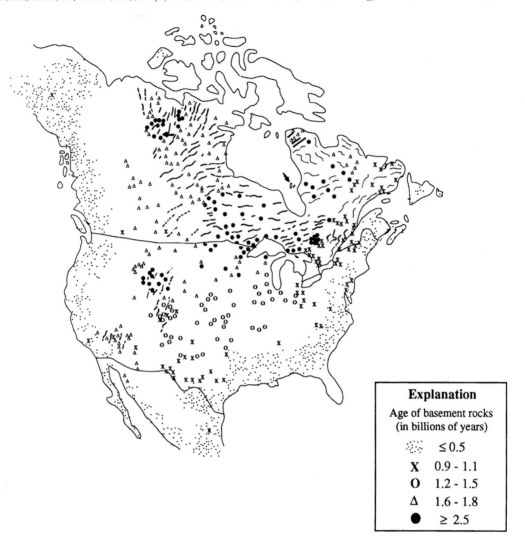

Explanation

Age of basement rocks
(in billions of years)

⬚	≤ 0.5
X	0.9 - 1.1
O	1.2 - 1.5
Δ	1.6 - 1.8
●	≥ 2.5

2. If a younger orogenic belt cuts across an older one, the radiometric "clocks" in the disturbed part of the older belt are reset, by heating and metamorphism, to the date of the younger belt. In view of this, what is the probable explanation for the isolated cluster of solid dots (2.5 billion years) in northwestern Canada?

_____

_____

_____

_____

_____

3. Now shade in with pencil the probable extent of the Superior Belt. If, as shown in Plate 1 (found after Chapter 20), orogenic belts are built at the edges of a

continental craton, where was the craton at whose edges the Superior Belt was built?

_____

_____

_____

_____

_____

Some geologists believe that the North American continent has grown as an isolated unit by the successive additions of orogenic belts to the margin of the Superior Belt or some previous nucleus. However, according to the hypothesis of sea-floor spreading (see Chapter 6 for review of sea-floor spreading), the American plate separated from the African and

Table 15.1 Precambrian Rocks of the Lake Superior Region

Era	Stratified Rocks	Intrusive Rocks
Paleozoic	Fossiliferous sandstone and limestone	
Middle Proterozoic		Duluth gabbro in Minnesota and Logan gabbro sills in Ontario. Age about 1.1 billion years.
	Keweenawan, "The Copper Sequence"	
	Upper: Red and green immature sandstones and shale; white quartz sandstone. About 15,000 feet thick (4572 m).	
	Middle: Mainly basaltic lavas, some conglomerate. Deposits of native copper in conglomerate and in porous part of some lava flows. About 20,000 feet thick (6096 m).	
	Lower: Sandstone and conglomerate, about 400 feet thick (122 m).	
Early Proterozoic		Intrusions of granitic rocks in Minnesota and Wisconsin. Age about 1.7 billion years.
	Animikian, "The Iron Sequence"	
	Upper: Shale and slate, graywacke, quartzite, volcanics, some iron formation. About 10,000 feet thick (3048 m).	
	Middle: Iron formation, quartzite, slate. About 2500 feet thick (762 m).	
	Lower: In Michigan, dolomite underlain by quartzite and a basal conglomerate. About 1500 feet thick (457 m).	
Archean		Intrusions of granitic rocks, ranging in age from about 2.4 to 2.6 billion years.
	Keewatin Sequence	
	Mainly graywacke and metamorphosed volcanic rocks (greenstones). Some iron formation. About 20,000 feet thick (6096 m).	
	Older rocks	

Eurasian plates about 150–200 Ma; thus, the present continents may not have evolved as separate units. In particular, small **microcontinents** of crustal material may have been moved in the same manner as the much larger continental plates and subsequently were accreted onto the margins of the craton. This concept is called **accretionary tectonics.**

4. What type of evidence would you look for in order to evaluate whether the material that was accreted to the craton was an orogenic belt or a microcontinent (also called a **microplate** or **allochthonous terrain**)?

_____

_____

_____

_____

_____

Evidence of life in these Precambrian rocks consists of stromatolites in the Steeprock Group of western Ontario (of Middle Proterozoic age) and microfossils in the black cherts of the Gunflint Formation exposed along the north shore of Lake Superior at Schrieber, Ontario. The Gunflint biota, approximately 1.9 billion years old, include the remains of cyanobacteria (blue-green algae), bacteria, and other microorganisms of complex structure and uncertain affinity.

PRECAMBRIAN HISTORY OF THE LAKE SUPERIOR REGION

The Lake Superior region is of particular interest because rocks ranging in age from the Archean to the Middle Proterozoic are well exposed there and, after decades of geological investigation, the history is well known in comparison with that of most Precambrian areas. In addition, the **iron ranges** of the region have been the primary domestic source of iron ore for many decades, and large quantities of copper have been mined from the Proterozoic rocks of Michigan. Among the major scenic attractions are the drive along the north shore of Lake Superior in Canada and Isle Royale National Park in Lake Superior, a wilderness park in which automobiles are not permitted.

As shown in Table 15.1, the Precambrian rocks of the Lake Superior region can be grouped into the **Keewatin** (Archean), **Animikian** (Early Proterozoic), and **Keweenawan** (Middle Proterozoic) sequences of stratified rocks, each separated by extensive intrusions of igneous rocks. Each sequence, particularly the Keewatin, records a long and complicated history, of which only the major events are considered here. The geologic map (Fig. 15.2) and the geologic cross section (Fig. 15.3) have been compiled mainly from Goldich (1961), Leith (1935), Ontario Dept. of Mines Map No.1958B, and Goodwin (1991). For a more extensive overview of this

Figure 15.2 Geologic map of the Lake Superior region.

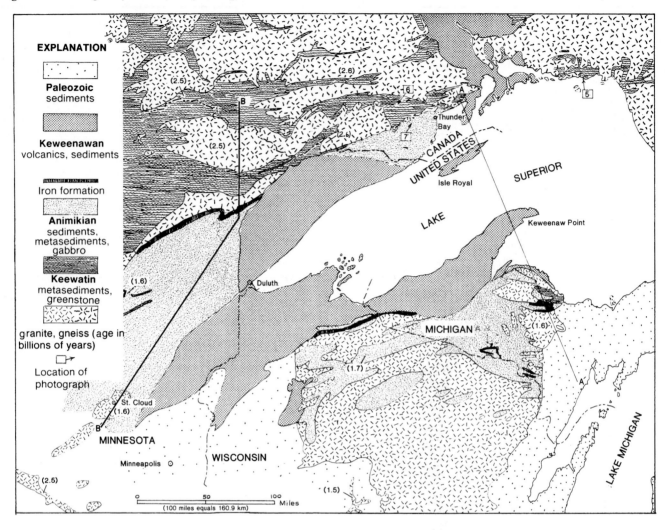

Figure 15.3 Geologic structure sections of the Lake Superior region.

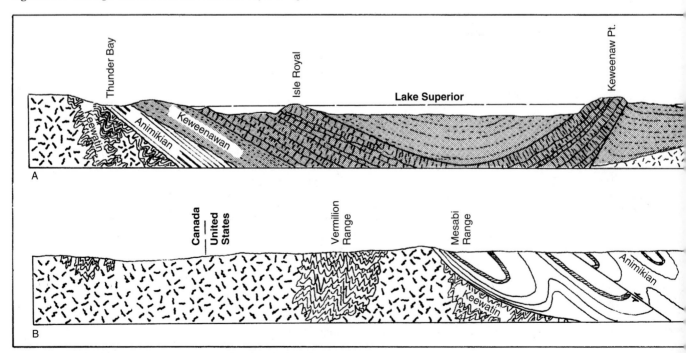

region, see also the Geologic Map of North America. Use Table 15.1 and Figures 15.2, 15.3, and 15.4 to answer the following questions about the geologic history of the Lake Superior region.

Study Questions

Archean

1. Rounded pebbles of granite have been reported from the lower strata of the Keewatin sequence. What does this imply about the contact between the underlying granite and the overlying Keewatin sequence? Explain your reasoning.

_____

_____

_____

_____

2. Judging from the trend directions (attitudes such as **strike** and **dip;** see Chapter 14 for a review of structural terms) of the present outcrops of Keewatin metasedimentary rocks, what was the trend of the mountains formed in connection with their deformation and intrusion?

_____

_____

_____

_____

Early Proterozoic

3. Figure 15.3 (structure section A-A′, see Fig. 15.2 for location) indicates that the Animikian strata were deposited unconformably on a nearly flat surface eroded across the Keewatin. On the basis of available radiometric ages (Fig. 15.1 and Table 15.1), make a reasonable estimate of the length, in millions of years, of this period of erosion, during which the Keewatin mountains were leveled. What type of unconformity does this represent? (Examine Figs. 15.3 and 15.4 and see Chapter 7 for review of unconformities).

_____

_____

_____

_____

4. In the Michigan area, beds of dolomite occur in the lower parts of the Animikian (see Table 15.1). Does this indicate the existence of an ocean in this region during the Early Proterozoic? What other evidence would you need to help substantiate the existence of an ocean at this time?

_____

_____

_____

_____

_____

Animikian

Paleozoic

A'

St. Cloud

Animikian

Increasing metamorphism ⟶

B'

Figure 15.4 Lithofacies of the Animikian at the Mesabi Iron Range.

Generalized from White (1954), Bull. **38,** Minnesota Geological Survey.

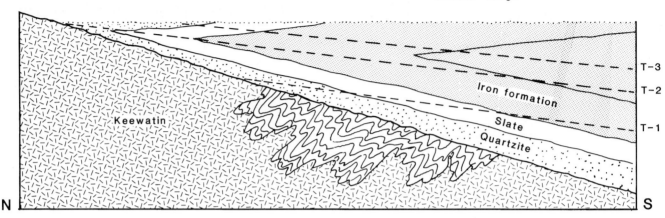

Mesabi range

5. The Animikian sandstones in Minnesota, Wisconsin, and Michigan are crossbedded throughout the strata and outcrop in many locations. In addition, the majority of the **foreset beds** of these strata (see Chapter 4 for a review of foreset beds) dip to the southeast. Where was the probable source area for this sand?

_____

_____

_____

_____

_____

_____

_____

6. The Upper Animikian sequence consists of thick units of graywacke and volcanic rocks. What change in tectonic setting probably occurred from the Lower to Upper Animikian sequence? With respect to the plate tectonic hypothesis, what type of interaction might have occurred?

_____

_____

_____

_____

_____

7. In northern Minnesota and Ontario, the Animikian sequence is nearly horizontal, very thin, and unmetamorphosed (see structure section A-A′ of Figs. 15.2 and 15.3). Southward in Minnesota, the thickness, degree of deformation, and degree of metamorphism increase in the direction of a body of intrusive granite near St. Cloud (see structure section Section B-B′, Fig. 15.3). What does this tell us about the location of the mountain belt formed in connection with the deformation and intrusion of the Animikian?

_____

_____

_____

_____

_____

_____

_____

8. What was the probable trend direction of the mountain belt formed in connection with the metamorphism and intrusion of the Animikian?

_____

_____

_____

_____

_____

9. The **iron formation** of the Animikian sequence (often called **taconite**) consists of chert and iron compounds (hematite and/or magnetite), is delicately banded, and is generally free of sand or silt. The Mesabi iron range (indicated on Fig. 15.4 by the band of Biwabik iron formation north of Duluth, and labeled on Section B-B′, Fig. 15.3) was nearly parallel to the shoreline of the sea in which the Animikian strata were deposited. The lithofacies of the Lower Animikian at the Mesabi range are shown diagrammatically in Figure 15.4. The shifting of a shoreline two billion years ago is recorded in these lithofacies.

   a. Assume that the dashed lines T-1, T-2, and T-3 (Fig. 15.4) are time lines representing the depositional surface at a particular time. (Compare with Fig. 9.3.) Draw arrows on the cross section indicating the position of the shoreline at T-1, T-2, and T-3. Was the shoreline advancing or retreating? How can you tell?

   _____

   _____

   _____

   _____

   _____

   _____

   b. Why would one expect the iron formation to be deposited farther offshore than the quartzite and slate? Explain your reasoning.

   _____

   _____

   _____

   _____

   _____

   c. Which of the tectonic settings in Figure 5.3 would most reasonably account for the properties of these Lower Animikian strata?

   _____

   _____

   _____

   _____

Middle Proterozoic

10. Keweenawan rocks were deposited on an erosion surface that cuts across all older rocks (see Fig. 15.3). On the basis of available radiometric dates (from Table 15.1), make a reasonable estimate of the length, in millions of years, of the period of erosion that preceded the deposition of the Keweenawan sequence. (The mountains formed in connection with the deformation of the Animikian sequence were leveled during this period.) What type of unconformity separates the Keweenawan sequence from the Animikian sequence?

   _____

   _____

   _____

   _____

   _____

   _____

   _____

   _____

   _____

   _____

11. About half the Keweenawan strata are sedimentary and half are volcanic (lava) flows. The sedimentary strata are mostly immature sandstones or conglomerates, and many are of reddish color. What is the likely sedimentary environment of the Keweenawan sequence (refer to Table 4.1)? What type of tectonic setting (and environment of deposition) might be envisioned (refer to Fig. 5.3)?

   _____

   _____

   _____

   _____

   _____

   _____

   _____

   _____

Figure 15.5 Outcrop of **greenstone** (slightly metamorphosed basalt) showing well-developed **pillow** structure, as exposed along the Trans-Canada Highway near Jackfish, Ontario. See numeral "5," Figure 15.2 for location.

Pye, 1969. Geology and Scenery, North Shore of Lake Superior; Ontario Department of Mines, Geological Guide Book 2, p. 148. © Queen's Printer for Ontario, 1969. Reproduced with permission.

1. To what rock sequence do these greenstones belong?

_____

_____

2. What is their minimum age, in billions of years?

_____

_____

3. Pillow structure in basalts results from submarine extrusion onto the sea floor. The upper part of the basaltic crust beneath the oceans has pillow structure. Judging from Figure 6.4, where were these basalts extruded?

_____

_____

4. Notice that the "pillows" have a flat upper surface, and their lower part projects into the space between the pillows beneath. Draw an arrow on the photo showing which way was up.

Figure 15.6 Interbedded slate (dark) and siltstone (light), exposed along the Trans-Canada Highway west of Thunder Bay. Location at "6" on Figure 15.2.

1. To what rock sequence do these beds belong?

_____

_____

_____

_____

2. What is their minimum age, in billions of years?

_____

_____

_____

_____

3. What aspect of the bedding suggests deposition by turbidity currents?

_____

_____

_____

_____

Figure 15.7 Shale at Kakabeka Falls, on the Trans-Canada Highway west of Thunder Bay. Location at "7" on Figure 15.2.

1. To what rock sequence does the shale belong?

_____

2. What is its minimum age?

_____

_____

3. Would you expect it to contain fossils?

_____

_____

4. Shales usually are not very resistant to running water. What reasons would you infer for the highly resistant nature of the shale at Kakabeka Falls?

_____

_____

12. As shown in the structure section A-A′ of Figure 15.3, the southwestern part of Lake Superior occupies a great syncline formed of Keweenawan rocks. The Keweenawan lavas and the Duluth **gabbro** represent a large volume of magma that must have come from somewhere. Suggest a possible connection between the rise of this magma and the development of the syncline. Evaluate this connection with regard to the plate tectonic hypothesis.

_____

_____

_____

_____

_____

_____

13. In the Lake Superior region, it is inferred that no further mountain-building episodes followed the deposition of the Keweenawan sequence. What is the basis for this inference?

_____

_____

_____

_____

14. As depicted in structure section A-A′ (Fig. 15.3), the Archean and Proterozoic rocks, including the Keweenawan sequence, are overlapped by nearly horizontal rocks of Paleozoic (Upper Cambrian) age. What length of time, approximately, is missing from the geologic record here? What might this deposition of Upper Cambrian rocks represent?

_____

_____

_____

_____

_____

PHOTO INTERPRETATION OF PRECAMBRIAN ROCKS IN THE BELCHER ISLANDS, CANADA

The Belcher Islands, in the southeastern part of Hudson Bay, are a group of stratified, Proterozoic-age rocks. Their excellent exposure in the field, and their lithologic characteristics, make this area a good region for an investigation of the distribution of strata. Use Figures 15.8 and 15.9 to answer the following questions.

Figure 15.8 Air oblique view, looking south, of stratified Precambrian rocks on one of the Belcher Islands (Tukarak Island), southeastern part of Hudson Bay. Location is indicated by an arrow in Figure 15.1; also refer to the Geologic Map of North America. Distance across foreground of photo is about 2.4 km; distance to extreme background is about 24 km.

Courtesy of the Royal Canadian Air Force.

Study Questions

1. What is the direction of dip of Rock Units 7 to 9? (Units are numbered on Fig. 15.9.)

_____

_____

_____

_____

2. What is the explanation for the cross-linear pattern shown by the prominent dolomite bed of Unit 7 (examine Fig. 15.8)?

_____

_____

_____

_____

3. Of all the rock units shown in Figure 15.8, which is youngest? What is the evidence?

_____

_____

_____

_____

4. In Figure 15.8, a single large-scale fold is shown. Is it an anticline or a syncline? What is its direction of plunge? Explain your reasoning (see Chapter 14 for review of fold terminology).

_____

_____

_____

_____

Figure 15.9 Sketch of Figure 15.8, showing boundaries of rock units. Water is indicated by horizontal ruling. Rock Units 4 through 7 consist mainly of dolomite, siltstone, slate, and quartzite. The prominent, light-colored belt in the foreground, which is part of Unit 7, consists mainly of dolomite. Unit 8 is iron formation and slate, and Unit 9 consists mainly of pillow basalts, fragmented or pyroclastic volcanic rocks, and slate.

Geology from Jackson (1960), Geologic Survey of Canada Paper 60-20 and Map 28-1960. Reproduced with the permission of the Minister of Public Works and Government Services Canada, 2007, and Courtesy of Natural Resources Canada, Geological Survey of Canada.

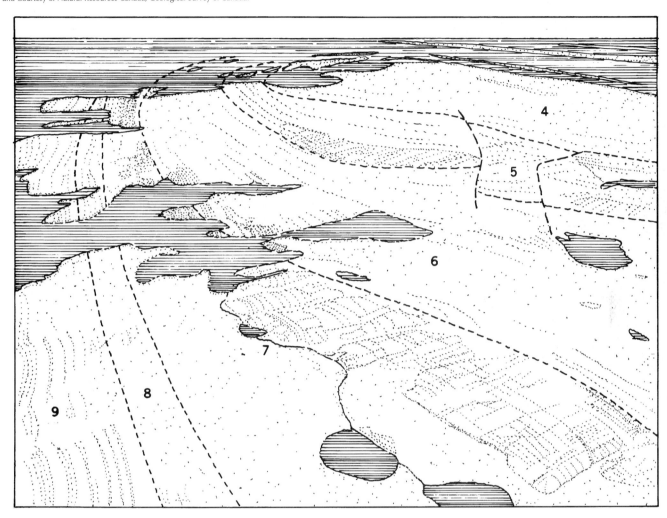

5. Which of the tectonic settings of Figure 5.3 would best account for the association of rock types in Figure 15.9? Which of the sedimentary environments (Table 4.1)?

_____

_____

_____

_____

6. On the basis of similarity in association of rock types, with which of the Precambrian sequences in the Lake Superior region (Table 15.1) would you correlate the Belcher Island strata?

_____

_____

_____

7. Although no radiometric ages are actually available from the Belcher Islands, let us assume that an age of 1.6 billion years has been obtained for these rocks. Would this tend to confirm your correlation? Explain.

_____

_____

_____

_____

8. Even if the association of rock types in the Belcher Islands was very similar to one of the Lake Superior sequences, correlation on this basis alone would be uncertain. What is the reason for this uncertainty?

_____

_____

_____

Figure 15.10 Geologic map of the area around Prosperous Lake, Yellowknife Gold District, Canada (1 mile = 1.609 km). Faults are indicated by rows of x's.

Map 868A, Prosperous Lake. District of Mackenzic, Northwest Territories. Reproduced with the permission of the Minister of Public Works and Government Services Canada, 2007, and Courtesy of Natural Resources Canada, Geological Survey of Canada.

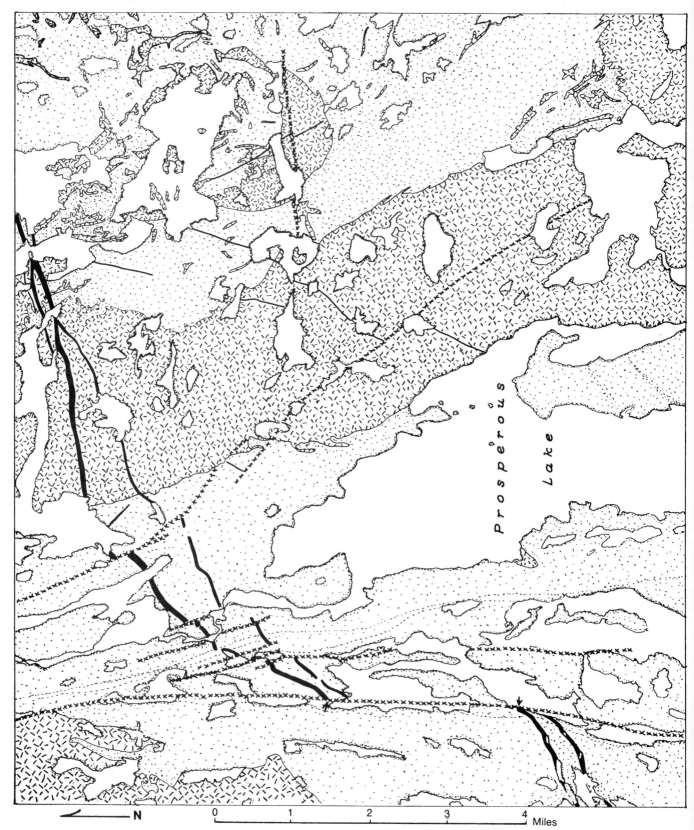

Figure 15.11 Vertical airphoto of the area around Prosperous Lake, Yellowknife Gold District, Canada.

THE YELLOWKNIFE GOLD DISTRICT, NORTHWEST TERRITORIES, CANADA

The Precambrian rocks of the Canadian Shield contain many valuable deposits of metallic minerals, including gold, silver, and uranium. Among the well-known gold-producing districts is the Yellowknife region, a small part of which is shown in Figures 15.10 and 15.11. Besides the presence of gold, the district is of interest for two other reasons. First, it is cross-cut by faults that rank with the largest known, steeply dipping displacements of the earth's crust. Total lateral displacement along a series of northward-trending parallel faults is about 11 miles (17.7 km). Second, about 230 miles (370 km) north of Yellowknife, one of the oldest known outcrops of the early crust of the earth was found recently. Radiometric dating of the tiny zircon mineral grains within these *felsic* rocks (rocks rich in quartz, muscovite, and feldspar minerals; see Chapter 19) indicate that they are 3.96 billion years old.

Figure 15.10 is a generalized geologic map in which three rock units are shown: the Yellowknife metasediments and volcanic rocks (dotted pattern), the Prosperous granite and **pegmatite** (geometric pattern), and gabbro dikes (solid black). Faults are indicated by rows of x's. The **metasediments** consist mostly of quartz + mica schist and phyllite, derived from slate and graywacke (see Chapter 20 for a summary of metamorphic minerals and rocks).

Study Questions

1. The area shown in Figures 15.10 and 15.11 is easily located on the Geologic Map of North America. First locate Great Slave Lake, and then find the word "Yellowknife" at the northern edge of the lake. Prosperous Lake, which looks very small on the map, is immediately above the "k" in Yellowknife. Remember that north is to the left in Figures 15.10 and 15.11. Both the major fault in the lower (western) part of the area and the body of Prosperous granite are shown on Figure 15.10.

   a. To what part of Precambrian time do the metasedimentary rocks belong (Archean or Proterozoic; Early, Middle, or Late)?

   b. What is their isotopic age? (See arrow on Fig. 15.1 for general location; do not include the one occurrence of 3.96-billion-year-old rock.)

   _____

   _____

   _____

   _____

2. By comparison with Figure 15.10, draw in the faults on the airphoto (Fig. 15.11). Next, draw in the gabbro dikes.

   a. Is the granite younger than the metasedimentary rocks? Give evidence in support of your answer.

   _____

   _____

   _____

   _____

   b. Are the dikes older or younger than the granite? Give evidence in support of your answer.

   _____

   _____

   _____

   _____

3. In the lower (western) part of the area, note that two prominent dikes (having a maximum width of 350 ft, or about 107 m), are offset along a system of parallel faults.

   a. What is the approximate horizontal displacement along the faults, as measured between the two arrows on Figure 15.10?

   _____

   _____

   _____

   _____

   b. The dikes are nearly vertical. Would you classify the faults as normal, reverse, or strike-slip? Explain.

   _____

   _____

   _____

   c. If the dikes had been emplaced after the period of faulting (rather than before), what evidence of this event would you expect to find?

   _____

   _____

   _____

4. The narrow, linear streaks of granite, outside of the main granite mass, are mostly pegmatitic dikes, and quartz veins branch out from some of these dikes. The gold and other metallic minerals (such as pyrite, galena, and sphalerite) are scattered through these veins, making up not more than a few percent of the volume of the vein. About four miles south of the area shown in the figures are two gold-producing mines, the Giant Yellowknife and the Ptarmigan.

   a. Is the mineralization of the quartz veins likely to have been associated with the emplacement of the gabbro dikes or with the granite?

   _____

   _____

   _____

   b. Some of the quartz veins are offset by faulting. If a geologist were tracing a gold-bearing vein and found it offset by a fault, would he expect to find the continuation of the vein to his left or his right, as he faced the fault?

   _____

   _____

   _____

TERMS

accretionary tectonics   Process whereby an allochthonous mass of continental or oceanic material is added to the margin of a craton by collision or welding (called suturing).

allochthonous   Formed or produced elsewhere than in its present location. Generally applied to a mass of rock that has been moved from its place of origin by tectonic processes such as thrust faulting, usually with kilometers of displacement.

Animikean sequence   Rocks of Early Proterozoic age exposed around the western shores of Lake Superior and famous for their iron ore deposits.

craton   That portion of a continent that has been tectonically stable for several hundred million years, which is underlain by Precambrian crystalline rocks. Cratons are commonly divided into areas of platforms or shields on the basis of the rocks that are exposed at the surface. **Platforms** are those areas of the craton in which flat-lying sedimentary rocks are exposed. **Shields** are those areas in which basement rocks are exposed at the surface.

dip   The maximum angle formed by the intersection of an inclined bed with a horizontal plane, measured perpendicular to **strike.**

epeirogeny   Tectonic activity of a primarily vertical, upward or downward, nature that affects large parts of the continent. Epeirogeny produces the larger features of continents and oceans, such as plateaus and basins, *in contrast* to the more localized process of orogeny, which forms mountain chains. Some epeirogenic and orogenic structures grade into each other in detail.

foreset bed   In a crossbedded unit, the inclined layers of material deposited on the relatively steep frontal slope. As these materials are deposited they cover the *bottomset bed* and in turn are covered (or truncated) by the *topset bed.*

gabbro   A phaneritic igneous rock composed mostly of calcic plagioclase and ferromagnesian minerals (such as amphibole or mica).

granite gneiss   A light-colored, foliated metamorphic rock where the quartz and feldspar minerals alternate in layers (bands) with more mafic minerals (such as micas and amphiboles).

greenstone   A metamorphosed (or altered) mafic igneous rock that owes its color to the presence of green minerals such as chlorite, epidote, and hornblende.

iron formation   A chemical sedimentary rock, typically thin-bedded and/or finely laminated, that contains at least 15% iron of sedimentary origin and commonly has layers of chert.

iron range   A term used in the Great Lakes region of the United States and Canada for a productive belt of iron formations. The term applies to a linear region rather than a topographic (*i.e.,* mountainous) elevation.

Keewatin sequence   Archean rocks (graywackes and greenstones) of the Lake Superior region.

Keweenawan sequence   Middle Proterozoic rocks (sandstones, conglomerates, some volcanic rocks, and deposits of native copper) of the Lake Superior region.

metasediments (or **metasedimentary rock**)   A sediment or sedimentary rock that shows evidence of having been subjected to some degree of metamorphism.

microcontinent (also called **allochthonous terrain,** *suspect terrain,* or a **microplate**).   A mass of rock, generally of continental crust, that has been postulated to be a fragment of a plate that has been accreted to a larger plate.

pegmatite   Very coarsely crystalline igneous rocks, most commonly of granitic composition, and often occurring in association with a large mass of plutonic rock of finer texture.

pillow lavas   Spheroidal or billowy masses of basaltic (or andesitic) lava formed by submarine extrusions.

strike   A direction indicated by the intersection of an inclined bed and a horizontal plane.

taconite   A local term used in the Lake Superior iron-bearing district for any bedded ferruginous chert or iron formation. The term is specifically applied to this type of rock when the iron content is at least 25%.

tonalite gneiss   A foliated metamorphic rock formed from an igneous rock (a form of **diorite** called **tonalite**) that contains at least 10% quartz.

16

Mountain Belts of North America

1. Demonstration specimens of igneous and metamorphic rocks: granite, basalt, schist, gneiss, greenstone, serpentine.

2. Geologic map of North America (instructor-provided) and Plate 1, Plate 2, and Plate 3 (following Chapter 20).

3. Additional map exercises, (instructor-provided) that may be used to supplement those in this chapter.

Colorplates

Appalachians

1–3 Athens Quad
4–6 Pine Grove Quad

Cordillera

7–9 Waucoba Spring Quad
10–12 White Mountain Peak Quad

Questions and mapping exercises for these colorplates are located at the end of the colorplate section (following Chapter 14).

HYPOTHESIS FOR THE ORIGIN OF MOUNTAIN BELTS

If new **oceanic crust** is being created by sea-floor spreading (Fig. 6.4), and if the earth remains the same size, something must be happening to the old oceanic crust. Hypotheses relating to the fate of old crust, and to the movements of lithospheric plates generally, have provided a rational explanation for mountain belts and for many other geological puzzles, such as the deep oceanic trenches and accompanying island arcs. As described in Chapter 6, the conceptual scheme relating to sea-floor spreading and the behavior of lithospheric plates is called **plate tectonics.** There are two basic premises of plate tectonics. (1) Lithospheric **plates,** such as the two spreading apart in Figure 6.4, are rigid and undergo little internal deformation. (2) Each plate is in motion relative to the others and may be deformed only at its boundaries. The boundaries of plates include *ridge axes* where plates are diverging and new oceanic crust is being

Figure 16.1 The trace of the San Andreas Fault in San Luis Obispo County, California can be seen in this vertical airphoto mosaic. The San Andreas is an example of a *transform fault.* It trends along the zone of slippage between the Pacific and American lithospheric plates (see Fig. 6.1 and Plate 1).

Courtesy U.S.G.S.

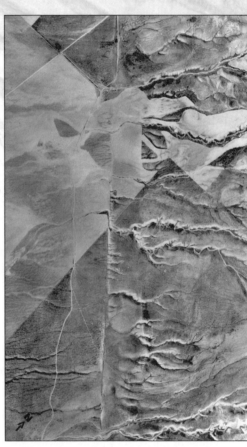

1. What features on the airphoto indicate the fault has been active relatively recently?

_____

_____

2. Strike-slip faults are often termed either **right lateral** or **left lateral.** To determine which term is appropriate, one may look across the fault trace to see if the opposite side moved to the right or the left. Is the San Andreas Fault right or left lateral?

_____

_____

_____

Figure 16.2 Hypothesis for the origin of mountain belts at a zone of subduction.

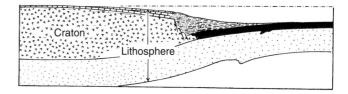

A

A. Sediments have accumulated at passive margin of craton that was rifted and separated from its adjoining block by sea-floor spreading (see Figure 6.4 and discussion on rifting). Carbonate rocks on the continental shelf grade seaward into the immature sandstones and shales of the continental rise, and these sediments grade into deep-sea sediments. Over time, a thick wedge of sedimentary material is deposited on the passive margin. Subduction begins by the reversal of plate motions (*convergent* rather than *divergent*) and the underthrusting of oceanic crust (solid black) lithosphere (note buckling on underside of the oceanic crust lithosphere).

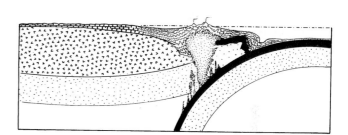

B

B. Lithospheric plate (at right), capped by oceanic crust, is subducted beneath lithospheric plate capped by continental crust and sediment. As the oceanic plate reaches depths of about 100 km, melting of the lithosphere generates basaltic-type magmas that rise toward the surface where submarine volcanic rocks are erupted onto the sea floor. A trench is formed at the zone of subduction, and sediments, as well as oceanic crust, are subducted.

C. Subduction continues. Granitic-and andesitic-type magmas, generated by partial melting of the oceanic crust in conjunction with continental crust (and sediments) subducted to depths of about 100 km, rise towards the surface. The rise and injection of these magmas, as well as the heating of the crustal rocks by the magmas, produces expansion and doming of the surface, and a mountain belt begins to rise. Sedimentary rocks adjacent to the magmas are metamorphosed (both contact and regional), and those on the flanks of the dome are folded and faulted. Characteristic metamorphic rocks of high pressure and low temperature (called *blueschists*) are formed and are evidence of subduction zone tectonics. Immature (orogenic) sediments accumulate both seaward and landward of the rising mountain belt.

C

D. Subduction continues. Further generation of granitic (and/or andesitic) magmas and the uplift of the mountain belts results in lateral spreading and thrust faulting toward the craton. Sheets of folded and metamorphosed immature sedimentary rocks are thrust over the carbonate rocks, which also become folded and thrust faulted (see Figure 16.4). New sediments continue to accumulate landward of the rising mountains. To seaward, the orogenic sediments accumulate and result in subsidence, allowing for great thicknesses of sediment to accumulate. The new granitic (or andesitic) and metamorphic rocks formed in this event become "welded" to the edge of the craton, increasing the area of crystalline cratonic basement. Mountain building (*orogenesis*) may continue intermediately for several geological periods, and many tens to hundreds of kilometers of old oceanic crust may be subducted and recycled.

D

Another craton or a volcanic island arc (similar to the Japanese islands), riding on the subducting plate, may be brought into collision with the craton creating a wide and complex orogenic belt. As the plate motions of these and other plates change with time, subduction ceases. These two plates may become *sutured* (welded) together, they may rift apart, or their junction may become a *transform fault* (Figure 16.1).

generated (Fig. 6.4); **transform faults** where plates slide past each other (Fig. 16.1); and **subduction zones** where plates converge and old oceanic crust is recycled as one plunges beneath the other.

Mountain belts consist of topographically mountainous, elongate zones of intensely deformed continental crust that generally have been associated with igneous activity and metamorphism. Mountain belts originate as **orogenic belts,** a term used for linear zones that are subjected to intense folding and deformation. Orogenic belts are localized along convergent plate boundaries and typically continue to be the locus of deformation for tens to hundreds of millions of years. There are two main types of orogenic belts: *marginal orogenic*

belts, formed along the margins of an over-riding plate above a subducting slab of lithosphere (Fig. 16.2); and *collisional* (or *alpine*) *orogenic belts,* formed along the suture zone of two continents (or fragments of continents) that have collided.

Marginal orogenic belts are of two types: *Andean-style* and *arc.* Andean-style orogenic belts are formed along a continental margin as a subduction slab slips underneath. This causes volcanic and seismic activity and results in mountain ranges such as the Andes Mountains of South America. Arc orogenic belts are the result of a volcanic island arc formed on oceanic **basalt,** which is then subducted beneath a continental margin. Unlike the marginal orogenic belt, the arc orogenic belt is separated from nearby continents by a

backarc basin. The islands of Japan provide a good example of this arrangement of island arc and backarc basin.

Marginal orogenic belts are characterized by chains of volcanoes and narrow zones of intense deformation. Partial melting of the continental crust and subducted oceanic crust (and the sediment riding on top of that crust) produces intermediate **(andesitic)** to felsic (granitic) magmas (Chapter 19). Arc volcanoes are generally andesitic to andesitic-basaltic in composition, whereas the Andean-style volcanoes are of a more felsic nature because they are formed from magmas that have more interaction with continental (silicic) crust. Andean-style orogenic belts have more intrusive activity than is characteristic of the arc belts. Normal faulting (Chapter 14) is more prominent in the arc belts, whereas thrust faulting in the Andean–style belts is common. Metamorphism, which may range from high pressure-low temperature (called **blueschists,** see Chapter 20) to low pressure-high temperature (*contact metamorphism*), is concentrated in the narrow zones of deformation and is not as predominant as the igneous activity that occurs in these marginal orogenic belts.

Collisional orogenic belts are characterized by intense deformation and regional metamorphism over a wider area than the marginal orogenic belts. This is easy to conceptualize as you realize that with the forces that form a collisional orogenic belt are two continental plates in collision. Where the two continental plates converge, the crust thickens to form great mountain ranges. The Himalaya Mountains are a result of the collision of the Indian and Eurasian plates in the Cenozoic. Generally, igneous activity, both intrusive and extrusive, is subordinate in collisional belts.

Along the **suture zone,** where the two continental plates are welded together, thrust faulting may have occurred and caused pieces of the oceanic crust caught between the two colliding plates to be **obducted** onto the continental crust. These pieces of oceanic crust (and possibly upper mantle) are called **ophiolites** and are composed of **mafic** and ultra-mafic rocks (Chapter 19), which are then metamorphosed to rocks that contain chlorite, **serpentine,** epidote, and plagioclase (albite composition). Ophiolites provide important physical evidence to the structure and materials of the oceanic crust.

THE APPALACHIAN OROGENIC BELT

Although we cannot determine how well the hypothesis depicted in Figure 16.2 accounts for the major features of our eastern or Appalachian orogenic belt, we will use it as a basis for discussion. The following questions concerning the Appalachian Mountain Belt relate to Figures 16.2, 16.3, and 16.4 and Plates 1, 2, and 3 (following Chapter 20).

Study Questions

1. Plate 1 is called a tectonic map because it emphasizes structural features, such as folding and faulting. It is not hypothetical, and it has not been influenced by the hypotheses of plate tectonics. Use it to answer the following questions.

   a. Judging from Plate 1, for what probable reason is the core of the Appalachian belt apparently wider in New England than it is farther south?

   _____

   _____

   _____

Figure 16.3 Diagrammatic structure sections across the northern part of the Appalachian Orogenic Belt. (A) Section from the Adirondack Mountains of New York across Vermont and New Hampshire. (B) Section from the Catskill Mountains of New York across Massachusetts. Note that the mountain building processes (*orogenesis*) for these sections is a result of two different events. The western portion (left side) of the sections was produced by Andean-style mountain building during the late Ordovician (Taconic orogeny). The eastern portion (right side) of the sections represents orogenesis related to continent—continent collision tectonics in the Devonian (Acadian orogeny). For scale conversion, 1 mile equals 1.609 km.

LEGEND: **p€:** Precambrian igneous and metamorphic rocks; **Pzs:** Paleozoic schists (mainly metamorphosed sandstone and shale and volcanic rocks of Cambrian, Ordovician, and Devonian age); **Pzi:** Paleozoic intrusive rocks (mainly granitic composition).

A. King, Philip B.; *The Evolution of North America.* © 1959 Princeton University Press, 1987 renewed PUP, 1977 revised edition, 2005 renewed PUP. Reprinted by permission of Princeton University Press.

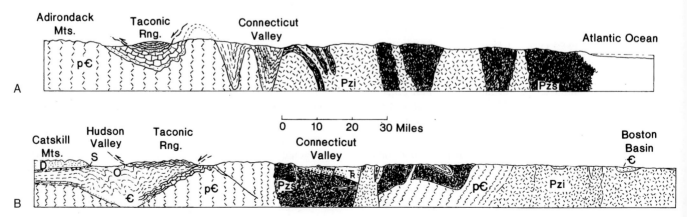

_____

_____

_____

b. According to the hypothesis depicted in Figure 16.2, why does the orogenic belt occur along the margins of the continents? What type of evidence can you find to support your answer?

_____

_____

_____

_____

_____

_____

c. Why are **granitic** bodies intruded into the core of the belt rather than in the marginal parts?

_____

_____

_____

_____

_____

_____

_____

_____

d. Why are the rocks of the marginal part not metamorphosed?

_____

_____

_____

_____

2. Figure 16.3A and B are geologic sections across the northern part of the Appalachian Belt in New England. (Note approximate location from the geography on Plate 2.)

a. Why do carbonate rocks occur in the western part of the region and immature sandstone, shales, and volcanic rocks (now metamorphosed) in the eastern part? Review margin orogenic belt formation.

_____

_____

b. In the Taconic Range, note the overthrusting of immature sedimentary rocks onto carbonate rocks and shales. According to the reconstructions above Section A, about what distance were these rocks overthrust? Where were the immature sandstones originally located?

_____

_____

_____

c. Note that the Cambrian and Ordovician rocks greatly decrease in thickness from the Hudson Valley to the Catskills. What is the explanation for this?

_____

_____

_____

_____

d. The Catskills are underlain by a great thickness of immature sedimentary rocks of Devonian age. What relation do these deposits have to the mountain building?

_____

_____

_____

e. In Section B, the Connecticut Valley is underlain by immature sandstone and conglomerate, predominantly of a reddish color and of Triassic age. Are these rocks accounted for by the hypothesis depicted in Figure 16.2?

_____

_____

_____

What type of depositional and tectonic environment would you expect for these Triassic strata?

_____

_____

_____

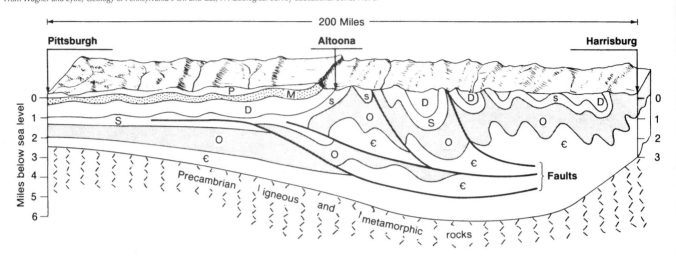

Figure 16.4 Section across the marginal part of the Appalachian Belt in Pennsylvania. Except for the Precambrian basement rocks, all rocks in this section are unmetamorphosed sedimentary rocks. For scale conversion, 1 mile equals 1.609 km.

LEGEND: **Є:** Cambrian; **O:** Ordovician; **D:** Devonian; **M:** Mississippian; **P:** Pennsylvanian.

From Wagner and Lytle, *Geology of Pennsylvania's Oil and Gas,* PA Geological Survey Educational Series No. 8.

If you do not think the Triassic rocks conform to the hypothesis of Figure 16.2, what subsequent events would be required to account for their present structural condition?

_____

_____

_____

_____

_____

3. Figure 16.4 is a structural section across the marginal part of the Appalachian Belt in Pennsylvania. (See also Plate 2.)

a. If a very deep well were drilled at Altoona, how many times would it pass through the same section of Ordovician rocks? Why does this occur? Explain.

_____

_____

_____

_____

_____

b. In the western part of the section, what is the approximate total thickness of sedimentary rocks above the Precambrian basement?

_____

_____

_____

_____

_____

c. Why do these rocks become thinner toward the west?

_____

_____

_____

_____

_____

d. Why are the rocks of the Appalachian Plateau (between Pittsburgh and Altoona) not folded and faulted?

_____

_____

_____

_____

_____

4. See Plate 3 (following Chapter 20) for a structure section across the Appalachian Belt in Virginia. Although covered, the core of the Appalachian Belt doubtless extends beneath the Coastal Plains sediments and to the edge of the craton.

a. Notice that the Piedmont, although in the core of the Appalachian Belt, is not mountainous and that the Mesozoic and Cenozoic sediments of the Coastal Plain are deposited on a nearly flat surface. What subsequent event, not included in the hypothesis of Figure 16.2, would account for this?

_____

_____

_____
_____
_____
_____
_____

b. According to the hypothesis, should Paleozoic sediments, derived from the rising mountain belt, be present along the edge of the craton? Explain your reasoning. (Note: The absence of these sediments has required an extension of the hypothesis. According to one current explanation, subduction brought the African and European cratons against the edge of the North American craton, forming a wide orogenic belt. This belt was then rifted in half, and sea-floor spreading moved the other half of it to its present position on the other side of the Atlantic.)

_____
_____
_____
_____
_____
_____

THE CORDILLERAN OROGENIC BELT: PACIFIC RANGES

Like the Appalachian Orogenic Belt, the Cordilleran Orogenic Belt of western North America has a core zone of igneous and metamorphic rocks and a marginal belt of folded and faulted rocks. However, the tectonic history of the Cordilleran is very different from the Appalachian tectonic history. While much of the tectonic history of the Appalachians was a result of continent-to-continent collision or subduction (with some arc-related subduction), the Cordilleran tectonics were dominated by collisions and subduction-related activity with island arcs and **accreted terranes** (see Chapter 15). These accreted terranes (often called *microcontinents*) often possessed geological formations, rock types, and structures that were very different from the rocks of the western North American craton. Thus, this diversity of rock types had a history prior to the formation of the Cordilleran. When they were subducted and sutured to the western North American craton, they produced similar types of features as those seen in the Appalachian orogenic belt; however, the origin of some of the structures and the rock types are controlled not by the tectonic

combination processes but by the original formation of the accreted terranes before they reached western North America.

Study Questions

1. Which orogenic belt (Appalachian or Cordilleran) has the widest marginal zone?

    _____
    _____
    _____

    Which orogenic belt has the greatest area of granitic intrusions? What is the distribution of the granitic intrusions (are they uniform or do they follow some pattern)? Why do you think they have this type of distribution (think about how they would form)?

2. Other differences between the Appalachian and Cordilleran Orogenic Belts can be examined on the structure section of Plate 3 (following Chapter 20).

    a. The core of the Sierra Nevada Mountains (Cordilleran Orogenic Belt; Plate 3; third structure section) is separated from the oceanic crust by a great thickness of immature sandstone, shale, and volcanic rocks that have been folded and faulted but not, for the most part, metamorphosed. What does this suggest about the tectonic history of these rocks (review ideas in Figure 16.2)? Where would the subduction zone be located? How could this interpretation change if the sedimentary and volcanic rocks were part of an accreted terrane?

    _____
    _____
    _____

    b. There is, at present, no trench or subduction zone off California. The plate junction between the Pacific Plate and the American Plate (western margin) is hypothesized to be a transform or strike-slip fault (see Chapter 14 for review). Notice the faults in the California Coast Ranges (Plate 3; third structure section). Which of these faults could be described in that manner? Is it possible that the faults are normal or reverse faults (see Chapter 14)? Explain.

    _____
    _____
    _____

    c. The granitic intrusions of the Sierra Nevada Mountains have been dated radiometrically and the ages range from 143-88 million years old

Figure 16.5 Structural cross section in the northern Cascade Range near Seattle, Washington. Note that this section shows some of the complex features of the tectonic history of the western North American craton. The Mesozoic granites represent subduction-related magmatism while the "flood" basalts represent mid-Tertiary rifting and the metamorphosed ultramafic rocks could represent obducted crust caught in an accreted terrane (Pm unit) collision. For scale conversion, 1 mile equals 1.609 km.

After Smith and Calkins (1906), USGS Geologic Atlas, Folio 139; see also King (1977), *Evolution of North America.*

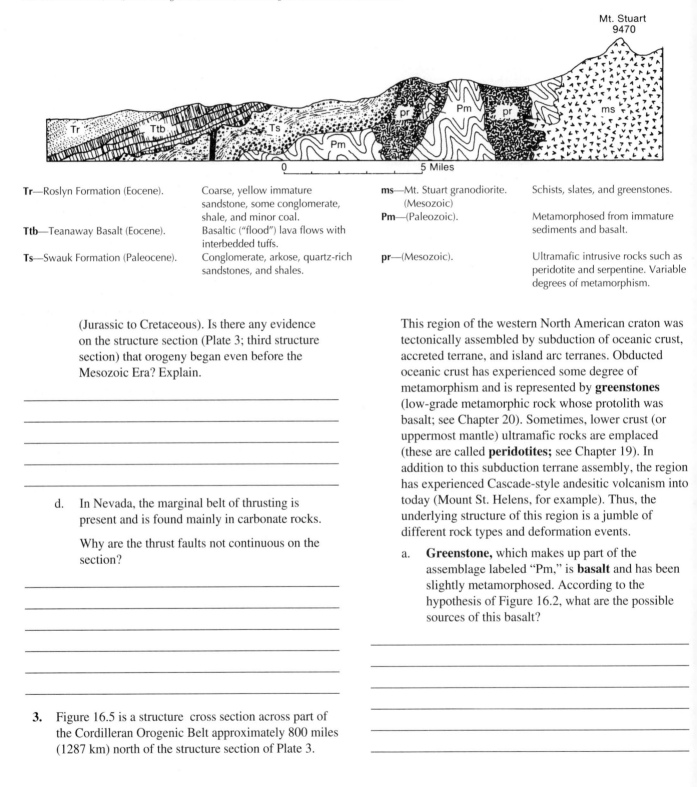

Tr—Roslyn Formation (Eocene).	Coarse, yellow immature sandstone, some conglomerate, shale, and minor coal.	**ms**—Mt. Stuart granodiorite. (Mesozoic)	Schists, slates, and greenstones.
Ttb—Teanaway Basalt (Eocene).	Basaltic ("flood") lava flows with interbedded tuffs.	**Pm**—(Paleozoic).	Metamorphosed from immature sediments and basalt.
Ts—Swauk Formation (Paleocene).	Conglomerate, arkose, quartz-rich sandstones, and shales.	**pr**—(Mesozoic).	Ultramafic intrusive rocks such as peridotite and serpentine. Variable degrees of metamorphism.

(Jurassic to Cretaceous). Is there any evidence on the structure section (Plate 3; third structure section) that orogeny began even before the Mesozoic Era? Explain.

_____

_____

_____

_____

_____

  d. In Nevada, the marginal belt of thrusting is present and is found mainly in carbonate rocks.

Why are the thrust faults not continuous on the section?

_____

_____

_____

_____

_____

3. Figure 16.5 is a structure cross section across part of the Cordilleran Orogenic Belt approximately 800 miles (1287 km) north of the structure section of Plate 3.

This region of the western North American craton was tectonically assembled by subduction of oceanic crust, accreted terrane, and island arc terranes. Obducted oceanic crust has experienced some degree of metamorphism and is represented by **greenstones** (low-grade metamorphic rock whose protolith was basalt; see Chapter 20). Sometimes, lower crust (or uppermost mantle) ultramafic rocks are emplaced (these are called **peridotites;** see Chapter 19). In addition to this subduction terrane assembly, the region has experienced Cascade-style andesitic volcanism into today (Mount St. Helens, for example). Thus, the underlying structure of this region is a jumble of different rock types and deformation events.

  a. **Greenstone,** which makes up part of the assemblage labeled "Pm," is **basalt** and has been slightly metamorphosed. According to the hypothesis of Figure 16.2, what are the possible sources of this basalt?

_____

_____

_____

_____

_____

_____

_____

b. **Peridotite** (or serpentine, which is altered peridotite) has the same composition as the lower lithosphere (pattern of crosses on Fig. 16.2). What plate tectonic process(es) would be required to allow part of the lithosphere to be crushed and squeezed by the rocks of an orogenic belt?

_____

_____

_____

_____

c. According to the structural cross section of Fig. 16.5, was the peridotite squeezed into place before or after intrusion of the granite? What is your reasoning? Be specific.

_____

_____

_____

_____

_____

d. What is the probable sedimentary environment of the Swauk Formation? What events preceded its deposition?

_____

_____

_____

_____

e. What evidence indicates that deformation of the region took place before Paleocene time?

_____

_____

_____

_____

After Paleocene time?

_____

_____

_____

_____

After Eocene time?

_____

_____

_____

_____

THE CORDILLERAN OROGENIC BELT: COLORADO AND WYOMING ROCKY MOUNTAINS

As shown in Plate 2 (following Chapter 20), the Colorado and Wyoming Rocky Mountains consist (mainly) of large isolated mountain ranges, of which the Big Horn Range (labeled BH on Plate 2), the Wind River Range (WR), the Colorado Front Range (FR), and the Uinta Range (U) are examples. Plate 1 suggests that the emplacement of these mountain ranges may have been related to tectonic uplift rather than thrust faulting. In addition, the uplift of these mountains does not seem related to emplacement of large intrusive bodies of granitic magma (called **batholiths;** see distribution on Plate 2 and correlate with position of these mountain ranges). This type of emplacement occurred in the Cenozoic as a result of the Laramide Orogeny and is called block faulting. This type of mountain range formation is unusual and not related to the two main types of orogenesis described earlier. Since these blocks were uplifted, they are mainly composed of basement crystalline rock and experienced very little magmatic activity.

Figure 16.6 Reconstructed stages in the development of the Medicine Bow Mountains, WY.

From S. H. Knight (1953), *Guidebook for 8th Annual Field Conf.,* Wyoming Geological Association.

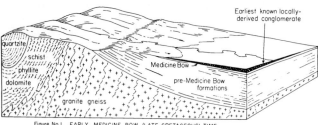

Figure No.1, EARLY MEDICINE BOW (LATE CRETACEOUS) TIME.

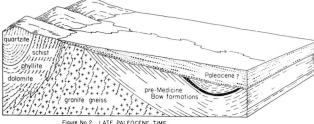

Figure No.2, LATE PALEOCENE TIME.

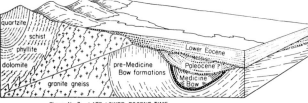

Figure No.3, LATE LOWER EOCENE TIME.

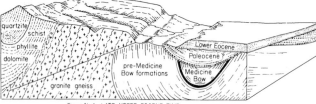

Figure No.4, LATE UPPER EOCENE TIME

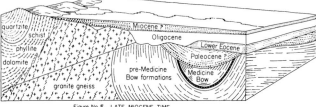

Figure No.5, LATE MIOCENE TIME.

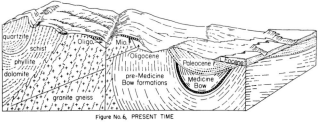

Figure No.6, PRESENT TIME

Study Questions

1. The history of the Medicine Bow Range (indicated by letters "MB" in Plate 2) is analogous to that of other ranges in Colorado and Wyoming. The basement rocks shown in Figure 16.6 (quartzite, granite gneiss) are of Precambrian age. The Medicine Bow Formation (Late Cretaceous) is composed of gravel derived from the range. The "pre-Medicine Bow Formations" are mainly marine (Paleozoic and Mesozoic) and reach a thickness of about 15,000 ft (4572 m) east of the range.

   a. In Figure 16.6, uppermost block, why do the pre-Medicine Bow Formations end at the margin of the range?

   _____
   _____
   _____
   _____
   _____
   _____

   b. During what period of geologic time did uplift of the range begin?

   _____
   _____
   _____
   _____

   c. On what evidence is thrust faulting of the range dated as prior to the lower Eocene?

   _____
   _____
   _____
   _____
   _____

   d. Why is it inferred that no uplift of the range took place during the middle and upper Eocene?

   _____
   _____

   e. Why is it inferred that renewed uplift, by normal faulting, took place in the late Miocene?

   _____
   _____
   _____

MAP INTERPRETATION OF MOUNTAIN BELTS

Major features of the mountain belts are well shown on the Geologic Map of North America. The meanings of the colors and symbols on the map are given on its *Explanation,* but it will not be necessary to refer to the explanation in detail.

First, notice the very large area of Precambrian rocks exposed in the Canadian Shield. These same colors and symbols are used for the smaller exposures of Precambrian rocks, found from place to place in the United States. Paleozoic sedimentary rocks are mostly in hues of purple and blue; Mesozoic sedimentary rocks are in hues of green, whereas Cenozoic sedimentary rocks are in yellow and orange. Igneous rocks are depicted mostly in hues of pink or red.

Figure 16.7 Map patterns and structures typical of mountain belts.

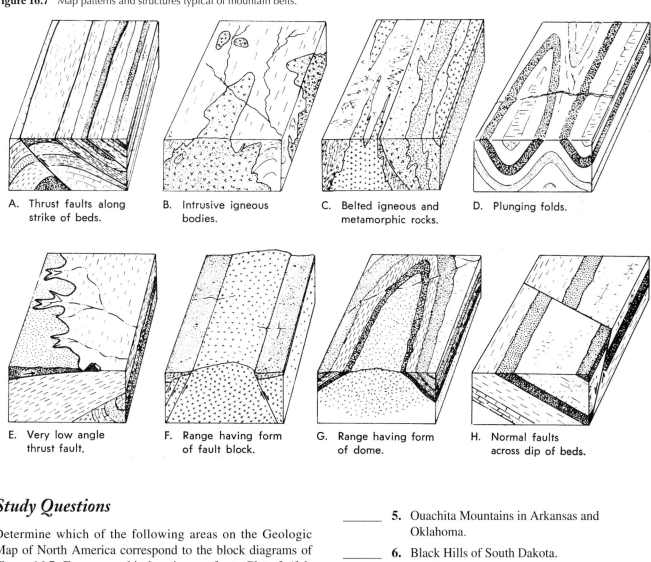

A. Thrust faults along strike of beds.

B. Intrusive igneous bodies.

C. Belted igneous and metamorphic rocks.

D. Plunging folds.

E. Very low angle thrust fault.

F. Range having form of fault block.

G. Range having form of dome.

H. Normal faults across dip of beds.

Study Questions

Determine which of the following areas on the Geologic Map of North America correspond to the block diagrams of Figure 16.7. For geographic locations, refer to Plate 2 (following Chapter 20).

_____ 1. New England.

_____ 2. Ridge and Valley Province in Pennsylvania.

_____ 3. Appalachian Piedmont.

_____ 4. Southern part of Ridge and Valley Province.

_____ 5. Ouachita Mountains in Arkansas and Oklahoma.

_____ 6. Black Hills of South Dakota.

_____ 7. Big Horn Range of Wyoming.

_____ 8. Uinta Range in Northern Utah.

_____ 9. Eastern part of Canadian Rocky Mountains (in Alberta).

_____ 10. Eastern part of Northern Rocky Mountains at international boundary.

TERMS

andesite   A dark-colored, fine-grained igneous rock composed of mafic minerals (biotite, pyroxene, hornblende), plagioclase (more Na-rich), and quartz (0 to 20% modal abundance); intermediate in composition between *rhyolite* and **basalt.** The name andesite comes from the Andes Mountains in South America.

backarc basin   Depositional region situated between a converging volcanic arc and a continent. Generally elongate, these basins fill with detritus from both the arc (volcaniclastic matter) and the continent (terrigenous sediment). The Sea of Japan is a good example of a backarc basin.

basalt   A dark-colored, fine-grained igneous extrusive rock composed mainly of rock glass, plagioclase feldspar, and mafic minerals such as pyroxene and olivine. A basalt is the common rock type of the ocean floor. An intrusive igneous rock with the same mineralogy as a basalt but larger grain size and no rock glass is called *gabbro.*

batholith   A large, generally discordant plutonic mass that has more than 100 km² of surface expression and no known floor.

blueschist   A schistose metamorphic rock with a blue color due to the presence of kyanite, sodic amphibole (glaucophane), and bluish-gray lawsonite. Formed in low temperature, high-pressure metamorphic conditions associated with subduction. Properly termed the lawsonite-glaucophane metamorphic facies.

granite   A light-colored, coarse-grained igneous intrusive rock composed mainly of quartz, plagioclase and potassium feldspar, and mafic mineral such as biotite mica or amphibole. Granite is a common rock found within the core of mountain ranges. An extrusive igneous rock with the same mineralogy (but with rock glass and a fine grain size) is called *rhyolite.*

greenstone   A metamorphosed (or altered) mafic igneous rock that owes its color to the presence of green minerals such as chlorite, epidote, and hornblende.

left lateral fault   A strike-slip fault along which displacement of the far block is to the left when viewed from the opposite block.

mafic   Term used to describe an igneous rock composed chiefly of one or more ferromagnesian (iron- and magnesium-rich), dark-colored minerals such as olivine, pyroxene, or amphibole. See Chapter 18 for further details.

magma   A naturally occurring mobile rock material, generated within the earth and capable of intrusion and extrusion, from which igneous rocks are thought to have been derived through solidification and related processes.

obduction   The overriding or overthrusting of oceanic crust onto the leading edges of continental lithospheric plates.

oceanic crust   The type of crust that forms the floor of the ocean. It is about 5 km thick, with a density of 3.0 g/cm³ and has a basaltic composition.

ophiolite   A group of mafic and ultramafic rocks (coarse-grained gabbros and diabase, volcanic rocks, usually overlain by radiolarian cherts), whose origin is thought to be a result of **obduction** of oceanic crust. Most ophiolites are variably metamorphosed and are difficult to identify.

orogenic belt (also called *fold* or *mountain belt*).   Linear regions of the earth's crust that are or have been subjected to intense deformation so as to produce elongate mountain ranges.

peridotite   A general term for a coarse-grained plutonic rock composed chiefly of olivine with (or without) other mafic minerals such as pyroxenes, amphiboles, or micas and containing little or no feldspar.

plate   A rigid, thin segment of the lithosphere, which may be assumed to move horizontally, and which adjoins other plates along zones of frequent earthquake activity.

plate tectonics   A theory of *global* tectonics where the lithosphere is divided into a number of rigid **plates** that interact with one another at their boundaries, causing seismic and tectonic activity.

right lateral fault   A strike-slip fault, along which displacement of the far block is to the right when viewed from the opposite block.

serpentine   A group of common rock-forming minerals having the formula $(Mg,Fe)_3Si_2O_5(OH)_4$. Serpentines are always secondary minerals, derived by alteration of magnesium-rich silicate minerals (especially olivine), and are found in both igneous and metamorphic rocks.

subduction zone   An inclined, planar zone formed by the descent of the leading edge of a lithospheric plate beneath an opposing plate.

suture zone   A region, often many kilometers in width and hundreds to thousands of kilometers in extent, where two continental plate margins collided (and were welded) together during a convergent plate interaction. This region is often identified by contrasting rock types, igneous intrusions, and various forms of faulting and deformation.

transform fault   A variety of strike-slip fault that is formed across the trends of divergent plate junctions and results in offsets along the divergent plate junctions.

17

The Interior Plains and Plateaus

WHAT YOU WILL NEED

1. Demonstration specimens of sedimentary rocks: limestone, fossiliferous, limestone, dolostone, sandstone (various maturities), conglomerate, breccia, shale, mudstone, and siltstone.

2. Geologic map of North America (instructor-provided) and Plate 1, Plate 2, and Plate 3 (following Chapter 20).

3. Additional map exercises, (instructor-provided) that may be used to supplement those in this chapter.

MAJOR STRUCTURAL FEATURES

As discussed in Chapter 15, the North American craton is composed of the Canadian Shield and the platform, which is the region of the Interior Plains and Plateaus (see Plate 2 for location). The strata of the platform (called the **cover**) are not strictly horizontal. The platform cover rocks, which overlie the Precambrian **basement rocks,** have been deformed by epeirogenic processes into broad **synclines, anticlines, basins,** and **domes** (or **arches;** see Chapter 14 for a review of geologic structural features). These are the major structural features of the region, and they were formed mainly during the time the sedimentary rocks were accumulating. The great width and breadth of these structural features is evident in that the limbs are very gently inclined, with dip generally measured in terms of meters of displacement per kilometers of traverse, rather than in degrees. The cover rocks vary in thickness and range from about 10,000 ft in basins to nonexistent (or very thin) on domes (or arches).

The outer parts of the region, which border the mountain belts, are called the **forelands.** The sedimentary strata of the forelands thicken toward the mountain belts and are uplifted along the belt margins.

The eastern and southern coastal areas of North America are formed by a wedge of Mesozoic and Cenozoic sedimentary rocks that thickens and dips gently toward the Atlantic Ocean and the Gulf of Mexico. The land surface of this wedge constitutes the Coastal Plains; the seaward extension forms the continental shelves. Although the Coastal Plains are somewhat different geologically from the interior region, they often are similar in surface aspect and will be treated briefly.

Figure 17.1 is a **structural contour map** in which the contours represent the depth of Precambrian basement below the present land surface. The contour lines also represent the thickness of the sedimentary cover overlying the basement. Even if you're not confident about reading contour maps, it will be apparent to you that the sedimentary cover increases in thickness toward the center of a basin and decreases in thickness toward the center of a dome.

Study Questions

1. Identify the following features by writing their names where they appear on Figure 17.1. Their geographic names indicate their general location (see Plate 2 for geographic information; following Chapter 20).

   Michigan Basin

   Illinois Basin

   Cincinnati Arch

   Nashville Dome

   Ozark Dome

   Dakota (Williston) Basin

   Foreland of the Appalachian Mountain Belt

   Foreland of the Cordilleran Mountain Belt

2. At what depth, to the nearest 1000 ft, would a well encounter the Precambrian basement at St. Louis?

   _____

   _____

   _____

   At Chicago?

   _____

   _____

   _____

   At the southwestern corner of Minnesota?

   _____

   _____

   _____

239

Figure 17.1 Structure contours on the Precambrian basement rock of the Interior Plains and Plateaus. Datum is present land surface.

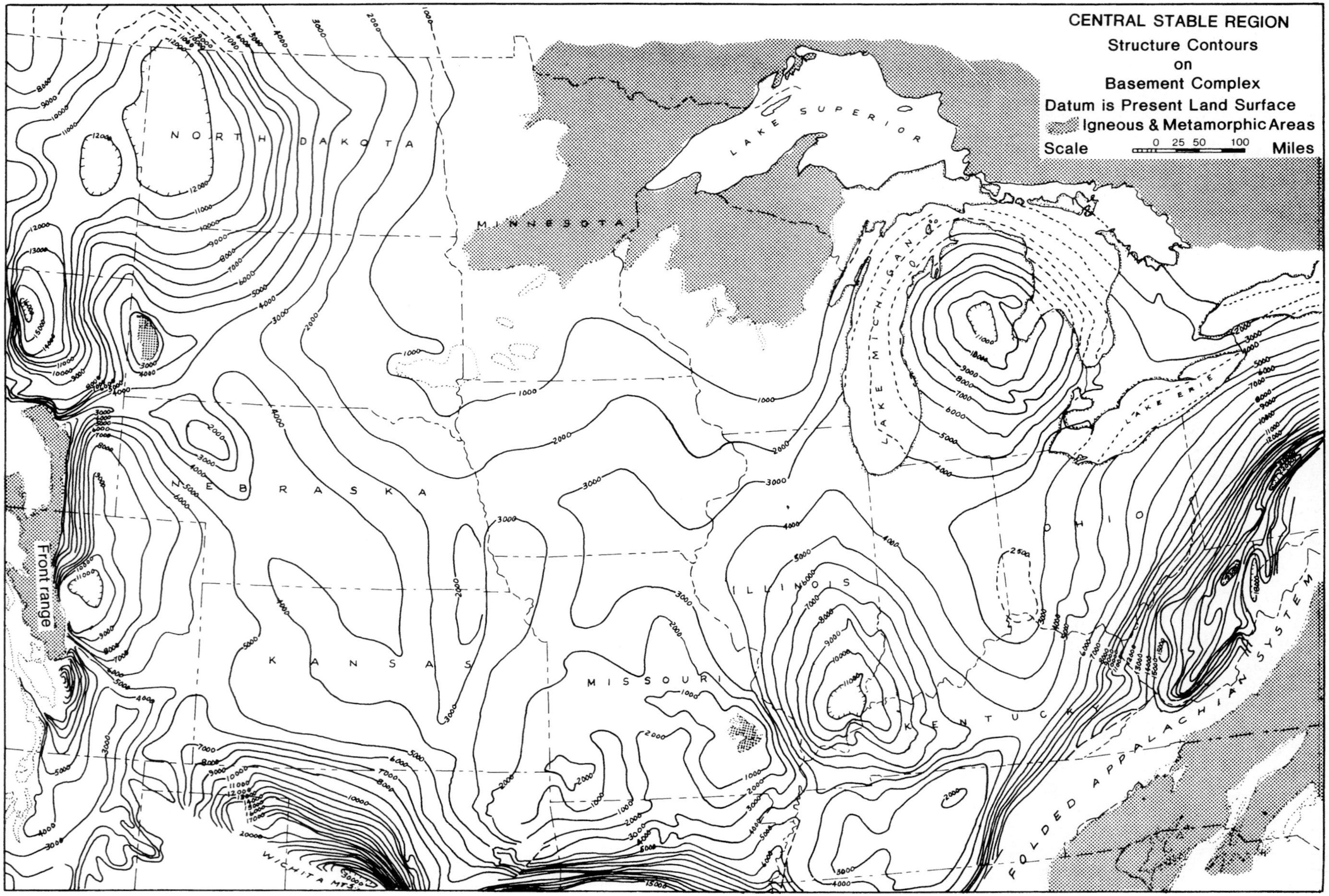

Figure 17.2 Map patterns and structural features of plain and plateau regions.

A. Dome or arch.

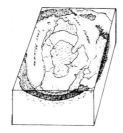

B. Basin.

C. Horizontal or gently dipping strata.

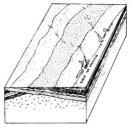

D. Dipping strata.

E. Unconformity— horizontal over dipping strata.

F. Erosional remnants of horizontal strata.

G. Unconformity— horizontal strata over crystalline rocks.

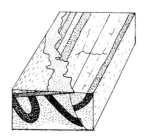

H. Unconformity— dipping strata over folds.

INTERPRETATION OF THE GEOLOGIC MAP OF NORTH AMERICA

Map patterns typical of major structural features in the region of plains and plateaus are shown by block diagrams in Figure 17.2. They may be helpful in interpreting the geologic map of North America that your instructor has provided (or the colorplates in this laboratory manual). In doing this interpretation, you will learn a great deal about the geology of the country. Use Plate 1 and Plate 2 (following Chapter 20) for aid in geographic locations.

Study Questions

1. Let's start with Hudson Bay in Canada. Notice the large area of purple Ordovician (O) and Silurian (S) strata (carbonate) along the south shore of the bay. What two alternative explanations account for the absence of the Cambrian?

_____

_____

_____

_____

_____

How would you account for the fact that the band of O is discontinuous beneath the S?

_____

_____

_____

_____

The Ordovician and Silurian strata also appear to the southwest, around Lake Winnipeg. Does it seem probable or improbable that the intervening belt of Precambrian was once covered by them? (Notice also the isolated patches of O and S in the northern part of Canada.)

_____

_____

_____

_____

2. In Wisconsin, just south of Lake Superior, a large, roundish patch of pink Precambrian is bordered by orange Cambrian. What is the structural form there?

_____

_____

_____

_____

Farther to the west, and shown well on Figure 17.1, a large prong of Precambrian rock extends southwest across Minnesota. This Precambrian high is sometimes called the "Continental Backbone."

3. Continuing southwest across the plains, notice the outcrop pattern of green and yellow in Kansas. To which of the blocks in Figure 17.2 does this most nearly correspond?

_____

_____

_____

_____

What is the direction of dip of the strata shown in green?

_____

_____

_____

What is their geologic age?

_____

_____

_____

Why does it seem probable that the green is everywhere present beneath the yellow?

_____

_____

_____

4. Farther to the south, in Texas, notice the relation between the dark green K1 and the underlying bands of blue. What is the age of the blue?

_____

_____

_____

_____

_____

What is the age of the dark green?

_____

_____

_____

_____

What is the explanation for the fact that the green cuts across the bands of blue?

_____

_____

_____

_____

Why does it seem probable that the green once covered a much larger area of blue?

_____

_____

_____

_____

_____

5. In fact, the hues of green seem to form a wide band under the plains, all the way from Mexico into northern Canada, and extending westward into the Rocky Mountain region where it is discontinuous because of mountainous uplifts. According to the map explanation, is the Cretaceous marine or nonmarine (continental)?

_____

_____

_____

_____

Could this band have been the edge of the continent in Cretaceous time? (Recall the age of the western Cordillera in California; see Chapter 16.)

_____

_____

_____

_____

If not, what must have been the location of the sea in which it was deposited, relative to the western Cordillera? (See also the block diagram of Fig. 5.3.)

_____

_____

_____

_____

6. Notice the bands of green, pale orange, and yellow that cross Texas diagonally, nearly parallel to the present coastline. Except in Texas, these bands form the Coastal Plain, whose inner margin begins with the green. In Texas, the inner margin is placed arbitrarily along the line of faults (like the ones near Austin), which make a prominent **escarpment** called the Balcones (see Plate 2). To which of the blocks in Figure 17.2 do these bands correspond?

_____

_____

_____

_____

Do the bands get younger or older toward the coastline?

_____

_____

_____

From this relation, do you infer that (a) the sea has been gradually regressing since Cretaceous time or (b) the Tertiary strata once extended inland and covered the Cretaceous but have since been removed by erosion?

_____

_____

_____

7. Along the Mississippi Valley, the Coastal Plain swings northward in a great **embayment.** Can you suggest a reason why the bands of green are missing from the western side of the embayment in Arkansas and Missouri? (Hint: What is the tectonic nature of the regions that border the embayment there?)

_____

_____

_____

Now think about the general outcrop pattern of the embayment. Does it most resemble a large plunging anticline or a syncline? What is the direction of **plunge?**

_____

_____

_____

_____

_____

_____

_____

8. Notice the striking pattern east of the embayment where the belts of green and orange swing across the Appalachian Orogenic Belt. Which of the blocks in Figure 17.2 does this resemble?

_____

_____

_____

_____

_____

_____

Does it seem likely that the Appalachian Belt continues indefinitely to the southwest beneath the Coastal Plain strata? (See Plate 1.)

_____

_____

_____

_____

_____

9. Now follow the band of green along the inner edge of the Atlantic Coastal Plain to the place where it disappears beneath the orange in North Carolina. Propose an explanation for this disappearance.

_____

_____

_____

_____

10. The whole of the Coastal Plain belt gradually becomes more narrow northeastward and disappears from view; its last remnants are at Long Island and Cape Cod. What could have happened to it? (Hint: Notice the trend of the continental shelf, whose edge is marked by the closely spaced blue contours of the continental slope; for example, south of Georges Bank. Recall also that the northern United States and Canada were depressed beneath the weight of the massive Pleistocene continental glaciers.)

_____

_____

_____

_____

_____

11. Now let's return for a moment to the domes and basins of the central states. Remember that the stratigraphic units get younger toward the center of a basin and older toward the center of a dome. The locations of domes and basins on the geologic map of North America can be seen by comparing it with Figure 17.1. The most symmetrical of these structures is the Michigan Basin. What is the age of the youngest rocks in its center?

_____

_____

_____

_____

Of the oldest stratified rocks on its northern margin?

_____

_____

_____

Do you see any connection between the basin and the shapes of Lakes Michigan and Huron?

_____

_____

_____

_____

12. What is the age of the oldest rocks in the center of the Cincinnati Arch and the Nashville Dome?

_____

_____

_____

To the east of the Cincinnati Arch are rocks (blue) of the Appalachian Foreland, whose geomorphic expression is the Appalachian Plateau. What is their apparent structural form?

_____

_____

_____

_____

13. To the west of the Cincinnati Arch is the Illinois Basin. What is the maximum thickness of stratified rocks in this basin?

_____

_____

_____

_____

Coal beds are associated with the Pennsylvanian strata, south of Michigan. Illinois has very large coal reserves, as does West Virginia. Would you also expect to find coal in Alabama? In Kansas? In Indiana? Explain your reasons.

_____

_____

_____

_____

14. To the southwest of the Illinois Basin is the broad Ozark Dome, whose geomorphic expression is the Ozark Plateau. What is the age of the oldest rocks in its center?

_____

_____

_____

_____

It is not a symmetrical dome. What is the direction of its elongation?

_____

_____

_____

_____

Notice the virtual absence of Silurian and Devonian strata in the Ozark Dome. Why are they missing?

_____

_____

_____

_____

INTERPRETATION OF GEOLOGIC STRUCTURE SECTIONS

The **geologic map** shows the strata where they are exposed at the earth's surface. Structure sections are also essential for geological purposes because they show what is beneath the surface. The following exercise examines a number of geologic features in the interior plains and plateaus region and asks questions that refer to the structure sections of Figure 17.3.

Michigan Basin, Illinois Basin, and Adjoining Domes

In Figure 17.3, Section A begins north of Lake Michigan and crosses the Michigan Basin in a southeasterly direction, ending just south of Lake Erie. Section B begins on the Ozark Dome and crosses the Illinois Basin in a northeasterly direction. Although deposition in the basins was not continuous, strata of all Paleozoic periods except the Permian are present.

Study Questions

1. Notice the thick salt beds (a maximum thickness of about 1800 ft) of Silurian age in the Michigan Basin. The deposition of thick accumulations of **evaporites** required a slowly subsiding basin and restricted circulation, which was provided by barrier reefs bordering the basin. With this type of arrangement, the evaporation of seawater was faster than the influx of fresh seawater, and the result was a net accumulation of evaporites. Why would the Michigan Basin be more favorable for this situation than the Illinois Basin?

    _____

    _____

    _____

2. In Section A, the Devonian strata (black) thicken toward the center of the Michigan Basin. What is an explanation for this?

    _____

    _____

    _____

    _____

    However, in Section B, the Devonian strata thin toward the Ozark Dome but thicken toward the Kankakee (Cincinnati) Arch. What is an explanation for this?

    _____

    _____

    _____

    _____

    _____

3. What evidence for an unconformity between Pennsylvanian and older strata can be seen in the sections? What is the type of unconformity?

    _____

    _____

    _____

    _____

    _____

4. The Pennsylvanian strata consist mainly of immature sandstone and shale, in part fluvial and deltaic, but the older Paleozoic strata are mainly carbonate. What was the probable source of the sand and silt that form the Pennsylvanian?

    _____

    _____

    _____

    _____

    _____

The Appalachian Foreland

Section C of Figure 17.3 begins at the Cincinnati (Findlay) Arch and crosses the Appalachian foreland in an easterly direction across West Virginia. It represents the strata as they are presently deformed, and Section D shows how they would look if they were straightened out and not eroded.

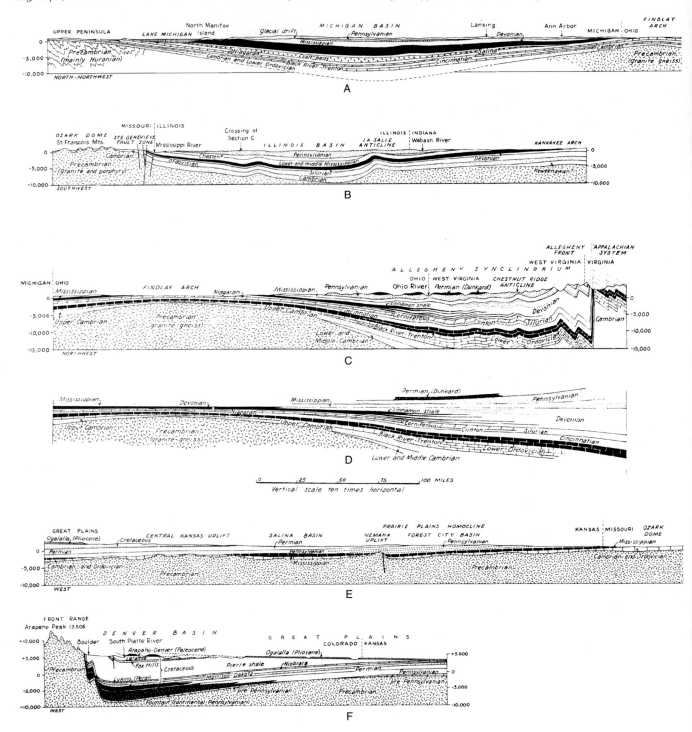

Study Questions

1. In Section D, notice that the thickest strata are carbonate rock, deposited from the Cambrian through the Devonian. What is their inferred tectonic setting of deposition? (Refer to Fig. 5.3.)

   _____

   _____

   Why do they thicken greatly toward the east?

   _____

   _____

   _____

2. What is the explanation for the lesser thickness of Mississippian strata and then the greatly increased thickness of the clastic Pennsylvanian?

   _____

   _____

   _____

   _____

Midcontinent Region and the Cordillera Foreland

Section E of Figure 17.3 begins (at right) on the Ozark Dome and crosses Kansas in a westerly direction. Section F is nearly continuous with Section E and continues westerly to Denver and the Colorado Front Range.

Study Questions

1. In Section E (and also in Plate 3), notice the thinness of the pre-Pennsylvanian Paleozoic strata. Why would they be thinner here than they are in the eastern United States?

   _____

   _____

   _____

   _____

2. The history of the Colorado Front Range is recorded in the rocks of the Cordilleran Foreland, Section F.

   a. Was an uplift present at the site of the Front Range when the Pennsylvanian strata there were deposited (red, immature sandstone and conglomerate)?

   _____

   _____

   _____

   b. Was an uplift present when the Cretaceous strata (marine shale) were deposited?

   _____

   _____

c. The Arapahoe and Denver Formations, named on Section F, are of very Late Cretaceous and Paleocene age. They contain conglomerate and immature sandstone. What do they tell us about the Colorado Front Range?

   _____

   _____

d. The Ogallala Formation, which appears on the geologic map as an extensive area of yellow mantling the plains, consists of fluvially deposited conglomerate and sandstone. What does it tell us about the Colorado Front Range, as well as other ranges of the Rockies?

   _____

   _____

3. Upper Cretaceous deposits of the Cordilleran Foreland resemble Devonian, Mississippian, and Pennsylvanian deposits of the Appalachian Foreland. What is the reason for this resemblance?

   _____

   _____

   _____

Submergence of the Ozark Dome

The following questions refer to Figures 17.4 and 17.5.

Study Questions

1. In Figure 17.4, the Precambrian rocks beneath the nearly horizontal strata appear to be vertically dipping, but the vertical planes are joints (fractures) in igneous rocks. Draw a line along the unconformity. What becomes of the lowermost strata as they are traced from right to left? What type of unconformity is this?

   _____

   _____

   _____

2. In Figure 17.5, notice that the Precambrian basement is hilly or even mountainous, with a local relief of more than 1000 ft. What evidence shows that this relief was present when the Cambrian sea invaded the region?

   _____

   _____

   _____

3. The lowermost formation in the section is the LaMotte Formation (Upper Cambrian), and the uppermost is the Gasconade Formation (Lower Ordovician). Thus, roughly one-third of the Cambrian Period was required to bury these hills beneath strata. About how many million years is this?

   _____

   _____

Figure 17.4 Dolomite and shale beds in the Upper Cambrian Bonneterre Fm., which unconformably overlies the Precambrian crystalline basement. This photograph shows the Taum Sauk Power Project at Proffit Mountain (Missouri) in the eastern part of the Missouri Ozark Mountains.

Courtesy of Union Electric Company.

Figure 17.5 Geologic structure section near locality of Figure 17.4.

From Dake (1930), Missouri Geological Survey, **23.**

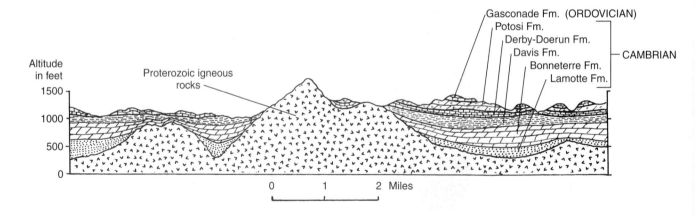

TERMS

alluvium   Recent sediments deposited on existing rocks. These unlithified materials may be considered as a mappable formation if abundant or pervasive. Also called **cover.**

anticline   A geologic structure in which strata are bent into an upfold where the limbs dip toward the fold's axis and where older rocks occur near the center of curvature (fold axis).

arch   A broad, open anticlinal fold on a regional scale.

basement rocks   The igneous and metamorphic rocks that lie unconformably beneath a covering of sedimentary strata. In the interior plains and plateau regions, these basement rocks are mostly Precambrian (Proterozoic) in age.

basin   A low area in the earth's crust, of tectonic origin, in which sediments have accumulated. The beds of a basin dip in toward the center.

cover   A package of recent sediments deposited on existing rocks. Also called **alluvium.**

dome   An uplift or anticlinal structure, either circular or elliptical in outline, in which the rocks dip away from the center in all directions.

embayment   A bay, either the deep indentation or recess of a shoreline or the large body of water (as an open bay) thus formed.

escarpment   A term generally applied to a cliff (or line of cliffs) formed by differential erosion.

evaporite   A nonclastic sedimentary rock composed primarily of minerals produced from a saline solution as a result of extensive or total evaporation. Examples of evaporites include gypsum, anhydrite, rock salt (halite), various nitrates, and borates.

forelands   The outer regions of the interior plains that border a mountain belt and that contain a wedge of sediment that thickens toward the mountain belt.

geologic map   A map on which the underlying geology (such as the rock type, strike and dip, foliation, and age relationships) is depicted.

plunge (of a fold)   The angle of inclination of the axis of a fold measured from a horizontal surface.

structure contour map   A map that depicts the configuration of the surface of a key bed or formation by lines (structure-contour lines) of equal elevation.

syncline   A downfold in which the rock strata dip toward the fold's axis and where younger rocks occur near the center of curvature (fold axis).

18

Identification of Minerals

Before coming to this laboratory session, read the chapter on minerals in your textbook. This study focuses on the physical properties used in identifying common minerals.

BASIC INFORMATION

There are a large number of physical and chemical properties used by geologists to identify minerals. Fortunately, however, most common minerals can be identified by a few of their easily recognized properties. Among these are *color, streak cleavage, hardness, crystal form, luster, specific gravity,* and *magnetism.*

Color

Color can be a useful property in the identification of minerals having a constant color. Almandite garnet, for example, is wine red. Olivine is usually green. However, many minerals can vary in color. Quartz, for example, can be colorless, white, pink, purple, green, or blue. The mineral corundum includes gem varieties that are red (we call them rubies) or blue (sapphire). For this reason, color should always be used in combination with other phystical properties.

The expression "light-colored" in the identification table refers not only to white or nearly white minerals but also to the "warm colors" (red, yellow, orange, brown, and light gray) Dark colors include black, dark gray, green, and blue. Because plagioclase varies from white to gray, it is included under both light and dark colors in the mineral identification charts.

Streak

The term streak refers to the color of the powdered mineral formed as the mineral is rubbed against the streak plate. Hematite, for example, has a red streak. Pyrite, which has a metallic luster, has a black streak.

Cleavage

Cleavage is the tendency of a mineral to break smoothly along certain directions. Your instructor may demonstrate cleavage by striking an easily cleavable mineral such as calcite with a small hammer. The surface along which the mineral breaks is called the cleavage plane. Depending on how smooth that surface is, cleavage can be described as good,

WHAT YOU WILL NEED

1. Two sets of common minerals. The first will be used as a demonstration set. At the beginning of the laboratory session, your instructor will help you identify the minerals in this set and demonstrate the physical properties used in identification. The second set of minerals constitute "unknowns" for you to identify on your own.

2. The *demonstration set* should include the following minerals:

   a. Single crystal of quartz (to demonstrate crystal form and luster).
   b. Broken piece of quartz (to demonstrate fracture, hardness, luster).
   c. Halite (to demonstrate cleavage, luster, hardness, taste).
   d. Magnetite (to demonstrate magnetism, streak).
   e. Hornblende (to demonstrate streak, hardness).
   f. Pyrite (to demonstrate metallic luster, streak, crystal form).
   g. Muscovite (to demonstrate cleavage, hardness, luster).
   h. Calcite (to demonstrate cleavage, luster, effervescence).

3. One or more specimens of the following minerals may be included in the *study set.*

Hematite	Orthoclase	Pyroxene
Magnetite	Plagioclase	Amphibole
Pyrite	Quartz	Olivine
Galena	Halite	Garnet
Graphite	Calcite	Biotite
Selenite gypsum	Fluorite	Muscovite
Satinspar gypsum	Sphalerite	Talc
Alabaster gypsum	Chlorite	Hematite
Limonite		

4. In addition to the mineral sets, students will need:
   a. Steel nail (for the hardness test).
   b. Streak plate for streak and hardness tests.
   c. Small magnet.
   d. Dropper bottle containing 10% hydrochloric acid (and paper towel for blotting).
   e. 10X magnification hand lens.

250

Figure 18.1 Muscovite (A) and biotite mica. These two mica mineral have perfect cleavage in one direction and cleave into thin, elastic, flexible sheets. Note the hexagonal shape of the biotite crystal.

A

B

Figure 18.3 Calcite exhibiting rhombohedral cleavage.

fair, or poor. Also, the number and direction of cleavage planes are aids to identification. A mineral like mica (Fig. 18.1) has good cleavage in one direction. If a crystal of halite is broken, it will produce three cleavage planes at right angles to each other (Fig. 18.2). Calcite has three perfect cleavage directions but at oblique angles so as to produce rhombohedral cleavage pieces (see Fig. 18.3).

A smooth surface on a mineral may also be a *crystal face* and not a cleavage plane. Cleavage planes, however, are produced when the mineral is broken, whereas crystal faces are planes along which ions were added to a growing crystal. A crystal of quartz exhibits good crystal faces (Fig. 18.4) but when broken has no cleavage at all. Instead, the breakage produces uneven surfaces simply called fractures.

Figure 18.4 Quartz crystal.

Figure 18.2 Halite exhibiting perfect cubic cleavage.

Crystal Faces and Cleavage Planes

Minerals grow by addition of ions to its surfaces from the surrounding liquid or gas. If the growing surfaces do not come into contact with other mineral grains, the mineral may acquire a characteristic *crystal form* with smooth *crystal faces* that parallel uniform planes of atoms within the crystal (see Fig. 18.4). Except when crystals are growing into an open cavity, they usually come into contact with each other or other mineral grains so that perfect crystals having all crystal faces developed are not produced. Because the more common occurrence is for crystals to interfere with each other during growth, many minerals form an interlocking mass of crystals called a *crystalline aggregate* (Fig. 18.5).

Hardness

Hardness is a measure of a mineral's resistance to being scratched by another substance of known hardness. Relative values of hardness have been assigned to ten minerals, which comprise Mohs scale of hardness (see Table 1.2). Although the minerals on the Mohs scale can be used to find the hardness of an unknown mineral, it is more convenient to use other substances of known hardness as indicated below.

Hardness of less than **2:** You can scratch the mineral with your fingernail (hardness 2 to 2.5).
Hardness of **2 to 3:** Mineral can be scratched with a copper penny (hardness of 3) but cannot be scratched with your fingernail.
Hardness of **3 to 5:** Mineral will scratch a copper penny but will not scratch a steel nail (hardness of 5).

Hardness of **5 to 7:** Mineral will scratch a steel nail but will not scratch a streak plate (hardness of 7).
Hardness of **greater than 7:** Mineral will scratch a streak plate.

Luster

Luster refers to the appearance of a mineral in reflected light. The two major categories of luster are *metallic* and *nonmetallic.* Nonmetallic minerals can be further grouped into adamantine (as in diamonds), *vitreous* (bright shiny), *resinous, waxy,* or simply *dull* or *earthy.*

Specific Gravity

As we pick up minerals from our specimen tray, we can tell that some are heavier than others, and this can help us to identify them. To assure uniformity in determining whether one mineral is heavier than another we use a property called *specific gravity.* Specific gravity is the number of times a mineral is as heavy as an equal volume of water. Pyrite (fool's gold) has a specific gravity of 5.0. It is therefore five times as heavy as an equal volume of water. Real gold has a specific gravity of 19.3 and is thus easily distinguished from pyrite.

Other Properties

There are many other properties distinctive to particular minerals: the salty taste of halite, the malleability of gold, the flexibility of mica, the soapy feeling of talc, and the earthy odor of clay minerals when moistened. A few minerals are attracted by a magnet or act as magnets themselves. Magnetite (Fig. 18.6) is such a mineral. Calcite is easily recognized by the way it effervesces (bubbles) when a drop of dilute hydrochloric acid is applied to its surface.

Figure 18.5 Crystals of olivine have interferred with one another during growth, forming a crystalline aggregate rather than perfectly complete crystals. (Viewed microscopically in thin section with crossed polarizers.)

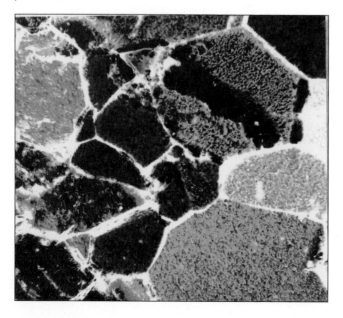

Figure 18.6 Magnetite can act as a natural magnet in attracting iron objects like these pins.

Exercise

By refering to the three **Mineral Description Tables** (Tables 18.1A, B, and C), enter the properties and name of each of the minerals in your study on the **Mineral Description Form**. Your instructor may correct your identifications in class or ask you to submit the completed form at the end of the laboratory session.

Table 18.1A   For Minerals Having a Metallic Luster

Steel-gray color in metallic variety (a dull earthy variety is also common). Brown-red streak. H 5–6.5, SG 5	HEMATITE* Fe_2O_3
Black, black streak, attracted to magnet. H 6, SG 5	MAGNETITE Fe_3O_4
Brassy yellow, streak green-black, occurs as well-formed crystals or granular masses. H 6–6.5, SG 5	PYRITE FeS_2
Silver-gray, dark gray streak, three planes of cleavage forming cubes, H 2.5, SG 7.5	GALENA PbS
Steel gray, soft, marks paper, greasy feel. H 2, SG 2	GRAPHITE C

*Steel gray metallic hematite has the variety name of specularite. On these tables, H = hardness and SG = specific gravity.

Table 18.1B   For Nonmetallic Light-Colored Minerals

Hardness 5–7	Good Cleavage	Cleavage in two directions at about 90°. Pearly to vitreous luster, white, gray, or pink. H 6–6.5, SG 2.5–2.7	ORTHOCLASE Potassium feldspar	
		Cleavage in two directions at about 90°. White to gray. Fine striations on one cleavage surface	PLAGIOCLASE NaAl Silicates to CaAl Silicates.	
	No Cleavage	Vitreous luster, varied colors, including opaque white, gray, and pink. Six-sided prismatic crystals. Also cryptocrystalline varieties like chert, flint, opal, agate, and jasper.	QUARTZ Silicon dioxide SiO_2	
Hardness less than 5 and good cleavage		Cubic cleavage, salty taste, transparent colorless and translucent white. H 2.1, SG 2.0–2.5	HALITE NaCl	
		Rhombohedral cleavage, transparent to opaque white or yellow. H 3, SG 2.7	CALCITE $CaCO_3$	
		Cleavage in one direction, pearly, vitreous, or dull; white, gray, or pink, tabular plates, fibrous, or granular. H 1–2.5, SG 2.2–2.4	GYPSUM Hydrous Calcium Sulfate. Three varieties:	*Selenite* (platy) *Satinspar* (fibrous) *Alabaster* (granular)
		Cleavage in four directions to form octahedrons, cubic crystals, vitreous, color white, pink, yellow, green, or purple;transparent to translucent. H 4, SG 3	FLUORITE CaF_2	
		Perfect cleavage in one direction, yielding flexible, elastic sheets: pearly to vitreous, colorless and transparent in thin sheets. H 2–2.5, SG 2.7–3	MUSCOVITE hydrous potassium aluminum silicate	
		Silky to greasy feel; green, white, or gray; occurs in foliated, fibrous, or compact masses. H 1–1.5, SG 1–2.5	TALC hydrous magnesium silicate	

Table 18.1C For Nonmetallic Dark-Colored Minerals

Hardness 5–7	Good Cleavage	Cleavage in two directions at nearly 90°, dark green to black, greenish gray to gray streak. H 6, SG 3.3	PYROXENE Ca, Mg, Fe, Al silicate
		Cleavage in two directions at 60° and 120°, dark green, black or brown. H 6, SG 3.35	AMPHIBOLE Na, Ca, Mg, Fe, Al silicate
		Cleavage in two directions at nearly 90°, dark gray to grayish white, striated on one cleavage face, vitreous, may show play of colors. H 6–6.5, SG 2.5–2.7	PLAGIOCLASE Mixture of Ca, Al, and Na, Al silicate.
	No Cleavage	Various shades of green, granular masses, vitreous. H 6.5–7, SG 3.2–3.6	OLIVINE Fe, Mg, silicate
		Red, brown, yellow, green, or pink, vitreous, often forms 12- or 24-sided crystals. H 6.5–7.5, SG 3.4–4.3	GARNET Ca, Mg, Fe, Al silicate
Hardness less than 5	Good Cleavage	Black, dark brown, perfect cleavage in one direction, forming thin sheets that split easily and are both flexible and elastic. H 2–2.5, SG 2.7–3.1	BIOTITE K, Al, Mg, Fe silicate
		Brown to brownish black with brownish yellow streak, resinous luster. H 3.5–4, SG 4	SPHALERITE ZnS
		Green color, pearly to vitreous luster, cleaves into fine scales, flexible but not elastic flakes. H 2–2.5, SG 2.8	CHLORITE hydrous Mg, Al silicate
	No Cleavage	Red or rust colored, earthy masses, red streak. H 1–4, SG 2.5–5	HEMATITE Fe_2O_3
		Yellow to brown, dull earthy masses, brownish-yellow streak, looks like clay. H 1.5–4, SG 3.6	LIMONITE hydrous iron oxide

Mineral Description Form

Record the properties of each mineral in your study set on the two forms provided. Use Table 18.1 to identify each specimen. Identification usually requires the use of several properties.

Sample No.	Metallic or Nonmetallic	Hardness	Cleavage	Color	Streak	Other Properties	Name

Record the properties of each mineral in your study set on the two forms provided. Use Table 18.1 to identify each specimen. Identification usually requires the use of several properties.

Sample No.	Metallic or Nonmetallic	Hardness	Cleavage	Color	Streak	Other Properties	Name

TERMS

cleavage   The tendency of certain minerals to break in preferred directions along bright, reflective, smooth surfaces.

crystal   A solid body composed of ordered, three-dimensional arrays of atoms or ions chemically linked together so as to produce a crystal form with smooth, flat surfaces.

crystal aggregate   A mass of interlocking crystals formed when developing crystals interfered with one another during growth.

hardness   The resistance of a mineral to being scratched as compared to the ten levels of hardness in Mohs hardness scale or objects that are equivalent to values on the Mohs scale.

luster   The appearance of the mineral under reflected light. The two major categories of luster are metallic and nonmetallic. Nonmetallic lusters include vitreous (bright and shiny like broken glass), resinous, waxy, pearly, greasy, or dull.

magnetism   Ability to be attracted by a magnet.

malleable   Capable of being pounded into different shapes without breaking as in gold.

streak   Color of the powdered mineral, shown by line of powder produced when mineral is drawn across a streak plate.

specific gravity   The relative weight of a substance compared to water. More specifically, the weight of a given volume of a mineral (or substance) divided by the weight of an equal volume of water at 4° C.

streak plate   A square piece of unglazed porcelain used to test streak (and also hardness).

19

Igneous Rocks

ORIGIN AND CHEMISTRY

Igneous rocks are rocks that solidify from a magma. **Magma** is a complex, high-temperature solution of rock-forming silicate minerals and **volatiles** (mainly carbon dioxide and water). Magma originates by the melting of preexisting rock at high temperatures and pressures deep within the earth. Because the density of the magma is less than that of the surrounding rock, it moves toward the surface of the earth. As it moves upward, either along fractures or faults within the earth or by the processes of **assimilation** or **stoping** of the overlying rocks, the magma slowly cools. If it cools and remains beneath the surface of the earth, it will solidify to form an **intrusive** or **plutonic** igneous body. If it reaches the surface, where it will quickly cool in contact with the atmosphere (or hydrosphere), it will form an **extrusive** igneous body.

The cooling of a magma follows a complicated chemical path where various silicate minerals are formed at particular temperatures in a definite sequence. This sequence, called a **reaction series,** was first recognized by N. L. Bowen in 1922 in his experimental studies of silicate melts. Its importance is that it shows how a wide range of igneous rocks can be developed from a single magma, depending on whether the early-formed crystals remain in the melt or are separated from it. This process of **differentiation** allows for the formation of rocks with a composition different from that of the original magma. If the crystals always remain in contact with the melt, chemical reactions between the melt and the crystals continue, causing a change in the chemical composition of the crystals. However, if the crystals are separated from the melt, the magma composition changes, depending on what type of crystals are separated from the melt.

Bowen's reaction series groups these chemical reactions between crystals and melt into two series (Fig. 19.1). In the **continuous series,** minerals such as the plagioclase feldspars change their chemical composition from calcium-rich at high temperatures to more sodium-rich at low temperatures but do not change their crystal structure. In the **discontinuous series,** early-formed crystals (formed at higher temperatures) react with the melt and recombine to form entirely new minerals with different crystal structures as the temperature decreases.

As shown in Table 19.1, six minerals/mineral groups make up the bulk of all common igneous rocks. The major elements contained in these minerals are silicon (Si), aluminum (Al), calcium (Ca), sodium (Na), potassium (K), magnesium (Mg), iron (Fe), hydrogen (H), and oxygen (O). These are, with the exception of hydrogen, the eight major elements from which the majority of rocks of the earth's crust are composed. The chemical data allows further classification of igneous rocks into two general types. **Mafic** igneous rocks, so called because of the large concentrations of the elements Mg and Fe, contain ferromagnesian minerals such as olivine, pyroxene, and amphibole. **Sialic** (or **felsic**) igneous rocks have high concentrations of the elements Si and Al and are composed of the minerals quartz, plagioclase feldspar, alkali feldspar, and mica (commonly biotite and/or muscovite).

In general, mafic rocks (with the minerals olivine, pyroxene, and calcic plagioclase) tend to form at higher temperatures while sialic rocks (with the minerals quartz, sodic plagioclase, mica, and alkali feldspar) form at lower temperatures. Examination of Figure 19.2 shows the relation of mafic to sialic rocks with regard to *color* (mafic rocks tend to be dark, felsic rocks light), *silica content* (expressed as the percent abundance of quartz), and the **modal abundance** (volume percent) of the minerals present in the rock.

OCCURRENCE OF IGNEOUS ROCKS

Intrusive Igneous Bodies

Intrusive igneous bodies (often called **plutons**) are classified according to their size, their shape, and the manner in which they are emplaced relative to the surrounding rock (**country**

Figure 19.1 Bowen's reaction series. The ferromagnesian minerals of the *discontinuous reaction series* form at specific temperatures. As the temperature decreases, previously formed minerals react with the remaining magma to form entirely new minerals. In the *continuous reaction series,* the earliest formed plagioclase also reacts with the remaining magma. But, instead of forming a new mineral, the plagioclase continuously changes its chemical composition from calcium-rich to sodium-rich.

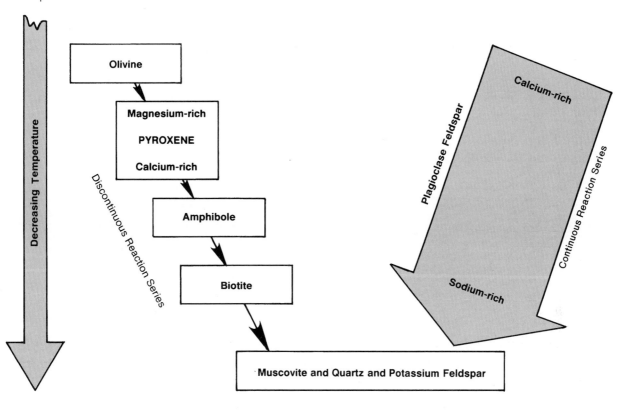

Table 19.1 Common Rock-Forming Silicate Minerals

Silicate Mineral	Composition	Physical Properties
Quartz	Silicon dioxide (SiO_2)	Hardness of 7; will not cleave (has conchoidal fracture); specific gravity: 2.65.
Feldspar group		
Potassium feldspar group	$KAlSi_3O_8$	Hardness 6.0–6.5; two directions of cleavage at 90°; specific gravity: 2.5–2.6; color: pink or white.
Plagioclase feldspar group	$Na(AlSi_3O_8)$-$Ca(Al_2Si_2O_8)$	Hardness 6.0–6.5; two directions of cleavage at 90°; may show striations on cleavage planes; specific gravity: 2.6–2.7; color: white or gray.
Mica group		
Muscovite mica	$KAl_3Si_3O_{10}(OH)_2$	Hardness 2–3; one direction of cleavage, yielding thin, flexible plates; colorless, transparent in thin sheets; specific gravity: 2.8–3.0.
Biotite mica	$K(Mg,Fe)_3AlSi_3O_{10}(OH)_2$	Hardness 2.5–3.0; one direction of cleavage, yielding thin, flexible plates; color: black to dark brown; specific gravity: 2.7–3.2.
Pyroxene group	$(Mg,Fe)Si_2O_6$	Hardness 5–6; two directions of cleavage at 90°; specific gravity: 3.1–3.5; color: black to dark green.
Amphibole group	$Ca_2(Mg,Fe)_5Si_8O_{22}(OH)_2$	Hardness 5–6; two directions of cleavage at 56° and 124°; specific gravity: 3.0–3.3; color: black to dark green.
Olivine group	$(Mg,Fe)_2SiO_4$	Hardness 6.5–7.0; no cleavage; specific gravity: 3.2–3.6; color: light green, transparent to translucent.

Figure 19.2 Guide to the classification of common igneous rocks. This approach is based on the *modal abundance* (amount of visible minerals present, expressed in volume percent), *color,* and *silica content* (weight percent of SiO$_2$ in the chemical composition of the rock). Note that a large modal abundance of the ferromagnesian minerals gives a darker color to the rock, and that the presence of feldspars (either plagioclase or alkali feldspar) and quartz gives lighter colors.

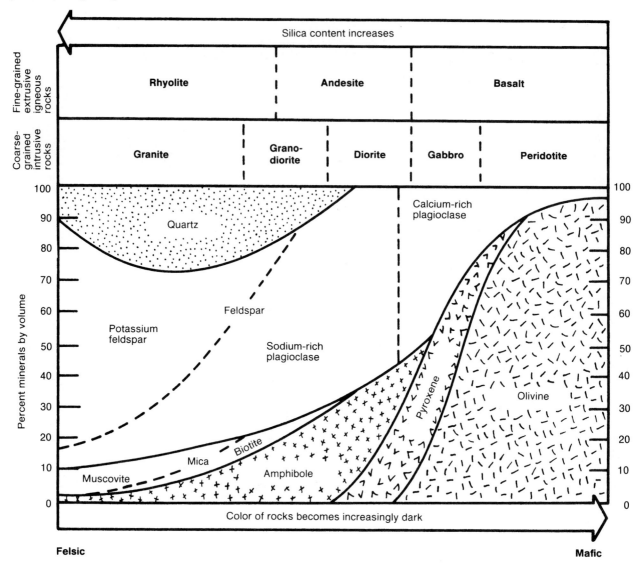

rock). If the country rock has a definite layering (as in a sedimentary rock), and the magma intrudes between the layers, the intrusion is called **concordant.** If the intrusion cuts across the layers, it is called **discordant.** Figure 19.3 shows a number of the various intrusive and extrusive igneous bodies. Concordant plutons, such as **sills** (Fig. 19.4) dome-shaped **laccoliths,** and the spoon-shaped **lopoliths,** are generally tabular and may be horizontal, inclined, or vertical, depending on the attitude of the rock in which they are found. Discordant plutons range in size from the relatively small, tabular, crosscutting **dikes** (Fig. 19.5) to massive **stocks** and **batholiths.** A discordant pluton is called a batholith if its surface expression is greater than 100 km^2 (or 36 square miles) and a stock if it is smaller.

One important concept associated with any pluton is its cooling history. If a large mass of magma is emplaced at depth, it will cool very slowly because the rock that surrounds the magma is a poor conductor of heat. Slow cooling allows ample time for the formation of large crystals and the chemical interactions theorized by the Bowen reaction series. Thus, the *texture* of plutonic igneous rocks is generally coarse grained. Often, crystals several centimeters (or more) long may develop. Slow cooling also provides sufficient time for the magma body to be moved about by tectonic forces, and thereby provides an opportunity for differentiation and development of more than one rock type from a particular magma.

Extrusive Igneous Materials

Extrusive magma may either erupt quietly as **lava** or, because of the dissolved gases (volatiles such as H$_2$O and CO$_2$) contained in the magma, erupt violently, spewing forth rock fragments called **pyroclastic debris** or **tephra.** Pyroclastic

Figure 19.3 Nomenclature for intrusive and extrusive igneous bodies. The concordant bodies include sills, laccoliths, and lopoliths. Discordant bodies include dikes, volcanic necks, stocks, and batholiths. Volcanoes can be formed from lava flows, volcanic ash flows or falls, or a combination of both.

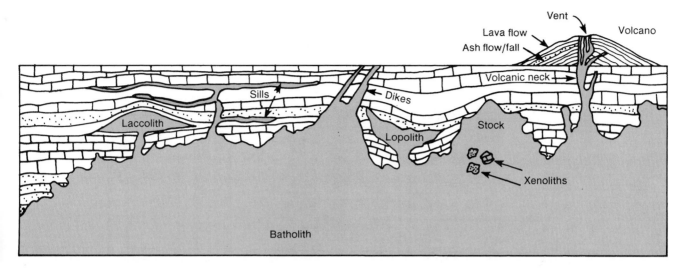

Figure 19.4 A dark colored basaltic sill of Precambrian age on Banks Island, Northwest Territories, Canada. To make room for itself, the magma forced the overlying rock upward and expanded a fracture where it changed level.

Courtesy of Geological Survey of Canada.

Figure 19.5 Basaltic dike crosscutting Rhyolite. Southeastern Missouri.

Figure 19.6 Volcanic bombs. Note the streamlined shape that formed by air resistance as the molten magma was erupted into the air.

Courtesy of Institute of Geological Sciences, London.

debris comes in all sizes and can range from huge volcanic bombs (Fig. 19.6) and blocks (some as large as a house) to fine volcanic ash (Fig. 19.7).

Pyroclastic debris and lava can find their way to the surface either through extensive **fissures** (or faults) in the earth's crust, producing laterally extensive lava flows such as the Columbia River basalt flows in Washington and Oregon, or through **vents,** around which the accumulation of lava and tephra may form volcanoes and cinder cones. In addition, pyroclastic debris, mainly volcanic ash, can be packed together (or agglomerated) to form a type of rock called **tuff.** When the volcanic ash composing a tuff is very hot, the ash particles may be welded together to form a dense, hard rock called **welded tuff.**

The cooling history of extrusive magma is usually very short and rapid. Contact of the hot lava or tephra with

Figure 19.7 (A) Large volcanic bomb (hammer for scale) in cross-bedded volcanic ash (tuff) deposit. Kilbourne Hole, New Mexico. (B) Detailed view of the cross-bedding in the volcanic ash (tuff) at Kilbourne Hole. Note the truncation of the topset beds and the graded bedding above the crossbeds.

A

B

air or water swiftly cools the material so that only very small crystals can form, or no crystals at all, as in the case of the volcanic glass called *obsidian* (Fig. 19.8). Gases dissolved in the hot lava may escape and result in the formation of gas **vesicles.** This process causes the formation of porous volcanic rocks such as *pumice* (Fig. 19.9) and *scoria*.

An important observation about extrusive volcanic material is that the pyroclastic debris can be reworked by wind and water soon after it is erupted and can be deposited in a manner similar to that of other common sediments. Thus, a volcanic eruption may produce igneous material that soon becomes sedimentary in nature due to the continuing processes of erosion and deposition on the surface of the earth.

Figure 19.9 Pumice. Note the gas vesicles as well as the frothy nature of the rock glass.

© The McGraw-Hill Companies, Inc./Jacques Cornell, photographer.

Figure 19.8 Obsidian. Note the glassy luster and conchoidal fracture.

Courtesy of Institute of Geological Sciences, London.

CLASSIFICATION OF IGNEOUS ROCKS

The classification of igneous rocks in a hand sample requires that you determine the identity of the major minerals (called the *composition*) and the *texture* of the rock. Table 19.1 displays the most common minerals found in igneous rocks and the physical properties that allow you to distinguish between them. Table 19.2 is a classification chart for separating the igneous rocks based upon the amount and identity of the minerals found. Since igneous magma may produce either an intrusive igneous rock or an extrusive igneous rock, the naming of the rock also depends upon the texture. For example, slow cooling of a mafic magma would produce large crystals of olivine, pyroxene, and calcium-rich plagioclase (called a **phaneritic** texture). This rock, based upon its mineral assemblage and texture, would be called a *gabbro*. However, if the same magma reached the earth's surface and cooled more rapidly, the crystals would be very fine-grained (called **aphanitic** texture). In addition, the extrusive igneous rock that formed would have the same olivine, pyroxene, and calcium plagioclase minerals. Because it has the same minerals as the gabbro but very fine-grained texture (and containing some rock glass), it would be called a *basalt*. Table 19.2 provides a key to the separation and identification of igneous rocks based upon composition and texture. Use these figures and tables to assist you in the identification of your samples.

Table 19.2 Key to Identification of Some Common Igneous Rocks

Texture	Rock Name	Composition	Occurrence	Comments
Phaneritic	**Granite** **Granodiorite** **Diorite**	K-feldspar + qtz with minor Na-plagioclase and biotite.	Continental crust as batholiths or stocks.	**Granite** is light-colored (pink, white, gray), because of qtz + K-feldspar. **Granodiorite** is gray. **Diorite** has little to no qtz, gray to gray-green color, feldspar and ferro- magnesian minerals in equal amounts.
	Gabbro	Ca-plagioclase + pyroxene with minor olivine, no quartz.	Layered plutons not very abundant. However, a rock type called **anorthosite** (formed only of Ca-plag) forms important economic deposits.	Dark green-black color because of ferromagnesian minerals (pyroxene and olivine).
	Peridotite	Olivine + pyroxene	Peridotite and dunite are found in lower crustal rocks and the cores of Alps, Urals, and Appalachian Mtns.	Dark green-black color like gabbro.
	Dunite	>95% olivine		Dunite is often green to yellow-green with a granular texture.
Aphanitic	**Rhyolite**	Same composition as granite, mostly rock glass.	Rhyolite domes and lavas.	High viscosity, syrupy lavas, may be flow banded, forms small highly explosive volcanic structures.
	Andesite	Same composition as diorite, abundant rock glass.	Andes Mtns., volcanic rocks associated with subduction zones, island arc volcanos.	Generally porphyritic, viscosity greater than basalt, explosive eruption style because of dissolved gases.
	Basalt	Same composition as gabbro, mostly rock glass.	Oceanic crust, Snake River plateau, mid-ocean ridges, Hawaiian islands.	Most abundant of all volcanic rock, may have porphyritic texture, glassy low viscosity lavas.
Glassy	**Obsidian**	Same composition as rhyolite and granite.	Rapid cooling (quenching), mainly rock glass, no vesicles.	Conchoidal fracture, usually jet black, may have crystal fragments or flow banding.
Vesicular	**Pumice**	Same composition as rhyolite and granite.	Highly vesicular, mainly rock glass as a froth.	Light colored, low density (due to vesicles; float on water).
	Scoria	Composition ranges from basaltic to andesitic.	Highly vesicular, mainly rock glass.	Dark colored, higher density, may have crystal fragments.
Pyroclastic	**Tuff**	Fragmental volcanic debris composed of rock glass, rock and crystal fragments, etc. Grain size varies.	Ash-fall or ash flow deposits associated with andesitic to rhyolitic volcanism.	Cross-bedding and other fluvial-style sedimentary features possible. **Welded tuffs** have flattened pumice fragments and may have flow banding.

Table 19.3 Igneous Rock Identification Form: Use with Tables 19.1 and 19.2

Sample Number	Texture	Composition		Other Features	Rock Name
		Minerals	*Modal Abundance*		

Student Name _____   Class/Section Number _____

Use Table 19.3 to record your observations concerning the mineral identity, modal abundance of minerals, texture, and rock name.

Composition

Minerals that make up igneous rocks are grouped into either *primary, accessory,* or *secondary* minerals. Primary and accessory minerals form during the **crystallization** of the magma. Secondary minerals are formed after the magma has crystallized and are usually the result of weathering and/or chemical alteration of the primary or accessory minerals. For our scheme of classification, secondary minerals are not considered important in the description of the rock.

The primary minerals of common igneous rocks are the six rock-forming minerals/mineral groups described in Table 19.1. The accessory minerals of common igneous rocks can include these minerals or others such as rutile (titanium oxide; TiO_2), magnetite (iron oxide; Fe_3O_4), and others. The distinction between a primary mineral and an accessory mineral is based on the **modal abundance** (volume percentage) of the mineral present in a particular igneous rock. The primary minerals in an igneous rock will generally compose more than 98% of the volume of the rock, whereas the accessory minerals present in that rock seldom make up more than 1% or 2%. In your examination of igneous rocks, identifying the primary minerals will allow you to name the rock (i.e., granite) while the accessory minerals are used to describe the rock name (i.e., muscovite granite).

Occasionally, during the processes that bring the magma toward the surface, pieces of the country rock are incorporated into the magma. These fragments are called **xenoliths** (Greek *xeno,* foreign, *lithos-,* rock). Their composition may be of sedimentary, metamorphic, or igneous origin and reflects the material through which the magma moved.

Texture

The size, shape, and arrangement of the interlocking mineral grains of igneous rocks are controlled mainly by the cooling history of the magma. These textural features allow the geologist to evaluate the conditions under which the rock originated. The following is a synopsis of the common textures found in both intrusive and extrusive igneous rocks.

Phaneritic texture　If an igneous rock is composed of crystals large enough to be recognized without the aid of a hand lens, the texture is referred to as **phaneritic** (Greek *phaneros,* visible). The grains are approximately equal in size and constitute an interlocking mosaic. This type of **equigranular** texture suggests a uniform rate of cooling, and the large crystal size implies that the cooling rate was slow. This observation usually implies that the magma cooled at depth, but other factors, such as the type and amount of dissolved volatiles and the chemical composition of the magma, can influence the size and uniformity of crystals.

Porphyritic texture　In some igneous rocks, two distinct sizes of crystals are apparent. The larger, well-formed crystals are called **phenocrysts,** and the smaller crystals are called the **matrix** (or **groundmass**). This type of **porphyritic** texture is thought to represent an example of two-stage cooling, in which the phenocrysts form by slow cooling in the magma and then, by some mechanism, the magma cools quickly (relative to the rate at which the phenocrysts formed) to form the finer-grained matrix. *A note of caution:* Sedimentary rocks can have a matrix that is composed of finer-grained materials (such as silt or clay) that surround the individual mineral grains or rock fragments. This matrix is deposited contemporaneously with the larger sediment grains. The matrix found in igneous rocks is formed from the remaining magma that surrounds the earlier-formed mineral grains. It forms by crystallization of the remaining molten material and is completely different from the matrix of a sedimentary rock.

Aphanitic texture　If the crystals of an igneous rock can be recognized only with the aid of a hand lens, the texture is called **aphanitic** (Greek *a,* not; *phaneros,* visible). These fine-grained rocks indicate rapid cooling and typically occur in shallow, near-surface intrusions and lavas.

Glassy texture　Glassy texture is produced by very rapid cooling of the magma. This rapid cooling, called **quenching,** takes place when the magma comes into contact with air or water. Few, if any, distinct crystals are visible, but it is possible to have numerous **vesicles** caused by the dissolved gases escaping during the rapid cooling. These vesicles give the characteristic appearance (and low density) of the volcanic rocks *pumice* or *scoria.* In volcanic glass, called *obsidian,* the fracture is conchoidal (curved), like that of common silica glass.

Pyroclastic texture　Pyroclastic texture results from the agglomeration of tephra fragments. Because the size of pyroclastic debris ranges from cobbles to very fine volcanic ash, the textural composition may resemble various sedimentary rocks. The presence of glass, welding caused by the hot volcanic material, flow banding and cross-bedding, and flattened pumice fragments indicates pyroclastic nature.

PEGMATITES AND HYDROTHERMAL ORE DEPOSITS

As a magma cools and crystals form in the manner indicated by Bowen's reaction series, some chemical elements in the magma are restricted from entering into the atomic structure of certain minerals because of their large ionic size or their characteristic electrical charge. These **incompatible elements** accumulate as the magma crystallizes and are concentrated in the late-stage, water-rich magmatic fluids. Because these fluids are both water-rich and hot (150 to 350° C), they are called **hydrothermal fluids.** When hydrothermal solutions crystallize, they may form very coarsely crystalline igneous rocks called **pegmatites.** Because they represent late-stage crystallization, the common minerals of pegmatites are the end members of the Bowen reaction series: potassium feldspar, quartz, and muscovite mica. In addition, because of the concentration of incompatible elements, these pegmatites may contain a variety of rare minerals such as lepidolite (a lithium-bearing mica), tourmaline and topaz (gem minerals), and pitchblende (a uranium-bearing mineral).

Hydrothermal ore deposits are closely related to pegmatites and contain valuable minerals and elements. Many of these deposits are thought to have occurred because the hydrothermal fluids escaped from the nearly crystallized magma and migrated to the surface through cracks, faults, and fissures. As these hydrothermal fluids cool, late-stage minerals such as quartz and mica often crystallize (forming a pegmatite). Sometimes, depending on the chemical composition of the magma from which the hydrothermal fluids were generated, economically important elements such as gold (Au), silver (Ag), lead (Pb), copper (Cu), and zinc (Zn) are often deposited along with the quartz or into the surrounding country rock.

TERMS

aphanitic A textural term to describe mineral grains that can be recognized only with the aid of a hand lens.

assimilation The incorporation and digestion of solid or fluid foreign material, such as wall rock, in an uprising magma.

batholith A large, generally discordant plutonic mass that has more than 100 km² of surface expression and no known floor.

concordant Term describing the orientation of the contact between an igneous intrusion and the country rock where the intrusion parallels the foliation, layering, or bedding planes of the country rock.

country rock A general term for any rock that surrounds a body of magma.

crystallization The process(es) by which elements combine to form minerals (and rock) from a liquid (or gaseous) substance such as a magma. Crystallization occurs in magmas as a result of cooling and can be modeled for igneous rocks by the use of the Bowen reaction series.

differentiation The process of developing more than one rock type from a common magma.

dike (or *dyke*) A tabular igneous intrusion that cuts across the bedding or foliation of the country rock.

discordant Term describing the orientation of the contact between an igneous intrusion and the country rock where the intrusion crosscuts the foliation or bedding planes of the country rock.

equigranular A textural term for an igneous rock in which the mineral grains are the same size and interlock with a polygonal junction.

extrusive An adjective describing an igneous rock that formed because magma was erupted onto the surface of the earth.

felsic Term applied to an igneous rock having abundant light-colored minerals such as plagioclase feldspar, alkali feldspar, and quartz. Similar term: **sialic.**

fissure A fracture or a crack in rock, along which there is a distinct separation. Fissures are generally long, linear breaks in the rock, as opposed to **vents.**

hydrothermal fluids A hot (150 to 350° C), water-rich fluid that may contain a variety of dissolved elements, which when precipitated during cooling (i.e., crystallization), may form either minerals or metal-rich rocks. These fluids may be of magmatic, metamorphic, or meteoric origin.

incompatible elements Elements that are not readily incorporated into the atomic or crystalline structure of certain minerals. Usually they are excluded from most minerals because of large ionic sizes or differences in electrical (ionic) charge.

intrusive An adjective describing an igneous rock that has formed by cooling of magma below the surface of the earth and within preexisting rock (i.e., country rock).

laccolith A concordant, dome-shaped igneous intrusion with a known or assumed flat floor and a postulated dike-like feeder commonly thought to be beneath its thickest point.

lava Magma that has reached the surface of the earth along fractures. Lavas are commonly extruded onto the surface through vents and fissures.

lopolith A large, concordant, typically layered igneous intrusion with a spoonlike shape caused by sagging of the underlying country rock.

mafic Term used to describe an igneous rock composed chiefly of one or more ferromagnesian (iron- and magnesium-rich), dark-colored minerals such as olivine, pyroxene, or amphibole.

magma   A naturally occurring mobile rock material, generated within the earth and capable of intrusion and extrusion, from which igneous rocks are thought to have been derived through solidification (i.e., crystallization) and related processes. It may or may not contain suspended solids, such as mineral crystals and rock fragments, and/or dissolved gas/liquid phases (called volatiles).

matrix (or **groundmass**)   The material between the phenocrysts of a porphyritic igneous rock. It is relatively finer grained than the phenocrysts and may be crystalline, glassy, or both.

modal abundance   A method of evaluating the composition of an igneous rock in terms of the relative amounts (volume percentage) of primary minerals present.

pegmatite   An exceptionally coarse-grained igneous rock, with interlocking crystals that may be greater than 1 cm in diameter. Pegmatites represent the last and most hydrous (fluid-rich) portion of a magma to crystallize. Some pegmatites may contain high concentrations of minerals (or elements) that are generally only present in trace amounts as accessory minerals. Pegmatites that are rich in valuable minerals or elements may result in economically important ore deposits (called **hydrothermal ore deposits**).

phaneritic   A textural term used to describe mineral grains that are recognized without the aid of a hand lens.

phenocrysts   A term used for a relatively large, conspicuous crystal in a porphyritic rock.

pluton   A general term for an igneous intrusion of any size.

porphyritic   A textural term for an igneous rock of any composition that contains conspicuous phenocrysts in a fine-grained groundmass.

pyroclastic   Pertaining to clastic rock material formed by a volcanic explosion or aerial explosion from a volcanic vent or fissure.

quenching   Cooling at a very rapid rate.

reaction series   A model for the formation of different minerals in a crystallizing magma. The model examines how the minerals change (or interact) as the magma cools and the composition of the magma changes. This interaction is called **differentiation** of the magma and shows how a variety of different igneous rocks can be formed from a single magma. There are two different reaction series: a **continuous reaction series,** which evaluates how the plagioclase feldspar minerals change in chemical composition as cooling continues, and a **discontinuous reaction series,** which examines how ferromagnesian minerals such as olivine, pyroxene, amphibole, and mica form during cooling.

sialic   A term used for rocks that are rich in the elements silicon (Si), oxygen (O), and aluminum (Al). Similar term: **felsic.**

sill   A tabular igneous intrusion that parallels the planar structure of the surrounding rocks.

stock   An igneous intrusion that is less than 100 km^2 in surface exposure. It is usually discordant and resembles a batholith except in size.

stoping   A process of magmatic emplacement or intrusion that involves detaching and engulfing pieces of the country rock. The engulfed material may be assimilated or retained as **xenoliths.**

tephra   A general term for all types of pyroclastic debris.

tuff   A general term for all consolidated pyroclastic rocks.

vent   The opening at the earth's surface through which volcanic materials (either lava or tephra) are erupted.

vesicle   A cavity of variable shape in a lava, formed by the entrapment of a gas bubble during the solidification of the lava.

volatile   A term for dissolved substances in a magma, such as water (H_2O) or carbon dioxide (CO_2), whose vapor pressures are sufficiently high for them to be concentrated as a gas phase.

welded tuff   A glass-rich pyroclastic rock that has been indurated by the welding together of the tephra particles under the combined action of the heat retained by the particles as well as the compression of the particles by the weight of the overlying material. A welded tuff is often flow banded and is usually felsic in composition.

xenolith   A foreign inclusion in an igneous rock. Commonly, xenoliths are fragments of the surrounding country rock through which the magma forces its way to the surface of the earth.

20

Metamorphic Rocks

METAMORPHISM AND METAMORPHIC ROCKS

Metamorphic rocks are formed from previously existing rocks as a result of pressure and/or temperature change and chemical interaction with fluids and gases. The process by which **metamorphic rocks** are formed is called **metamorphism.** The parent material, called the **protolith,** could have been an **igneous,** a sedimentary, or even a previously formed metamorphic rock that responded to the change in conditions to produce a new texture and/or assemblage of minerals that are stable under the new conditions. The formation of a new mineral assemblage restores the status quo of chemical equilibrium. In attempting to attain **equilibrium,** the effects of metamorphism may include (1) the **deformation** and/or rotation of the constituent mineral grains; (2) chemical rearrangement and growth of new minerals, with or without the addition or removal of chemical elements via circulating fluids or gases; and (3) the **recrystallization** of minerals into larger grains. The result is a rock exhibiting textural and compositional differences that reflect the particular variations in surrounding conditions.

According to our models for the origin of mountains (Chapter 16), many metamorphic rocks are formed in the cores of orogenic belts during orogeny. Where large areas of metamorphic rocks occur at the earth's surface, or unconformably beneath younger strata, we infer that a mountain belt had been formed at some time in the past and subsequently has been deeply eroded. From detailed studies of the texture and mineral composition of metamorphic rocks, geologists make inferences as to the direction of the orogenic forces and the conditions of temperature and pressure under which metamorphism took place. The rock type from which a metamorphic rock is derived is another inference of historical significance. For example, if an Archean-age schist (approximately 2.5 billion years old) is derived from a sedimentary rock, this tells us that conditions necessary for sedimentation occurred early in the history of the earth.

TYPES OF METAMORPHISM

The characteristics of metamorphic rocks depend on the variety and intensity of the three metamorphic agents (pressure, temperature, and chemically reactive volatiles) acting on the protolith, as well as the protolith's chemical composition. As we review the two most common types of metamorphism, **contact** and **regional,** keep in mind that metamorphic effects do not depend entirely on the three agents of metamorphism. The length of time over which metamorphic activity occurs is also important. Therefore, before deciding that a particular mineral assemblage indicates a specific metamorphic event, one must ask if the same mineral assemblage can be caused by dissimilar conditions operating over a different time span.

Contact Metamorphism

When a **magma** is intruded in the crust, the rock surrounding the **intrusion** is altered by heat from the magma and the chemical interactions between volatiles (either from the magma or trapped in the surrounding rock) and the rock that is intruded. This alteration, called **contact metamorphism,** (also called **thermal metamorphism**), forms new mineral assemblages, which vary depending on the amount of **volatiles** (fluids or gases such as water or carbon dioxide), the temperature of the intruding magma, and the composition of the surrounding rock (called the **country rock**). Contact metamorphism, as shown in Figure 20.1, occurs in zones called **aureoles** (or halos), which may range in width from several millimeters to several hundred meters. These aureoles record the intensity and extent of the metamorphic event.

The effect of heat from the magma on the country rock is sufficient alone to produce new kinds of rocks and minerals. Marble may be formed from limestone, and diopside

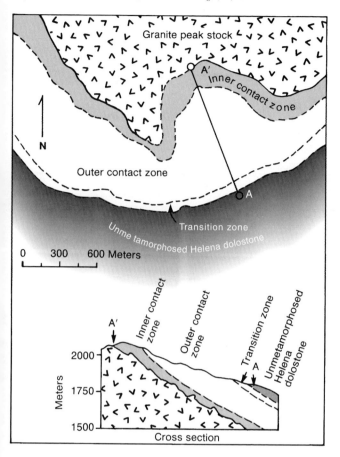

Figure 20.1 The Granite Peak aureole. This metamorphic aureole is developed around a granite stock that has intruded into dolostone country rock. Particular assemblages of metamorphic minerals occur within each contact metamorphic zone.

Simplified from Melson, W. G. (1966), *American Mineralogist,* **51,** 404.

Table 20.1 Regional Metamorphic Minerals, Zones, and Grades

T	Zone	Metamorphic Grade	Minerals
150° C	Chlorite	Low	Chlorite, muscovite, quartz
	Biotite	Low	Biotite, chlorite, muscovite, quartz
	Garnet	Middle	Garnet, biotite, muscovite, quartz
	Staurolite	Middle	Staurolite, garnet, biotite, quartz
	Kyanite	Middle	Kyanite, garnet, biotite, quartz
700° C	Sillimanite	High	Sillimanite, garnet, biotite, quartz, orthoclase, muscovite

tive intensity of metamorphism, evaluated by the difference between what is thought to be the original parent rock (protolith) and the metamorphic rock. This relative ranking by zone and grade has allowed geologists to draw maps of the regional metamorphism of entire areas and follow the mineralogic change from low-grade to high-grade metamorphic mineral assemblages (Fig. 20.2). However, caution must be used in assigning zones and grades, as variations in volatiles and protolith composition can produce an *apparent* high-grade metamorphic mineral assemblage at a much lower pressure and temperature.

STRUCTURES AND TEXTURES OF METAMORPHIC ROCKS

Foliation (from Latin *folium,* leaf) is a descriptive term applied to the parallel texture of a metamorphic rock, which is due mainly to a parallel arrangement of the mineral constituents that often causes the rock to split into nearly flat pieces. Conspicuously foliated rocks usually contain abundant flat minerals such as the micas (muscovite or biotite), but foliation may also result from the parallel arrangement of needle-shaped minerals such as hornblende (Fig. 20.3). Imperfect foliation may result from the segregation of granular minerals, such as quartz and feldspar, into streaks or bands. In a few metamorphic rocks, of which marble is the most common example, no foliation or directed arrangement of mineral constituents is observable.

The grains that compose metamorphic rocks do not interlock, as do the grains of typical igneous rocks (Chapter 19), nor are they in random contact, as are the grains of clastic sedimentary rocks (Chapters 1 and 2). Instead, grains of metamorphic rocks fit neatly together, like a wall built of well-fitted irregular stones, or in some metamorphic rocks, like the individual tiles

(a pyroxene mineral) from the chemical interaction of the minerals quartz and dolomite. Heat, however, also stimulates circulation of groundwater and other fluids. These **hydrothermal** (hot water) solutions are highly reactive. They alter the country rock and carry in solution metallic ions that may form valuable ore minerals when deposited. Iron deposits in the Urals (Russia), copper deposits in Arizona and Utah, and zinc deposits in New Mexico and Ontario are examples.

Regional Metamorphism

In contrast to contact metamorphism, which is a relatively localized event, **regional metamorphism** (also called **dynamothermal metamorphism**) may occur across regions that are thousands of square kilometers in area. Because of the large lateral extent, metamorphic rocks formed by regional metamorphism may exhibit distinct **metamorphic zones,** which are characterized by minerals (called **metamorphic index minerals**) that developed in response to particular conditions of pressure and temperature (see Table 20.1). Another concept important in studies of regional metamorphism is that of **metamorphic grade,** which provides a measure of the rela-

Figure 20.2 Changes in the mineralogic composition of a terrain originally underlain by shales following regional metamorphism.

From Levin, H. L., *Contemporary Physical Geology,* 2nd ed. Philadelphia, Saunders College Publishing, 1986.

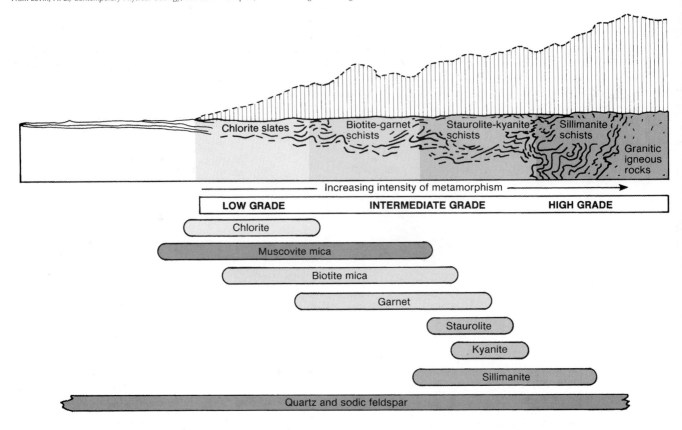

Figure 20.3 Well-developed foliation resulting from parallel alignment of biotite grains. The rock type is biotite schist and occurs as a xenolith in granite (the surrounding lighter-colored rock).

From an exposure at Wadi al 'Araban, Jordan; courtesy of F. Bender and the U.S. Geological Survey.

in a mosaic. Such metamorphic texture is evidence that the different minerals have grown at the same time in the solid state, unlike the igneous minerals, which develop in sequence from a magma, or the clastic sediments, which consist of grains piled (or cemented) together. In many metamorphic rocks, crystals considerably larger than the average grain size may develop; these large crystals are termed **porphyroblasts.** If the average grain size of a metamorphic rock is smaller than that of ordinary granulated sugar, the texture is described as fine-grained; if larger, as medium- to coarse-grained.

The dominant mineral constituents of metamorphic rocks are quartz, feldspar (plagioclase or alkali feldspar), mica, and amphibole. These are also dominant constituents of other rocks, but mica and amphibole are, in general, more abundant constituents of metamorphic rocks than of other rocks. Minerals such as garnet and chlorite are more common (and conspicuous) in metamorphic rocks than in other rocks (see Table 20.1).

COMMON KINDS OF FOLIATION IN METAMORPHIC ROCKS

Figure 20.4 shows the common types of foliation that are found in most metamorphic rocks. Remember that although foliation is common in a large number of metamorphic rocks, it is possible to have a metamorphic rock, such as marble, that does not have any type of foliation.

1. Foliation due mainly to the orientation of platy minerals, such as micas and chlorite, is displayed in Figure 20.4A. The sequence of metamorphic rocks from slate to

Figure 20.4 Examples of common metamorphic rocks and their mineral compositions and textures.

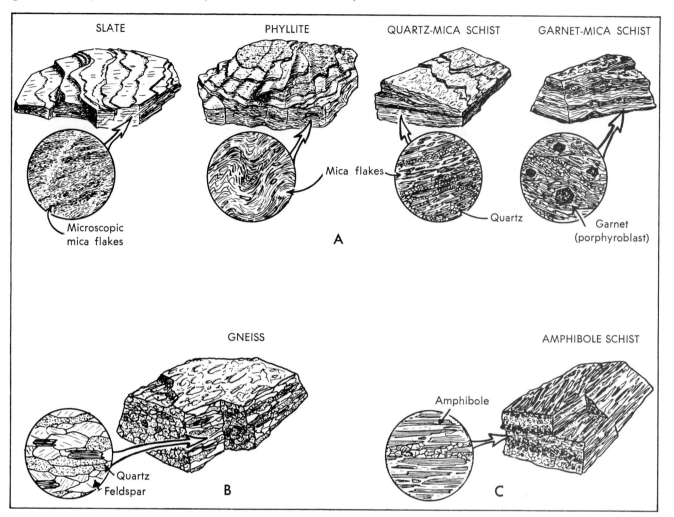

phyllite to mica schist represents successively higher grades of metamorphism and a successive increase in the size of constituent platy minerals. In *slate,* the flakes of mica and chlorite are microscopic; in *phyllite,* the flakes have grown larger, and this increase is attended by an increase in luster; in *mica schist,* the platy minerals have reached a size that is easily visible. Schists that are formed at higher metamorphic grade commonly contain porphyroblasts, around which the mica flakes are wrapped. The groundmass of schists is typically quartz or a mixture of quartz and feldspar. Also, in schists, the foliation due to orientation of platy minerals may be accompanied by foliation caused by the segregation of the different minerals into bands.

2. Foliation due mainly to the segregation of minerals into bands, lenses, or streaks is shown in Figure 20.4B. In this figure, quartz is segregated into irregular bands, while the feldspar and amphibole are segregated into other bands.

3. Foliation caused by the layered arrangement of needlelike minerals is portrayed in Figure 20.4C. In this figure, needles of amphibole are arranged in layers, alternating with layers rich in feldspar. Within a given layer, the needles are approximately parallel.

IDENTIFICATION OF METAMORPHIC ROCKS

The identification of common metamorphic rocks is based primarily on mineral composition and type of foliation (if present). In your laboratory examination of metamorphic rocks, use the following descriptions of common metamorphic rocks and Table 20.2 to record your observations concerning mineral composition, foliation, texture, and rock name.

Table 20.2 Metamorphic Rock Identification Form: Use with Table 20.1 and Fig. 20.4

| Sample Number | Grain Size | Foliation | | Identifiable Minerals | Rock Name |
		Degree of foliation? (i.e., non-foliated, weakly, moderately, or strongly)	Cause of foliation? (orientation of platy or needlelike minerals, segregation of minerals into layers, or combination of above)		

Student Name _____   Class/Section Number _____

Figure 20.5 These layers of slate from the Kodiak Peninsula of Alaska exhibit well-developed slaty cleavage. Note that the cleavage planes are vertical and are not parallel to bedding as in the fissility of shale.

Courtesy of U.S. Geological Survey.

Foliated

Slate Very fine-grained texture, commonly gray, black, green, or purple. Smooth foliation planes have weak luster, which is derived from tiny flakes of mica or chlorites. Plane of foliation may lie at any angle to plane of original bedding. If original bedding intersects plane

of foliation, the intersection may be marked by a line. As a result of the parallel growth of the minute flakes of mica in slate, the rock characteristically will split into thin slabs. This trait is called *slaty cleavage* (Fig. 20.5). Although the planes of slaty cleavage have no relation to the bedding planes in the shales from which the slate is derived, measurements of the orientation of slaty cleavage in outcrops can be used in reconstructing the deformational history of a metamorphic region.

Phyllite Fine-grained texture, although some grains (of mica, garnet, or quartz) may be visible. Surfaces commonly wrinkled or sharply bent. More lustrous than slate because of larger grains of mica or chlorite. Represents an intermediate degree of metamorphism between slate and schist. Derived from regional metamorphism of shale.

Schist Fine- to medium-grained texture, distinctly foliated, in which grains of platy minerals (mica or chlorite) or of needlelike minerals (commonly amphibole) are visible. Surfaces of specimen may or may not be wrinkled. The different minerals are commonly segregated into lenses or layers. The main varieties of schist are described below:

Mica schist Generally gray, composed of muscovite, biotite, and quartz in varying proportions. If quartz is a prominent constituent in a specimen, it is called quartz-mica schist; if garnet or some other mineral is conspicuous, the name is similarly qualified (i.e., as garnet-mica schist; Fig. 20.6). Derived from regional metamorphism of shale or high-silica volcanic rock.

Chlorite schist Generally dark green; soft, greasy feel. Composed of chlorite, epidote, and feldspar, but in hand specimen, only chlorite may be identifiable. Derived from regional metamorphism of low-silica volcanic rocks.

Amphibole schist Generally dark green, composed mainly of amphibole and feldspar, accompanied by garnet, quartz, or biotite in varying proportions. Foliation results from arrangement of amphibole needles, which lie flat in the plane of foliation; the needles may or may not point in the same direction. Derived from low-silica igneous rocks, represents a higher grade of regional metamorphism than a chlorite schist.

Gneiss Commonly coarse-grained and evenly granular, composed mainly of quartz and feldspar. Mica and amphibole are not abundant. Foliation caused by segregation of minerals into streaks or bands rich in quartz, feldspar, hornblende, or biotite. In general, the foliation is less distinct than that of schist. Derived from high-grade regional metamorphism of high-silica igneous rocks, sandstones, or aluminum-rich pelitic sedimentary rocks (Fig. 20.7).

Not Foliated or Weakly Foliated

Marble Fine- to coarse-grained, composed of calcite or dolomite and therefore relatively soft (easily scratched with steel). Many marbles have a variegated (marbled) appearance; others are brecciated. Derived from limestone or dolomite.

Figure 20.6 Mica schist with large porphyroblasts blasts of garnet.

Figure 20.7 Coarse foliation developed in a gneiss. The lens-like lighter-colored areas are feldspar and quartz. The darker bands are composed mainly of ferromagnesian minerals such as biotite and amphibole.

Courtesy of Wards Natural Science Establishment, Inc., Rochester, New York.

Quartzite   Commonly sugary-textured, composed of quartz and therefore hard (not scratched with steel). Generally light-colored, but can be almost any color. May be difficult to distinguish from quartz-cemented sandstone, but under the microscope, the grains that constitute quartzite are seen to have very irregular (sutured) boundaries. Derived from quartz sandstone.

Greenstone   Fine-grained, dark green color, and commonly has scattered dark porphyroblasts. Represents a lower degree of metamorphism than does chlorite schist; both are derived from low-silica volcanic rocks.

Amphibolite   Commonly green, coarse-grained, consists of amphibole and plagioclase. Similar to amphibole schist, except that little or no parallel orientation of amphibole needles is visible in hand specimen.

CONODONTS AND BURIAL METAMORPHISM

Because of the destructive effects of metamorphism (primarily due to the high temperatures and recrystallization), fossils are rarely encountered in metamorphic rocks. However, the hard mouth-parts of marine microchordate organisms known as **conodonts** (Fig. 20.8) are an exception. Conodonts are composed of calcium phosphate and are found most often in sedimentary rocks ranging from Late Proterozoic to Late Triassic in age. Their overall appearance did not attract much attention until geologists noticed that conodonts of the same age but from different localities exhibited color variation. The range of colors, from pale yellow to black, suggested to

Figure 20.8   Scanning electron microscope photographs of conodonts. Magnification 30X.

Courtesy of Karl M. Chauff.

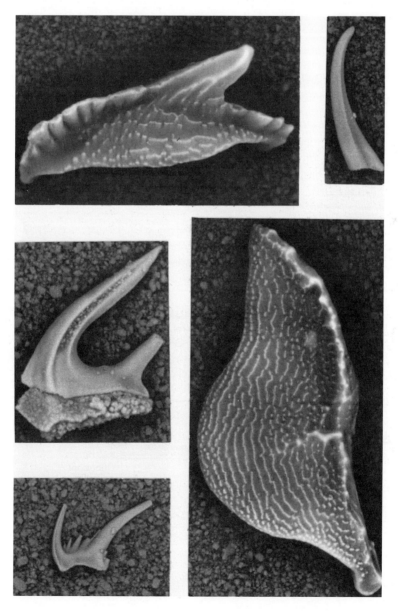

several members of the U.S. Geological Survey that the color change was a function of time and temperature. By testing this hypothesis, they found that the color of heat-treated conodonts could be compared to the colors of untreated conodonts collected in the Appalachian Mountains from rocks that had been buried to various depths and for different lengths of time. When the location and color of the conodonts is plotted on maps (Fig. 20.9), the color variation allows the geologist to determine the relative temperatures and burial depths that the conodont remains have experienced.

This gauge of the effects of **burial metamorphism** allows us to hypothesize how the rocks in which the conodonts are found correlate with metamorphic mineral assemblages.

In addition to using conodont color as an index to time and temperature relations in metamorphic terrains, geologists have observed that rocks containing black conodonts, which have experienced high temperatures, are less likely to yield commercial quantities of gas and oil. Thus, conodont color is another way to evaluate an area for potential commercial quantities of oil and gas.

Figure 20.9 Map showing the west-to-east gradation in conodont color resulting from eastward increases in the temperatures to which the conodont-bearing strata were subjected. In general, the darker-colored conodonts indicate higher temperatures. Conodont alteration colors range from black (numbered 5), through dark brown (numbered 3), to pale brown (numbered 1.5).

From Levin, H. L., *Contemporary Physical Geology,* 2nd ed. Philadelphia, Saunders College Publishing, 1986. Simplified from Epstein, A. G., Epstein, J. B., and Harris, L. D., 1977, *Conodont Color Alteration—An Index to Organic Metamorphism,* U.S. Geological Survey Professional Paper 995; conodont drawings added to illustrate concept.

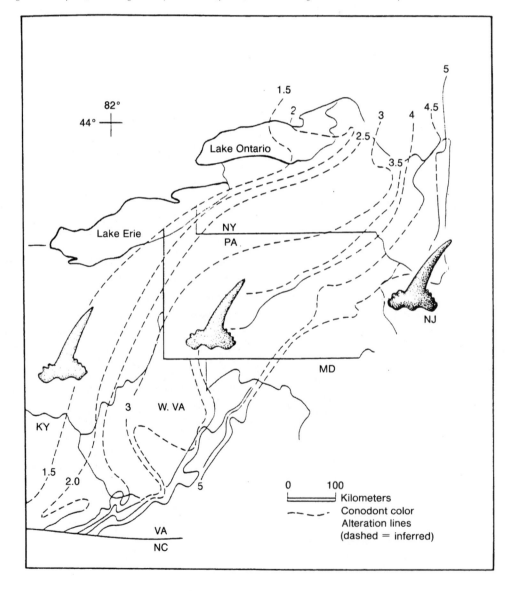

TERMS

burial metamorphism Type of low-grade regional metamorphism caused by heat and pressure resulting from deep burial.

conodont One of a large number of small fossil elements, phosphatic in composition, and commonly toothlike in form but not in function. Geologic range: Late Proterozoic to Late Triassic. Commonly abundant, widespread, and useful for biostratigraphy.

contact metamorphism A common type of metamorphism related to the intrusion of magmas into rocks at or near the earth's surface. Metamorphic changes in the surrounding rock (**country rock**) are affected by the heat and chemically reactive **volatiles** (fluids and gases) emanating from the magma or cooked from the country rock. Also called **thermal metamorphism.**

country rock A general term for any rock that surrounds the magma.

deformation A general term for the process of folding, faulting, shearing, compression, or extension of rock as a result of various forces.

equilibrium A state of balance (or stability) between two or more substances such that they retain their same properties (or characteristics) with the passage of time.

foliation A textural term that describes metamorphic rocks in which laminae develop by growth or alignment of minerals into a parallel orientation.

hydrothermal fluid Subsurface fluid (mainly water) whose temperatures are high enough to allow them to aid in chemical reactions. These fluids can carry a wide variety of ionic (elemental) species and assist in mineralogic changes in neighboring rocks.

igneous Said of a rock or mineral that solidifies from molten or partly molten material.

intrusion The process of emplacement of magma into preexisting rock.

magma Naturally occurring mobile, molten rock material, generated within the earth and capable of intrusion and extrusion, from which igneous rocks are thought to have been derived by solidification (i.e., crystallization) and related processes.

metamorphic aureole A zone surrounding an igneous intrusion in which the country rock shows the effects of contact metamorphism.

metamorphic grade The intensity of metamorphism measured by the amount or degree of difference between the original parent rock (protolith) and the metamorphic rock.

metamorphic index mineral A mineral developed under a particular set of temperature and pressure conditions, characterizing a particular degree of metamorphism.

metamorphic rocks Rocks resulting from the transformation of preexisting rocks as a result of the action of heat, pressure, and chemically reactive solutions.

metamorphic zoning The development of areas of metamorphosed rocks that may exhibit zones in which a particular mineral or suite of minerals is predominant or characteristic, reflecting the original rock composition, the pressure and temperature of formation, the duration of metamorphism, and whether or not material was added or removed during the event.

metamorphism A process whereby rocks undergo physical or chemical changes (or both) to achieve equilibrium with conditions other than those under which they were originally formed. (Weathering is excluded from this definition.)

porphyroblast A large crystal in a metamorphic rock that is enclosed in a matrix of smaller crystals.

protolith The unmetamorphosed rock from which a given metamorphic rock was formed.

recrystallization The formation, in the solid state, of new crystalline mineral grains in a rock. The new grains are generally larger than the original grains and may have the same *or* different mineralogic composition.

regional metamorphism A common type of metamorphism involving the effects of a wide range of confining pressures and temperatures. It is related both geographically and genetically to large orogenic belts, and hence is regional in character. Also called **dynamo-thermal metamorphism.**

volatiles Substances, such as water (H_2O) or carbon dioxide (CO_2), whose vapor pressures are sufficiently high for them to be concentrated in any gaseous phase.

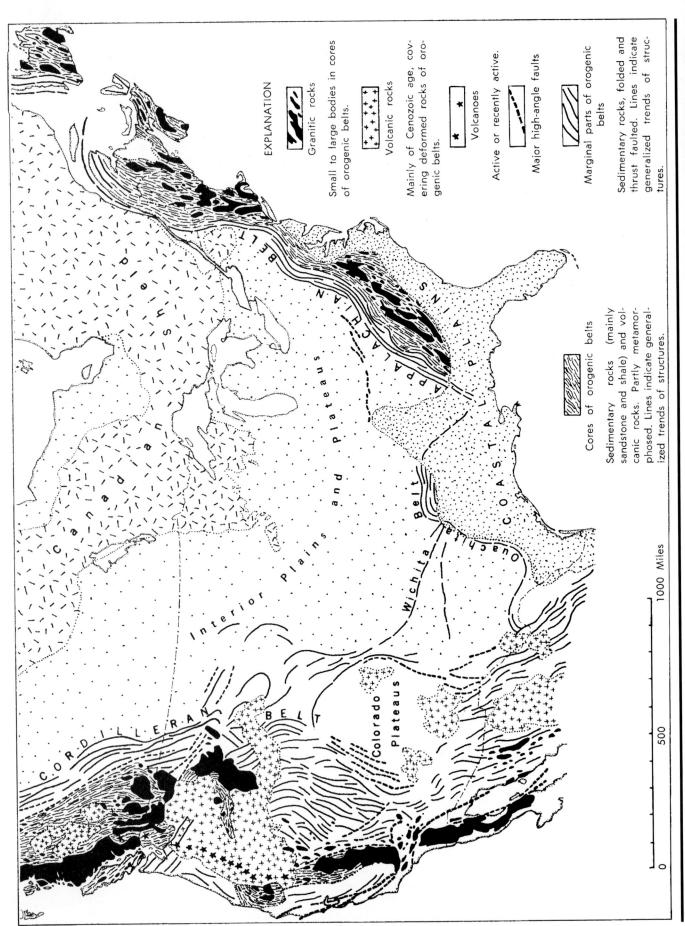

PLATE 1 — Tectonic map of the United States. (King, Philip B.; *The Evolution of North America*. © 1959 Princeton University Press, 1987 renewed PUP, 1977 revised edition, 2005 renewed PUP. Reprinted by permission of Princeton University Press.)

EXPLANATION

Granitic rocks

Small to large bodies in cores of orogenic belts.

Volcanic rocks

Mainly of Cenozoic age, covering deformed rocks of orogenic belts.

Volcanoes

Active or recently active.

Major high-angle faults

Marginal parts of orogenic belts

Sedimentary rocks, folded and thrust faulted. Lines indicate generalized trends of structures.

Cores of orogenic belts

Sedimentary rocks (mainly sandstone and shale) and volcanic rocks. Partly metamorphosed. Lines indicate generalized trends of structures.

C a n a d i a n   S h i e l d

I n t e r i o r   P l a i n s   a n d   P l a t e a u s

A P P A L A C H I A N   B E L T

C O A S T A L   P L A I N S

Ouachita Belt

Wichita Belt

C O R D I L L E R A N   B E L T

Colorado Plateaus

0   500   1000 Miles

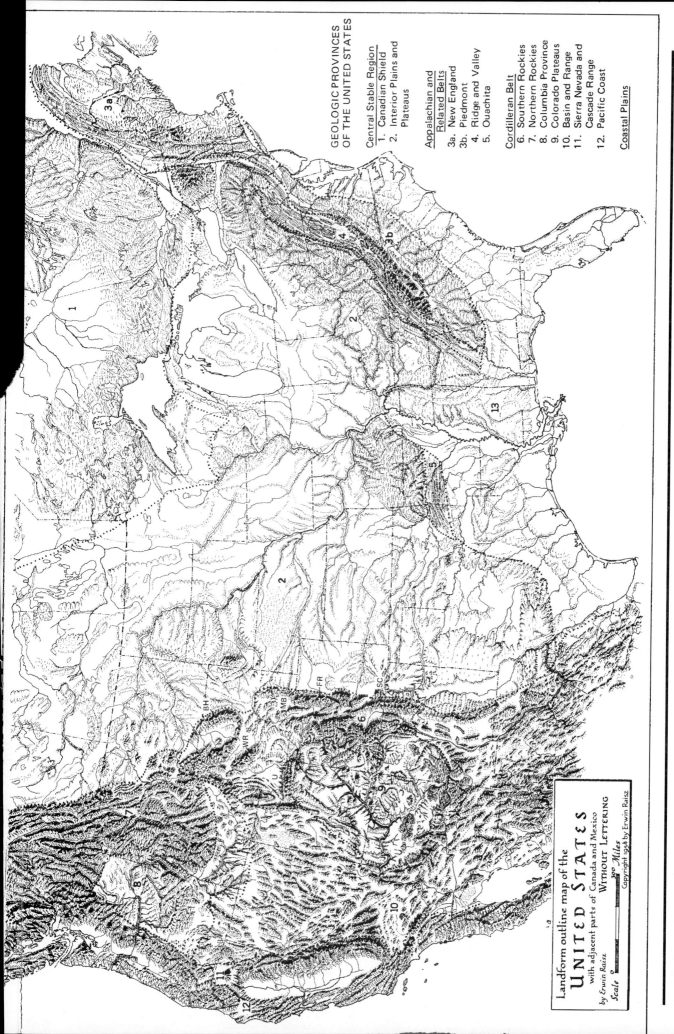

GEOLOGIC PROVINCES
OF THE UNITED STATES

Central Stable Region
1. Canadian Shield
2. Interior Plains and
   Plateaus

Appalachian and
Related Belts
3a. New England
3b. Piedmont
4. Ridge and Valley
5. Ouachita

Cordilleran Belt
6. Southern Rockies
7. Northern Rockies
8. Columbia Province
9. Colorado Plateaus
10. Basin and Range
11. Sierra Nevada and
    Cascade Range
12. Pacific Coast

Coastal Plains

Landform outline map of the
UNITED STATES
with adjacent parts of Canada and Mexico
by Erwin Raisz   WITHOUT LETTERING
Scale 0   300 Miles
                   Copyright 1954 by Erwin Raisz

PLATE 2   Geologic provinces of the United States.